EAA Series

EAA series is successor of the EAA Lecture Notes and supported by the European Actuarial Academy (EAA GmbH), founded on the 29 August, 2005 in Cologne (Germany) by the Actuarial Associations of Austria, Germany, the Netherlands and Switzerland. EAA offers actuarial education including examination, permanent education for certified actuaries and consulting on actuarial education.

actuarial-academy.com

For further titles published in this series, please go to
http://www.springer.com/series/7879

Marcus Kriele · Jochen Wolf

Value-Oriented Risk Management of Insurance Companies

Marcus Kriele
Hoboken, NJ, USA

Jochen Wolf
Fachbereich Mathematik und Technik
Hochschule Koblenz
Remagen, Germany

Translation from the German language edition:
'Wertorientiertes Risikomanagement von Versicherungsunternehmen'
by Marcus Kriele and Jochen Wolf

Additional material to this book can be downloaded from http://extras.springer.com

ISSN 1869-6929 ISSN 1869-6937 (electronic)
EAA Series
ISBN 978-1-4471-6304-6 ISBN 978-1-4471-6305-3 (eBook)
DOI 10.1007/978-1-4471-6305-3
Springer London Heidelberg New York Dordrecht

Library of Congress Control Number: 2014930223

Mathematics Subject Classification: 91B30, 91B70

Springer is part of Springer Science+Business Media (www.springer.com)

Preface to the English Edition

The English edition differs from the German original in that we have changed parts of the presentation in order to better address international readers. We have corrected any errors we are aware of. Some sections have been somewhat expanded, most notably the section on the Swiss Solvency Test (SST). The SST is a working example of a modern, economic capital based implementation for regulatory capital requirements.

This book contains example scripts using the statistical language R. These files can be downloaded from

http://extras.springer.com,

using the ISBN of this book.

We would like to thank Springer-Verlag for suggesting an English version of our book and for having the German text translated. We would also like thank our translator, Patrick Ion, who not only translated the whole text but who, on many occasions, also explained to us subtleties of the English language.

Hoboken, NJ, USA
Remagen, Germany
September 2013

Marcus Kriele
Jochen Wolf

Contents

Chapter 1
The Process of Risk Management

1.1 Risks and Opportunities

In ordinary language, the concept of risk is often associated with the danger of negative events or effects. From the point of view of economics, in contrast, a risk presents the possibility, resulting from the unpredictability of the future, of a deviation from a budgeted value or expected target value. For example the profit of an insurance company could be higher or lower than predicted by the company's planning. If one considers exclusively the variation from the budgeted figure in one direction, then one speaks of *one-sided risk*. However, if the discrepancies from the budgeted figure in both directions, for example higher or lower profit, are taken into consideration then one speaks of a *two-sided risk*.

Good risk management considers not only adverse deviations, but is integrated with value-oriented management. Therefore we focus in this section on the risk-opportunity profile as a basis for evaluations and decisions in management. Risk-opportunity profiles are described mathematically by probability distributions. By analyzing them risk management provides a basis for reliable and transparent corporate planning which is, with adequate estimates and control of risks, focused on the opportunities for the enterprise. The goal of entrepreneurial behavior is not to avoid risk. An insurance company, in particular, generates its profits by taking on risks. The goal of risk management is therefore the optimization of the risk-opportunity profile. In practice, this is often achieve through the combination of risk measures and profit measures.

For example, two strategies can be compared using the profit measure "expected profit" P and the measure of risk the *risk capital* C, i.e. the capital needed to sustain large unexpected losses.[1] The choice between two strategies can be made on the basis of a performance indicator that combines the risk and profit measures, e.g., using the risk-adjusted profit $P - hC$, where h is management's (relative) profit

[1] We will further develop the concept of risk capital in Chap. 2 from a mathematical point of view and in Chap. 4 from an economic point of view. An example of such a risk measure would be the "maximum loss, which may be exceeded with a probability of at most 0.5 %".

M. Kriele, J. Wolf, *Value-Oriented Risk Management of Insurance Companies*,
EAA Series, DOI 10.1007/978-1-4471-6305-3_1, © Springer-Verlag London 2014

expectation. If the company wants to decide whether it should invest into a certain line of business, it can calculate the $P - hC$ for this line of business. This provides management with a basis for deciding whether it is worthwhile to invest in this line of business.

Another example of a combination of the risk and profit measures is the valuation of an enterprise using the sum of the discounted future profits, where the discount rate depends on the risk. The more uncertain the future profits, the higher the discount rate, i.e., the lower the value of the enterprise.

The risk management process is closely interlocked with management. In order to see risk as a deviation from an intended target value, risk management requires transparent and informed corporate planning. First all relevant internal and external risks must be identified, evaluated and aggregated according to their dependence on one another. So risk management provides, on the one hand, feedback for strategic corporate orientation, on the other hand, the basis for concrete actions to optimize the risk-opportunity profile and thus for increasing the value of the business. Amongst such actions to manage risk belong risk avoidance, risk reduction and risk transfer. The development of risks must continually be evaluated over time, which necessitates a corresponding organization of control by assignment of responsibilities, clear communication structures, and reporting requirements. In many jurisdictions, the interlocking of risk management and value-oriented management is a central element in the legal or regulatory requirements. For instance, in Germany they follow from the KonTraG,[2] Solvency II[3] and MaRisk[4] directives and are illustrated in Fig. 1.1.

The process at the heart of the function "risk control"[5] comprises the steps from risk identification to risk monitoring, and thus fulfills the common regulatory requirements that risks be recognized early and suitable measures be taken. Risk control also calculates the risk capital. Risk capital allocation, the allocation of the risk capital to individual business areas and products, allows offsetting potential profits and opportunities against the cost of capital needed for taking on the corresponding risks. On this basis management can take decisions about products, and provide risk control with targets for the optimization of the risk profile.

[2]KonTraG stands for "Gesetz zur Kontrolle und Transparenz im Unternehmensbereich" (Control and Transparency in Business Act), a German law introduced on the 1st of May, 1998. The main consequence of KonTraG is the mandatory use of early warning and risk management systems. http://www.grc-resource.com/?page_id=21.

[3]The Solvency II Directive 2009/138/EC is an European Union (EU) Directive that codifies and harmonizes the EU insurance regulation. Primarily this concerns the amount of capital that EU insurance companies must hold to reduce the risk of insolvency. http://en.wikipedia.org/wiki/Solvency_II_Directive.

[4]MaRisk stands for "Mindestanforderungen an das Risikomanagement" (minimum requirements for risk management), a German ordinance introduced in December 2005. http://www.bafin.de/SharedDocs/Veroeffentlichungen/EN/Fachartikel/fa_bj_2012-08_marisk_en.html.

[5]Risk control may not be a separate function or department but could instead be a collection of tasks performed by the risk management department.

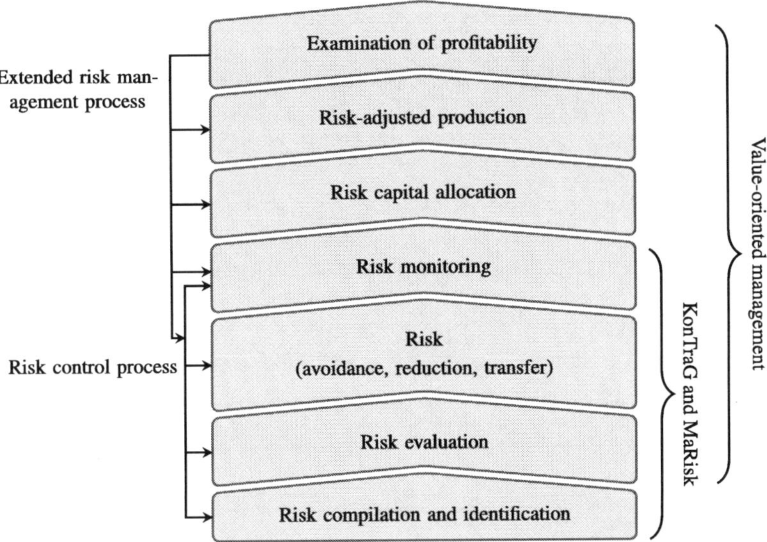

Fig. 1.1 The process of risk control and management (slightly modified form of a figure due to Bernd Heistermann, 2005)

While risk control is focused on deviations from target figures in the negative direction, management reaches decisions under uncertainty and therefore needs complete information about the probability distribution. So risk control is actually more closely attuned to the notion of one-sided risk. To calculate the risk capital C, usually risk measures that reflect the one-sided risk of negative discrepancies are utilized. Management also uses in addition to the risk measure the information from a profit measure such as the expected profit P. Thus, e.g., the indicator *Return On Risk-Adjusted Capital*, $\text{RORAC} = P/C$, expresses the profit-risk ratio.

In the remaining sections of this chapter we will describe the risk control process. The extended risk management process will be described in detail in the following chapters.

1.2 Compilation and Identification of Risks

The goal of *risk identification* is to compile all essential risks by a systematic up-to-date analysis of an insurance business and its economic environment.

A systematic analysis is needed to ensure that, on the one hand, all relevant material risks are recognized and, on the other, that measures to cope with risks are focussed on the relevant risks to be controlled, without being obscured by a flood of information about inessential risks.

Risk identification should always include the present exposure of the company to risk since early recognition of risks enables more efficient responses to risks.

One distinguishes between *systematic risks* which affect a great number of insurance companies, and *company-specific risks*. Amongst the systematic risks belong business cycles, movements in the financial markets, mortality trends, natural catastrophes, epidemics, and changes in the legal, regulatory and political environment as well as exogenous shocks such as oil crises or terrorist acts. Insurance companies cannot influence the systematic risks, but should take account of their exposure to such risks and respond to them. Enterprise-specific risks, which can be controlled by a particular company, include errors in strategy, management risk, risks to reputation (e.g., resulting from the sale of unsuitable product), bad liquidity planning, IT failures, cases of fraud. They also contain such components of market, credit and insurance risk as can be influenced by the company itself, such as, for example, the structure of its investments, its choice of reinsurers, premium risk (e.g., resulting from bad underwriting) and reserve risk (for instance, from using an inappropriate model or from inadequate claims settlement practice).

There are many ways to classify the risks faced by insurance companies. We will use the following classification of risk as our basis: strategic risk, market risk, credit risk, liquidity risk, insurance risk, operational risk (including management risk), reputational risk and concentration risk. How fine a classification of risk is used should always be adjusted to the individual risk profile of a company.

The individual risk classes can be influenced both by systematic and company-specific effects. For example, market risk depends on the movement of the financial markets and on the particular company's investments.

Strategic business planning sets the strategic direction in individual business areas. In doing this, the potential for success must be recognized, the position of the business in relation to the competition and trends in the market and society must be analyzed, and core competencies developed. Core competencies should thereby contribute significantly to what customers appreciate as their benefits, be important for many business areas, and ideally be difficult for competitors to copy. In addition, one must examine the cost structures and changes in risk profile resulting from the decision as to which services in the supply chain should be carried out by the company itself and which should be outsourced.

It is the job of strategic risk management to identify the factors influencing the achievement of strategic goals and to analyze the deviations from the strategic targets. Because strategic goals are often difficult to describe in terms of performance indices, *strategic risk* can usually only be judged qualitatively.

The Sharma Report [3] analysed crisis situations in the insurance industry and concluded that although crises could often be related to some external triggering event, the central cause was really a manifestation of management risk. Under the rubric *management risk* we collect all dangers that have to do with the internal organization and leadership of the business (Corporate Governance). We count lack of professional qualifications amongst employees, lack of clarity over distribution of competencies, and deficiencies in organizational and operational structure, as well as in the communication and reporting structures.[6] Quantitative evaluation of management risk is difficult and of lesser importance. The task of risk management in the first place is to ensure that all employees are involved in the risk management process.

The core business of insurance companies is assuming insurance risk. *Insurance risk* shows itself in variations from the underlying biometric probabilities, damage and claim frequency and rate distributions, as well as in the behavior of those insured (e.g., lapses, selection). Insurance risk is often divided into three components: random risk, risk of error and risk of change. While random risk describes the natural fluctuations of claims and insurance payments with respect to the underlying assumptions, risk of error depends on the incompleteness of information about the true nature of what is being insured and reflects the dangers of false assumptions. Error of change expresses the possible changes in risk characteristics over time (e.g., trends, structural breakdowns). As an additional component, catastrophe risk has to be considered. It describes extreme scenarios such as a pandemic or a serious accident in the chemical industry.

Especially in property insurance, insurance risk is divided into premium risk and reserve risk. *Premium risk* results from the fact that the premiums received may not be enough in the current business year, or future periods, to cover the insurance payouts and to establish the necessary technical provisions (e.g.,"Münchner Hagel",[7] "Wiehltalbrücke",[8] natural disasters). *Reserve risk* is the risk that the technical provisions which have been established during past periods prove to be insufficient. An example of this in non-life insurance would be the strengthening of technical provisions for asbestos claims. An example in life insurance would be the strengthening of technical provisions for longevity risk due to improved life expectancy.

Market risk is triggered by changes in prices in the finance markets. Oscillations in market prices result from changes in share prices (e.g, the 2002 crisis and the 2008 financial market crisis), interest rates (losses in fixed income securities resulting from increases in interest rates, increased spreads as a result of sub-prime or financial market crises), exchange rates (e.g., large deviations in the exchange rate of the Euro to the Dollar), volatility (e.g, volatility in spreads), real estate prices, and from other changes. The oscillations in market prices manifest themselves in changes in the value of assets and insurance liabilities.

Credit risk comprises default and decline in credit-worthiness of business partners (e.g., "Hypo Real Estate" as a consequence of the financial crisis in 2008).

---

[6]In 2008 the transactions made by Jérôme Kerviell cost the Société Générale a loss of 5 billion Euros. Though the loss was primarily the result of a crisis in the market, its origin was a manifestation of management risk: breakdown of the 4-eyes principle (double signatures), by-passing control mechanisms by using forged email and by forgoing vacation.

[7]Munich Hail: In 1984 a hailstorm in Munich led to economic damage to the extent of 3 billion DM (Deutschmark), of which 1.5 billion DM were insured.

[8]Wiehl Valley Bridge: In 2004, after a collision with a car whose driver was under the influence of drugs and had no driver's license, a tanker semi-trailer loaded with 32,000 liters of gasoline (petrol) fell from the bridge over the Wiehl Valley. The bridge was badly damaged by the resulting fire and had to be closed. The damages were of the order of 30 million Euros.

In the area of asset management this danger extends to the failure of creditors or counterparties in derivative financial instruments, as well as to reduction in value of bonds as a result of loss of credit-worthiness on the part of their issuers. A second important source of credit risk for insurance businesses consists in the possibility that the reinsurer is unable to fulfill its contractual obligations. Since the obligations of the primary insurer to the policy holder are unaffected by any reinsurance contract, if there is a major catastrophic event the failure of a reinsurer can quickly become a threat to the very existence of the primary insurer. In contrast to the default of a credit debtor, the failure of a reinsurer represents a secondary risk that may become obvious only when the loss has occurred.[9] Additional sources of credit risk come from the possible losses due to bad debt on the part of the insured (e.g., outstanding premiums), or tied agents and insurance brokers (claims for repayment of commissions).

Liquidity risk designates the possibility that an insurer cannot fully meet its obligations to pay when they are due. Even if furnishing the liquidity needed for the insurance business is usually easy, liquidity risk can be amplified by interaction with other risks, for instance, as a result of the downgrading of a rating, or when an increase in the lapse rates necessitates the realization of unrealized losses which in turn could trigger further lapses. Therefore liquidity risk is something to be considered not only on the basis of the cost of providing liquidity, but also as part of risk management. Finally, in extreme scenarios (massive catastrophes in property insurance, pandemics, or sudden increases in lapses as a result of changed conditions in the financial market, or collapse of the market for subordinated bank debt), it can endanger the existence of the company.

Operational risk is often considered as a residual category in the literature, which includes a broad spectrum of risks that cannot be put into the categories of market, credit or insurance risk. The definition according to Basel II[10] [1], which is also used in article 101 of the Solvency II Directive[11] [2], describes operational risk as

> "the risk of loss resulting from inadequate or failed internal processes, people, systems or from external events. This definition includes legal, but excludes strategic and reputational risk."

Operational risk thus includes

- company-internal and external criminal activities ("fraudulent acts"),
- political, legal and societal risks,

---

[9] In practice the failure of a reinsurer is more likely if a catastrophe has produced very high losses for a number of primary insurers. The primary insurance risk and the secondary risk are thus not independent.

[10] Basel II is the second of the Basel Accords. Basel II comprises recommendations on banking laws and regulations originally issued by the Basel Committee on Banking Supervision in June 2004. It is now extended and partially superseded by Basel III.

[11] The Solvency II Directive 2009/138/EC codifies and harmonizes the European Union insurance regulation. Solvency II represents a holistic approach to solvency regulation with an emphasis on the risk-based calculation of solvency capital, on risk management, and on public disclosure.

- damage to business assets, interruptions of business, and system failures (IT),
- operational mistakes by employees,
- losses resulting from disturbances in work flow, communication structures and weaknesses in organizational structure, as well as the already separately considered
- management risk.

Since operational risk is essentially characterized by rare events with very high losses, the data available to an insurance company may well be insufficient to quantify them. In this case the data available can be extended by pooling data from several companies, or by using commercial data bases. Bayesian models make it possible to combine historical data from an individual company with external data or with expert knowledge. In addition, so-called "near losses", i.e., losses avoided in a timely manner, can lead to a more informed estimate, provided loss severities and loss frequencies are modeled separately.

Further, it can be difficult to distinguish operational risk from other risk classes. While the distinction is irrelevant to the total risk, it is important for response to risk. For example, unexpectedly high claim payments can be the results of an insurance risk but also of a deficient business process, inadequate controls or the exploitation of vulnerabilities of the technological system used by the insurer. Even delimiting explicitly excluded strategic risk seems to be difficult. Strategic risk can essentially be distinguished from management risk only by its longer time horizon.

Against the background of these difficulties scenario-based methods offer a solution. On the one hand, one can investigate the effects of occurrences for which there are no historical data by using expert assessments of hypothetical scenarios. On the other hand, scenarios can capture the interactions between different risk categories and so alleviate the demarcation problem. Scenario-based models can be combined with models based on loss data bases.

Concentration risks arise if an insurance business is subject to a large exposure in one risk category, or undertakes strongly correlated risks. High exposure can, for example, be in relation to particular debtors, particular stocks, particular reinsurers or a particular region (e.g, the Hurricane Katrina). If an insurance company holds a significant portion of the shares in a single business whose whole fleet of vehicles it insures, there is a concentration extending over risk categories, which is reflected in an unfavorable total risk.

Reputation risk shows itself in the loss of reputation of a company triggered by a negative impression in the public, e.g., amongst customers (sale of unsuitable products), amongst business partners (badly communicated new forms of the distribution structure), amongst shareholders (losses as a result of inadequate management of risk) or amongst public authorities. Reputation risk can mostly only be judged qualitatively. It occurs often in connection with the realization of other risks (such as IT failures), but can occur as a risk on its own.

The result of risk identification is a complete inventory of the risks that provides a foundation for the further steps to be taken in the process of risk management.

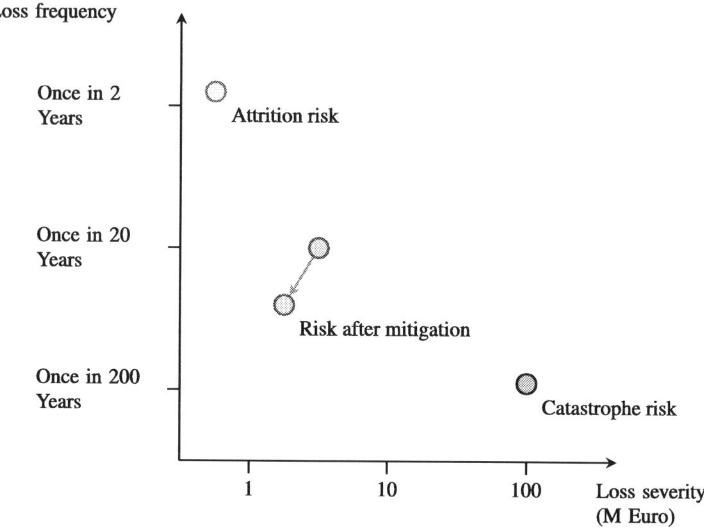

Fig. 1.2 Example of a risk matrix. Note the logarithmic scales, which allows a simultaneous overview of the catastrophe and attrition risks

1.3 Evaluating Risks

Having identified the risks faced by a company they can be evaluated in two stages, first a qualitative evaluation and then a quantitative measurement using a risk model.

The *qualitative evaluation* serves to estimate the relevance of risks. The individual risks can be arranged into a risk scale with levels from "insignificant" to "threatening our existence". One may also add qualitative estimates of their probability of occurring, their likely effects, and of their possible maximum losses.

The boundary between risk identification and qualitative evaluation cannot always be sharply drawn, since negligible risks are in possibility of not being included. Such risks could gain in relevance over time (e.g., destruction of an administrative building by an airplane crash after a change in the flight paths).

Ordering into a relevance scale reduces the complexity of a risk inventory and shows the relative importance of individual risks, which makes for easier communication about possible responses depending on the ramifications of the risks.

There are various ways of representing qualitative evaluations:

- A *risk matrix* is a two-dimensional representation, were the axes show the probability of occurrence and the severity of potential losses. Responses to cope with risks can be represented by moving risks within the risk matrix. Risk matrices are also used to exhibit the consequences of risk mitigation. See Fig. 1.2.
- *Risk trees* illustrate the classification of risk and its division into partial risks.
- *Check lists* can be provided for the relevant risk factors, cause and effect connections, and the environmental factors that influence risk factors, also called

influence factors. These check lists are to be filled in by the risk owners during structured workshops.
- An *Analysis of scenarios* explains the potential effects of risks under given constellations of influence factors.
- An *Analysis of dependencies* examines the interdependence of influencing and risk factors with an eye on accumulation and amplification effects.
- *SWOT (strengths-weaknesses-opportunities-threats)* analyses are made up of two partial analyses, analysis of strengths and weaknesses and analysis of opportunities and threats. In strength-weakness analysis on the basis of business data, estimates and expert knowledge, the current position of the business is investigated and its strengths and weaknesses compared to the competition. In a opportunity-threat analysis one establishes, with scenarios for the relevant influence factors, the effects of market developments on the business. By combining both analyses promising business strategies are then put together.

Quantitative evaluation of risks is done using risk models. First, partial models are developed for the individual risks. These partial models, which can be calibrated against the business's data using statistical methods, provide probability distributions for the individual risks, which can, using the methods of Chap. 3, be aggregated to a total risk distribution for the whole business. Risk measures (see Chap. 2) are then employed for assessment of the complex information of the probability distribution in terms of the key risk indicator (KRI).

1.4 Response to Risk

Following upon evaluation of risks, the responses to risks serve to optimize the risk-opportunity profile in concert with strategic business planning.

1.4.1 Avoiding Risks

Risk avoidance is intended to prevent the occurrence of certain risks and thus to exclude certain deviations from company goals. There are the following methods for risk avoidance:

- A restrictive underwriting policy does not accept certain risks, e.g., on the basis of a medical or financial examination of the risk. Contracts already in force with no longer acceptable risk-opportunity profiles are canceled, if possible.
- In developing a product, clauses can be included in the contract which exclude certain aspects of risk, such as war, acts of terror, civil strife, or storm and tempest damage. Additional contract provisions can completely transfer certain risks such as, for example, interest and currency risks, to the insured.
- A general or case-by-case selection of risks prevents the buildup of extraordinary levels of claims or of accumulated risks.

- Certain risky forms of investments can be excluded.
- In the event of insufficient risk capacity, a bad result of the SWOT analysis or bad risk-adjusted profit indicators, the company can give up a line of business.

Risk avoidance can nonetheless not play a central role in the risk response strategies of an insurance business, since taking on risk, above all insurance risk, represents its core profitable activity and is necessary for the long-term development of core competencies.

1.4.2 Reduction of Risks

Measures for the *reduction of risks* aim for a cause-oriented reduction of claim frequency, an outcome-oriented reduction of claim severity, or for improved diversification or hedging within the portfolio. There are many methods for risk reduction, and for the associated improvement in the risk-opportunity profile, open to insurance businesses.

- In underwriting, examination of risks can lead, in addition to the refusal of unacceptable risks and exclusion of opportunities, to a risk-sensitive differentiation in premiums. Limit systems can limit or prevent accumulated risks (e.g., exposure in respect of a major client, large insurance amounts for occupational disability, regional concentration in non-life insurance).
- Control mechanisms in checking payments and claims settlements can avoid malpractice and fraud, and thus reduce insurance payments.
- In developing products there is the possibility of drafting clauses in the contracts that oblige the policy holder to undertake measures to prevent damage (for example, installing burglar-proofing). Elements of the sharing of risks and fates with policy holders such as deductibles, forms of profit sharing and bonus-penalty systems depending on individual loss histories give concrete motivation for policy holders to avoid damages or at least to keep their effects small. Premium adjustment clauses and cancellation rights open possibilities for the insurer in reaction to worsening of the insured risks.
- Investment risks can be reduced by diversification over different investment classes and issuers as well as by hedging using derivative instruments.
- Risks on the liability side can be lessened by an insurer through diversification in lines of business, products, regions and sales organizations. Further, it can try by targeted marketing to have an influence on the composition of its portfolio of insurance contracts. For example, the influence of falling mortality rates on a pension population can be partly compensated for by their influence in the opposite direction on term insurance.
- Asset liability management (ALM) can also have risk-reducing effects, for example, through the reduction of duration gaps between assets and liabilities, thus reducing the risk from changes in interest rates.
- Simultaneous planning and balancing asset and liability cash flows additionally counteracts liquidity risk.

- Measures to reduce operational risk are amongst others business continuity planning, security mechanisms in the IT systems, the two-man rule (4-eyes principle, or requirement for two signatures), and internal limit systems to avoid fraudulent practices.

1.4.3 Transfer of Risks

One speaks of *risk transfer* if a risk is partly or wholly carried over to another economic entity. There is a wide spectrum of possibilities for risk transfer open to insurance companies.

To start with an insurance risk can be shared with the insured. Suitable arrangements can be made in agreeing over retentions and liability limits. Another possibility is profit sharing, in which, for example, the insured benefits from a favorable claims experience through premium rebates or participates in the run-off results through accounts linked to his or her claims experience. Furthermore, some risks such as investment risk in unit linked products can be shifted to the insured completely.

In *co-insurance* an insurance risk is carried by several insurers. Each company takes on a portion of the risk. The company that undertakes the technical administration gets a commission as a reward. In *open co-insurance* the insured has a contract with each participating company, while in *concealed co-insurance* the insured knows nothing of the sharing of risk with other insurers. Examples for co-insurance occur for major risks in liability insurance and for consortial agreements in life insurance. In contrast to an insurance pool, the risk per insurance volume is not lessened in this way.

In an *insurance pool* several insurers band together to take on risks. The motivation is to be able to insure major risks or risks that have not previously been viewed as insurable (e.g., risks from terrorist acts). The pool itself is not a risk taker but just organizes the risk taking. A pool contract fixes which insurer can, or must, bring in to the pool which risks in which form, and in what form the risks are divided amongst the members of the pool. The portion of the overall business that a pool member brings in is called the signing quota, the portion of the pool risks that it takes on is called the pool quote.[12] If the insured parties are known to all the pool members then one speaks of a *co-insurance pool*. Typical examples are the nuclear pools which have been set up in several countries in order to insure nuclear energy plants. Another example is the Pensions-Sicherungs-Verein (PSVaG), a German mutual that takes on occupational pension contracts from companies that have become insolvent, thus safeguarding the retirement of employees with occupational pensions, regardless of the financial future of their employer. While the PSVaG operates as a private company, membership in it is required by law. The PSVaG operates

[12] While the signing quota deals with the shifting of risks to the pool, the pool quota quantifies the portion of the risk borne by a pool member.

in both Germany and Luxembourg. If an insured is always under contract to a single pool participant, then one speaks of a *reinsurance pool*. Advantages of a pool are the broadened spread of risks within a bigger collective, a reduction of the administrative costs, and the widened statistical basis. An insurance pool is distinguished by high transparency, and thus makes the control of accumulation risk easier. An example of a reinsurance pool is the German pharmaceutical pool which reinsures liability risk of pharmaceutical companies.

In addition to the forms of risk sharing already mentioned there are other methods for risk transfer, which can be better tailored to the individual needs of insurers. Risk transfer within the insurance market is offered in the form of reinsurance, risk transfer within the finance market can be achieved with derivative instruments, and there are also "alternative" risk transfer offerings between the finance and insurance markets.

In classical reinsurance the primary insurer transfers insurance risk to a reinsurer. From the point of view of risk analysis one distinguishes between proportional and non-proportional reinsurance.

In *proportional reinsurance* the risk is split in a given ratio between the primary insurer and the reinsurer. Forms of proportional reinsurance are quota share reinsurance, surplus insurance, or a combination of both.

In *quota share reinsurance* the reinsurer fixes a percentage for each policy involved in the reinsurance contract. Thus the losses R_i of the i-th policy,

$$R_i = cR_i + (1-c)R_i,$$

is split into a reinsured portion $(1-c)R_i$ and the portion cR_i that remains with the primary insurer, where $c \in]0,1[$. Quota share reinsurance reduces the absolute size of the liability of the primary insurer. However, it cannot homogenize the portfolio. Important risk indicators such as the coefficient of variation and risk adjusted profit indicators, such as RORAC, are not affected by it.

In *surplus reinsurance* the reinsurer only participates in those policies i where the sum insured, v_i, exceeds an absolute deductible v_0 assigned to the primary insurer. Of the losses R_i of a policy whose sum insured exceeds the deductible, the primary insurer is only liable for the portion $c_i R_i$, which corresponds to the ratio of the deductible and the sum insured:

$$R_i = c_i R_i + (1-c_i)R_i, \quad c_i = \min\left(\frac{v_0}{v_i}, 1\right).$$

Surplus reinsurance serves to relieve the primary insurer of peak claims and leads to the homogenization of a portfolio by reducing its variance. However, surplus reinsurance is not very effective against accumulation risk.

Non-proportional reinsurance describes reinsurance contracts in which the payments by the reinsurer are not proportional to the claim payments by the primary insurer.

In *excess of loss reinsurance* the reinsurer takes on the part the cedant's portion of each individual claim X that exceeds the *priority* a (also termed the *retention*) up to an agreed limit of liability h, that is it pays

$$\min\bigl(h, \max(X - a, 0)\bigr).$$

Excess of loss offers an effective protection against large claims and reduces the variance in the distribution of claims. It is relatively easy to manage. However, it does only partially protect against growing frequencies of claims, since for each claim the priority must be carried by the cedant. Small and medium claims remain the cedant's responsibility.

Accumulation excess of loss reinsurance is suitable for cases where a single loss event causes many individual claims: The reinsurer pays whatever exceeds the priority portion of the total claim. Accumulation excess of loss reinsurance can supplement other types of reinsurance. Naturally, an additional excess of loss reinsurance has an effect only if its priority is below the priority of the accumulation excess of loss reinsurance.

Stop-loss reinsurance (annual excess of loss reinsurance) pays for that portion of the total claims S in a year that exceeds the agreed priority a of the ceding primary insurer: $\max(S - a, 0)$. Stop-loss effectively protects the balance sheet and smooths annual results, since it guards against all types of insurance risk. It is nonetheless associated with much moral risk on the part of the primary insurer, which could change its underwriting policy or could engage in less careful claims handling, either of which could lead to a significant increase of the number of claims above the threshold a. Therefore stop-loss is usually only agreed along with other reinsurance contracts and with a proportional deductible for the excess claims of the cedant.

The reduction of the insurance risk by reinsurance is also to be balanced by the *risk of insolvency* on the part of the reinsurer. Since interruption of payments from the reinsurer can have grave consequences for the primary insurer, the financial health of the reinsurer is of great importance to the cedant. In order to limit this type of credit risk for the primary insurer, the reinsurer may deposit the reinsurance premium or technical provisions at the primary insurer. It may also provide a letter of credit with which a bank guarantees the solvency of the reinsurer.

The transition between traditional reinsurance and *financial reinsurance* is fluid. In financial reinsurance the financial, balance sheet oriented or regulatory goals are in the foreground, while the transfer of insurance risk is only limited. Typical goals for financial reinsurance are increase of capital, help in financing in the case of strong growth, stabilization of business results, and improvement of performance indicators. Financial reinsurance uses the classic forms of reinsurance contracts. In addition to limited transfer of risk, it is distinguished by the long terms of the contracts, extensive setting of commissions, and the taking into account of investment yields in setting the price for the reinsurance. Typical examples in life insurance are quota share reinsurance contracts, where the reinsurer pays a large commission to the primary insurer.[13]

[13] Presently the German supervisory authority requires a significant transfer of risk since otherwise it considers the reinsurance contract as a loan, which is not allowed in this context.

For the transfer of financial risks on the assets side *derivative instruments in the financial market* can be used. Examples are puts, calls, forwards and futures to hedge against share price risks, or swaps to hedge against interest and currency risks.

If financial risks are transferred to the insurance market or insurance risks to the financial market, one speaks of *alternative risk transfer* (ART). For example, a reinsurer can take on the interest risk that is induced by corresponding guarantees in life insurance. Conversely, one may want to transfer insurance risk to the financial markets in order to overcome capacity shortages in the reinsurance market or in order to obtain tailored transfer methods within the framework of integrated risk management.

Insurance-linked securities (ILS) are bonds whose coupon and/or payments depend on the manifestation of an insurance risk. Examples are catastrophe bonds, whose coupon or principal would not be fully paid, if a given number of hurricanes in a particular region arise, or bonds which would not be paid back, if there is a sudden increase of mortality over a predefined threshold. The latter bonds can be employed to manage the financial risk of a pandemic. Furthermore the so-called longevity bonds insure against the risk of long life for a writer of annuities. For catastrophe, epidemic and longevity risks there is often only a very restricted capacity in the reinsurance market.

Since the nominal value of the ILS is invested in risk-free investments through a special-purpose vehicle, there is no credit risk, in contrast to traditional reinsurance. Since bonds have to be largely standardized on cost grounds and to avoid moral risk, the definition of the insured risks is often not coupled to the development of claims within the emitting insurance company but instead to a claim index. Thus for the insurance company there arises a basis risk, since the index develops differently from their own claim history.

ILS also comprise, in addition to bonds, futures and derivative instruments such as swaps.

Tailored offerings for risk management are in addition to be seen in multiple-trigger products, that hedge for a desired combination of several financial or insurance risks, or in contingent capital solutions. For example, put options on the issue of new shares could be linked to the event that a certain claim amount is exceeded. This would represent a hedge in the case that a large claim causes both a decrease in the share price of the insurer and the necessity to strengthen the equity base of the insurer.

1.5 Monitoring Risk

Risk control goes along with the process of risk management, and thereby supports both risk owners and company management. It ensures steady monitoring of risks, and that the provisions of strategic business planning and the risk policies of the executive board are carried out, and that the company abides by its risk limits and underwriting rules.

Ongoing attention paid to risks should ensure that new risks, and also changes in already identified risks, can be recognized early. To this end the responsibilities for the separate risk areas, and a regimen of monitoring and reporting requirements, are to be spelled out. In an effective organization, large changes in the risk structure lead to ad hoc notifications of risk control and of the superiors of the risk owner. Risks that endanger company existence are immediately reported to the executive board. In Germany the practices to be followed for existence-threatening risks are prescribed by law (KonTraG).

Risk control examines the results of the reports from the risk owners as to their plausibility, and evaluates the risk indicators and profit indicators. Controlling is backed by, amongst others, the following insurance activities:

- regular checking of the reasonableness of the valuation basis,
- recalculation and analysis in the face of changes in the valuation basis,
- calculation of technical provisions according to actuarial principles,
- calculation of projections to determine important cash flows,
- profit testing in various scenarios, in order to recognize risks in certain market conditions (such as selective lapse at a favorable point in time),
- asset liability management (ALM),
- creation and development of internal risk capital models,
- analysis of the suitability of the reinsurance structure.

1.6 The Role of the Appointed Actuary in the Risk Management Process

In many countries the actuarial duties in risk management are legally established.

For example in Germany, the institution of the appointed actuary is described in the VAG.[14] In this jurisdiction, the appointed actuary of a life insurance company has to ensure that premiums and technical provisions are calculated in accordance with legal requirements and recognized actuarial principles. Because of the importance of the technical provisions, the appointed actuary plays an essential role in risk identification and evaluation. This role is not limited to insurance risks, since the appointed actuary must also evaluate the financial situation of the business with an eye on the lasting satisfiability of its obligations and the necessity of ensuring solvability margins on an ongoing basis. In addition, the requirements of the actuary's report (which is required by law in Germany) have evolved from checking the basis of the calculations to a comprehensive report on risks, in which the appointed actuary must, as may be necessary, suggest suitable measures to improve the risk situation. Thus he is also implicated in the response to risk. Finally, the results from actuarial control support value-oriented management.

Even if the appointed actuary plays an important role in the process of risk management, his function is different from that of the *Chief Risk Officer* (CRO). The

[14] VAG stands for "Versicherungsaufsichtsgesetz" and is the German Insurance Supervisory Act.

CRO safeguards the structural and procedural organization of risk management and translates the strategic targets of the business into risk management.

In contrast to the CRO, in Germany the appointed actuary is liable with his personal assets.

References

1. Basel Committee on Banking Supervision, International convergence of capital measurement and capital standards. A revised framework comprehensive version, June 2006 (p. 6)
2. Commission of the European Communities, Directive of the European Parliament and the Council on the taking-up and pursuit of the business of insurance and reinsurance, Solvency II (recast), November 2009. PE-CONS 3643/6/09 REV 6, approved by the European Parliament on 2009-04-22 (p. 6)
3. Conference of Insurance Supervisory Services of the Member States of the European Union, Prudential supervision of insurance undertakings, December 2002. Sharma-Report (p. 4)

Chapter 2
Risk Measures

2.1 The Notion of a Risk Measure

In ordinary language, risk is simply understood as the possibility that "unfavorable events" occur. Deviations toward the positive are as a rule ignored. When one tries to capture "risk" quantitatively it turns out that risk is very much a many-sided phenomenon.

One way to describe risk mathematically consists in identifying risk in general with fluctuations (for example monetary fluctuations). In that way both "favorable" and "unfavorable" variations are taken into account. Such an approach is taken, for example, when one chooses as a measure of risk the standard deviation (see below).

Another focus would be to identify financial risks with an amount of money, which gives an indication of how much one can lose in the event of the risk occurring. This will be the approach that we will mainly follow. For this, various measures are suitable according to the situation. Very popular are measures whose results can be viewed operationally as the amount of capital that the company must put by, according to its level of risk aversion, in order to go about its business. The result from such a measure is referred to as the *risk capital*.

Let $(\Omega, \mathscr{A}, \mathbf{P})$ be a probability space with a σ-algebra $\mathscr{A}$ and probability measure $\mathbf{P}$. We denote by $\mathscr{M}_{\mathscr{B}}(\Omega, \mathbb{R}^k)$ the space of $\mathbb{R}^k$-valued random variables

$$X : \Omega \to \mathbb{R}^k, \quad \omega \mapsto X(\omega),$$

that is the maps measurable with respect to $\mathscr{A}$ and the Borel σ-algebra. If we wish to emphasize the σ-algebra $\mathscr{A}$, we will also say *$\mathscr{A}$-measurable maps* or the *maps measurable with respect to $\mathscr{A}$*.

Definition 2.1 A *risk measure* is a map

$$\rho : \mathscr{M}(\Omega, \mathbb{R}) \to \mathbb{R}, \quad X \mapsto \rho(X),$$

where $\mathscr{M}(\Omega, \mathbb{R}) \subseteq \mathscr{M}_{\mathscr{B}}(\Omega, \mathbb{R})$ is suitable vector subspace (depending on ρ).

M. Kriele, J. Wolf, *Value-Oriented Risk Management of Insurance Companies*,
EAA Series, DOI 10.1007/978-1-4471-6305-3_2, © Springer-Verlag London 2014

Remark 2.1 The restriction to a subspace is necessary since the interesting risk measures are often not defined on all of $\mathcal{M}_{\mathcal{B}}(\Omega, \mathbb{R}^k)$. If, in the following, we use the notation $\mathcal{M}(\Omega, \mathbb{R}^k)$, it is always intended in the context of an obviously suitable subspace of $\mathcal{M}_{\mathcal{B}}(\Omega, \mathbb{R}^k)$.

2.2 Examples of Risk Measures

Let Y be a random variable that describes an uncertain financial outcome. Then $X = -Y$ expresses the possible loss. Many risk measures contain a parameter $\alpha \in]0, 1[$, through which the (intuitive) safety levels described by this measure are fixed. Here we shall call this parameter a *confidence level* and reserve the notion of *safety level* for its intuitive meaning. Safety level is made mathematically concrete by giving a risk measure, a confidence level, and the time horizon, to which the profit and loss amounts refer. The terminology in the literature is however fairly confused, so that the actual meaning is only clear in context.

2.2.1 Measures Based on Moments

2.2.1.1 Measures Based on the Standard Deviation

A mathematically very simple measure of risk is the standard deviation,

$$\sigma(X) = \sqrt{\mathrm{E}\big((X - \mathrm{E}(X))^2\big)} = \sqrt{\mathrm{E}\big((Y - \mathrm{E}(Y))^2\big)} = \sqrt{\mathrm{var}(X)} = \sqrt{\mathrm{var}(Y)}.$$

It expresses how far on average the results differ from the expected value, where the "difference measure" is simply borrowed from Euclidean geometry. As a measure of risk, standard deviation is also used in the form

$$\rho(X) = a\,\mathrm{E}(X) + b\sigma(X) \tag{2.1}$$

where $a, b > 0$ are given parameters. A traditional area of application of this measure is in fixing premiums. A related principle applied in fixing premiums is the *variance principle* with the measure of risk

$$\rho(X) = a\,\mathrm{E}(X) + b\sigma^2(X). \tag{2.2}$$

In the variance principle it is to be noted that the variance does not represent, as the expected value does, an amount of money, but rather a quadratic amount of money, so that the sum of $a\,\mathrm{E}(X)$ and $b\sigma^2(X)$ is difficult to interpret.

The risk measure (2.1) has the unpleasant property that positive variations in the standard deviation have the same effect as negative variations. It is thus insensitive to whether an event is "favorable" or "unfavorable". To avoid this problem one can consider only the losses that exceed the expected value by using the one-sided standard deviation $\sigma_+ = \sqrt{\mathrm{E}(\max(0, X - \mathrm{E}(X)))^2}$.

2.2.1.2 Risk Measures Based on Higher Moments

Risk measures that are based only on the mean and standard deviation do not take into account that loss distributions are in general very asymmetric. Examples of this are the claim amount distributions in property insurance and surplus sharing in life insurance with guaranteed interest. This asymmetry can be taken into account by including higher moments in the risk measure.

2.2.1.3 Shortfall Measures

The danger of exceeding a given loss threshold a is measured by so-called shortfall measures. The higher and lower partial moments weight the discrepancy for this with a power function.

For loss amounts one considers the *upper partial moments*:

$$\mathrm{UPM}_{(h,a)}(X) = \begin{cases} \mathrm{E}(\max(0, X-a)^h) & \text{for } h > 0 \\ \mathbf{P}(X \geq a) & \text{for } h = 0. \end{cases}$$

Special cases are the probability of exceeding the critical boundary a $(h = 0)$, the mean excess $(h = 1)$ and the semi-variance $(h = 2)$.

For profit amounts there are analogously the *lower partial moments*:

$$\mathrm{LPM}_{(h,a)}(Y) = \begin{cases} \mathrm{E}(\max(0, a-Y)^h) & \text{for } h > 0 \\ \mathbf{P}(Y \leq a) & \text{for } h = 0. \end{cases}$$

2.2.1.4 General Problems with Moment-Based Measures

The biggest problem with moment-based measures is the fact that they can be interpreted financially only with difficulty. Most convincingly the standard deviation can be seen as the "average distance to the mean". However, the Euclidean metric, though a good measure of distance, is no natural measure of financial risks.

For many of the distributions used in the insurance industry higher moments do not exist. In modeling operational risks using the GPD (generalized Pareto distribution) with values of parameters that may occur in practice not even the mean is defined.[1] Moment-based measures cannot be applied to such distributions.

[1] In such a case the company cannot last indefinitely, if the management of operational risks is not significantly improved.

2.2.2 Value at Risk

In contrast, the value at risk is a direct and simple mathematical financial measure. It describes the amount that, with a given probability α, one may at most "lose".

Definition 2.2 The *value at risk* (or *VaR* for short) $\mathrm{VaR}_\alpha(X)$ is given by the formula

$$\mathrm{VaR}_\alpha(X) = \inf\{x \in \mathbb{R} : F_X(x) \geq \alpha\},$$

where F_X is the distribution function of X.

The value at risk, $\mathrm{VaR}_\alpha(X)$, is the *minimal* loss, that occurs in $100(1-\alpha)$ % of the worst scenarios for the portfolio (see Fig. 2.1).

In other words, if a company does not wish to consume, with the probability α, all its equity capital in a time period, then that equity capital must amount to *at least* $\mathrm{VaR}_\alpha(X)$, where X denotes the loss in this time interval. This measure is therefore suitable for a shareholder who is only liable for the money he has invested. For internal risk management, where one is also interested in higher risks beyond the quantile $\mathrm{VaR}_\alpha(X)$, this measure is not always suitable.

Remark 2.2 In exceptional cases $\mathrm{VaR}_\alpha(X)$ can be negative for large α. This value then would correspond to a profit not a loss.

In the language of statistics, value at risk represents the lower α-quantile of the distribution of X. In the special case that F_X is invertible, we have $\mathrm{VaR}_\alpha(X) = F_X^{-1}(\alpha)$.

Lemma 2.1 *For all* $\alpha \in]0,1[$ *we have* $F_X(\mathrm{VaR}_\alpha(X)) = \alpha$.

Proof This follows directly from the right continuity of the distribution function. □

The following two lemmas make clear that value at risk can be considered as "pseudo-inverse" of the distribution function of X.

Lemma 2.2 *If* F_X *is the distribution function of* X, *then* $\mathrm{VaR}_{F(X)}(X) = X$ *a.e.*

Proof Because of the monotony of F_X there holds

$$Y := \mathrm{VaR}_{F_X \circ X}(X) = \inf\{x \in \mathbb{R} : F_X(x) \geq F_X \circ X\} \leq X \quad \text{a.e.}$$

From Lemma 2.1 follows in addition that $F_X(Y(\omega)) = F_X(X(\omega))$ for all $\omega \in \Omega$. Hence F_X is constant on each interval $[Y(\omega_0), X(\omega_0)[$ where $\omega_0 \in \{\omega : Y(\omega) < X(\omega)\}$. As a result, $\mathbf{P}(Y < X) = 0$. □

Lemma 2.3 *Let* U *be a random variable with* $\mathbf{P}(U \leq u) = u$ *for all* $u \in]0,1[$. *Then the random variable* $\mathrm{VaR}_{U(\cdot)}(X)$ *has the same distribution function as* X.

Proof Let $\omega \in \Omega$ with $U(\omega) \le F_X(x)$. Then obviously

$$y_0 = \inf\{y : U(\omega) \le F_X(y)\} \le x,$$

since x itself satisfies the condition for y. Conversely it follows from the right continuity of F_X that the equation $U(\omega) \le F_X(y_0)$ is also satisfied for the infimum y_0. Thus we have shown

$$\{\omega \in \Omega : U \le F_X(x)\} = \{\omega \in \Omega : \inf\{y : U \le F_X(y)\} \le x\},$$

and it follows that

$$\begin{aligned}\mathbf{P}\big(\mathrm{VaR}_{U(\cdot)}(X) \le x\big) &= \mathbf{P}\big(\inf\{y : F_X(y) \ge U\} \le x\big)\\ &= \mathbf{P}\big(U \le F_X(x)\big)\\ &= F_X(x) = \mathbf{P}(X \le x).\end{aligned}$$

□

Lemma 2.4 *Let* $\mathscr{M}(\Omega, \mathbb{R})$ *and* $\alpha \in]0, 1[$. *Then*

$$\mathbf{P}\big(X < \mathrm{VaR}_\alpha(X)\big) \le \alpha \le \mathbf{P}\big(X \le \mathrm{VaR}_\alpha(X)\big).$$

If it is also true that $\mathbf{P}(X = \mathrm{VaR}_\alpha(X)) = 0$, *then, in particular, it follows that* $\alpha = \mathbf{P}(X \le \mathrm{VaR}_\alpha(X))$.

Proof Let U be a random variable with $\mathbf{P}(U \le u) = u$ for all $u \in]0, 1[$. Since the value at risk grows monotonely with the confidence level

$$\begin{aligned}\{\omega : \mathrm{VaR}_{U(\omega)}(X) < \mathrm{VaR}_\alpha(X)\} &\subseteq \{\omega : U(\omega) < \alpha\}\\ &\subseteq \{\omega : \mathrm{VaR}_{U(\omega)}(X) \le \mathrm{VaR}_\alpha(X)\}.\end{aligned}$$

From Lemma 2.3 now follows

$$\begin{aligned}\mathbf{P}\big(X < \mathrm{VaR}_\alpha(X)\big) &= \mathbf{P}\big(\mathrm{VaR}_{U(\cdot)}(X) < \mathrm{VaR}_\alpha(X)\big)\\ &\le \overbrace{\mathbf{P}\big(U(\cdot) < \alpha\big)}^{=\alpha}\\ &\le \mathbf{P}\big(\mathrm{VaR}_{U(\cdot)}(X) \le \mathrm{VaR}_\alpha(X)\big) = \mathbf{P}\big(X \le \mathrm{VaR}_\alpha(X)\big).\end{aligned}$$

Under the additional assumption that $\mathbf{P}(X = \mathrm{VaR}_\alpha(X)) = 0$, the inequalities degenerate into equalities, since $\mathbf{P}(X < \mathrm{VaR}_\alpha(X)) = \mathbf{P}(X \le \mathrm{VaR}_\alpha(X))$ holds. □

For the important class of normal random variables the value at risk can be given directly:

Proposition 2.1 *Let* $X : \Omega \to \mathbb{R}$ *be a normally distributed random variable with mean* m *and standard deviation* s. *If* $\Phi_{0,1}$ *is the standard normal distribution and*

$f: X(\Omega) \to \mathbb{R}$ is a strongly monotone increasing map, then

$$\mathrm{VaR}_\alpha(f \circ X) = f\big(m + s{\Phi_{0,1}}^{-1}(\alpha)\big).$$

Proof Since $F_{f \circ X}$ is strongly monotone increasing the value at risk is uniquely determined by $F_{f \circ X}(\mathrm{VaR}_\alpha(f \circ X)) = \alpha$. The assertion thus follows from

$$\begin{aligned}\mathbf{P}\big(f \circ X \le f\big(m + s{\Phi_{0,1}}^{-1}(\alpha)\big)\big) &= \mathbf{P}\big(X \le m + s{\Phi_{0,1}}^{-1}(\alpha)\big) \\ &= \mathbf{P}\left(\frac{X - m}{s} \le {\Phi_{0,1}}^{-1}(\alpha)\right) \\ &= \Phi_{0,1}\big({\Phi_{0,1}}^{-1}(\alpha)\big) = \alpha,\end{aligned}$$

where we have used that f is invertible on $X(\Omega)$ and $\omega \mapsto \frac{X-m}{s}$ has a standard normal distribution. □

Example 2.1 If X is log-normally distributed with the parameters m and s^2, then there holds $\mathrm{VaR}_\alpha(X) = \exp(m + s\Phi_{0,1}^{-1}(\alpha))$.

2.2.3 *Tail Value at Risk and Expected Shortfall*

The tail value at risk, in contrast to the value at risk, also weights higher losses.

Definition 2.3 The *tail value at risk* is given by the conditional expectation

$$\mathrm{TailVaR}_\alpha(X) = \mathrm{E}\big(X \mid X > \mathrm{VaR}_\alpha(X)\big).$$

It thus delivers, from the internal risk management point of view, more interesting information, namely the *expected* loss of the $100(1-\alpha)\%$ worst scenarios. It is clear that the tail value at risk for the same confidence level α is always bigger than (or in the extreme case equal to) the value at risk. See Figs. 2.1 and 2.2.

The tail value at risk has an economic interpretation. For continuous distributions X_1, X_2 it also has, as we shall see later, the important property of subadditivity

$$\mathrm{TailVaR}_\alpha(X_1 + X_2) \le \mathrm{TailVaR}_\alpha(X_1) + \mathrm{TailVaR}_\alpha(X_2),$$

which intuitively expresses that the risk in a diversified collective is less than the sum of the individual risks. This property does not hold in general for random variables X_1, X_2 with distribution functions that have discontinuities (jumps). In contrast, the closely related 'expected shortfall' shows subadditivity for all random variables (see Sect. 2.3).

Definition 2.4 The *expected shortfall* is given by the formula

$$\mathrm{ES}_\alpha(X) = \frac{1}{1-\alpha}\int_\alpha^1 \mathrm{VaR}_z(X)\,\mathrm{d}z.$$

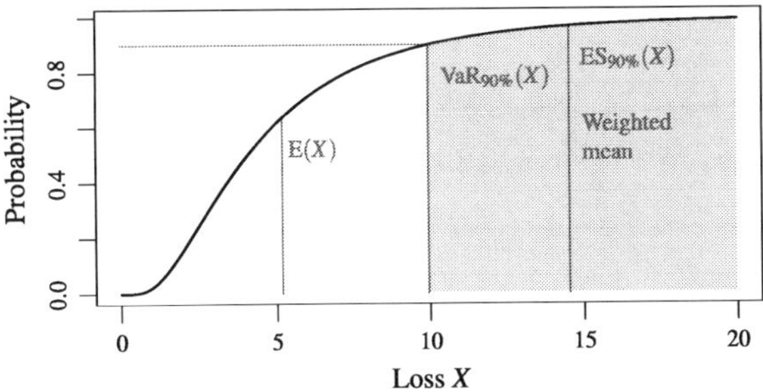

Fig. 2.1 Value at risk and tail value at risk from the perspective of the distribution function

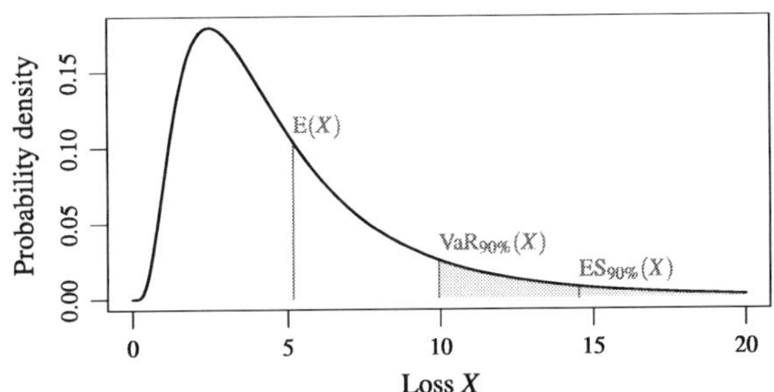

Fig. 2.2 Value at risk and tail value at risk from the perspective of the density

In the literature, the expected shortfall is sometimes called the *average value at risk*.

We will now derive an alternative formula for $\mathrm{ES}_\alpha(X)$ which shows that for continuous distribution functions $\mathrm{ES}_\alpha(X)$ coincides with $\mathrm{TailVaR}_\alpha(X)$.

Lemma 2.5 *Let $X : \Omega \to \mathbb{R}$ be a random variable and assume $x \in \mathbb{R}$. We put*

$$1_{X,x,\alpha} = 1_{\{X>x\}} + \beta_{X,\alpha}(x) 1_{\{X=x\}},$$

where

$$\beta_{X,\alpha}(x) = \begin{cases} \frac{\mathbf{P}(X\le x)-\alpha}{\mathbf{P}(X=x)} & \textit{if } \mathbf{P}(X=x)>0 \\ 0 & \textit{otherwise.} \end{cases}$$

Then we have

(i) $1_{X,\mathrm{VaR}_\alpha(X),\alpha}(\omega) \in [0,1]$ *for all* $\omega \in \Omega$,
(ii) $\mathrm{E}(1_{X,\mathrm{VaR}_\alpha(X),\alpha}) = 1-\alpha$,
(iii) $\mathrm{E}(X 1_{X,\mathrm{VaR}_\alpha(X),\alpha}) = (1-\alpha)\,\mathrm{ES}_\alpha(X)$.

Proof Some quantities used in the proof are illustrated in Fig. 2.3. (i) The assertion is obvious in the special cases $\mathbf{P}(X=\mathrm{VaR}_\alpha(X))=0$ and $\omega \notin \{X=\mathrm{VaR}_\alpha(X)\}$. By applying Lemma 2.4 twice, we obtain

$$\begin{aligned} 0 &\le \mathbf{P}\big(X \le \mathrm{VaR}_\alpha(X)\big) - \alpha \\ &= \mathbf{P}\big(X = \mathrm{VaR}_\alpha(X)\big) + \mathbf{P}\big(X < \mathrm{VaR}_\alpha(X)\big) - \alpha \end{aligned}$$

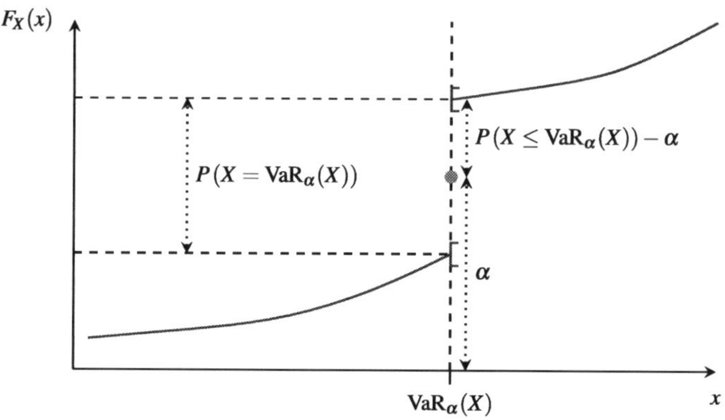

Fig. 2.3 For the proof of Lemma 2.5

$$\leq \mathbf{P}\big(X = \mathrm{VaR}_\alpha(X)\big).$$

If $\mathbf{P}(X = \mathrm{VaR}_\alpha(X)) > 0$, it follows that for $\omega \in \{X = \mathrm{VaR}_\alpha(X)\}$

$$1_{X,\mathrm{VaR}_\alpha(X),\alpha}(\omega) = \frac{\mathbf{P}(X \leq \mathrm{VaR}_\alpha(X)) - \alpha}{\mathbf{P}(X = \mathrm{VaR}_\alpha(X))} \in [0, 1].$$

(ii) We first consider the case $\mathbf{P}(X = \mathrm{VaR}_\alpha(X)) = 0$. Then Lemma 2.4 implies

$$\begin{aligned}
\mathrm{E}(1_{X,\mathrm{VaR}_\alpha(X),\alpha}) &= \mathrm{E}(1_{\{X > \mathrm{VaR}_\alpha(X)\}}) \\
&= \mathbf{P}\big(\{X > \mathrm{VaR}_\alpha(X)\}\big) \\
&= 1 - \mathbf{P}\big(\{X \leq \mathrm{VaR}_\alpha(X)\}\big) = 1 - \alpha.
\end{aligned}$$

For the case $\mathbf{P}(X = \mathrm{VaR}_\alpha(X)) > 0$ we obtain

$$\begin{aligned}
\mathrm{E}(1_{X,\mathrm{VaR}_\alpha(X),\alpha}) &= \mathrm{E}\bigg(1_{\{X > \mathrm{VaR}_\alpha(X)\}} + \frac{\mathbf{P}(X \leq \mathrm{VaR}_\alpha(X)) - \alpha}{\mathbf{P}(X = \mathrm{VaR}_\alpha(X))} 1_{\{X = \mathrm{VaR}_\alpha(X)\}}\bigg) \\
&= \mathbf{P}\big(X > \mathrm{VaR}_\alpha(X)\big) + \frac{\mathbf{P}(X \leq \mathrm{VaR}_\alpha(X)) - \alpha}{\mathbf{P}(X = \mathrm{VaR}_\alpha(X))} \mathbf{P}\big(X = \mathrm{VaR}_\alpha(X)\big) \\
&= \mathbf{P}\big(X > \mathrm{VaR}_\alpha(X)\big) + \mathbf{P}\big(X \leq \mathrm{VaR}_\alpha(X)\big) - \alpha = 1 - \alpha.
\end{aligned}$$

(iii) Let U be a random variable with $\mathbf{P}(U \leq u) = u$ for all $u \in]0, 1[$. Since $u \mapsto \mathrm{VaR}_u(X)$ is monotone increasing, we have

$$\{U \geq \alpha\} \subseteq \big\{\mathrm{VaR}_{U(\cdot)}(X) \geq \mathrm{VaR}_\alpha(X)\big\}.$$

If $U(\omega) < \alpha$ and $\mathrm{VaR}_{U(\omega)}(X) \geq \mathrm{VaR}_\alpha(X)$, then we must have (likewise because of monotony) $\mathrm{VaR}_{U(\omega)} = \mathrm{VaR}_\alpha(X)$. Thus we also obtain the relationship

$$\{U < \alpha\} \cap \big\{\mathrm{VaR}_{U(\cdot)}(X) \geq \mathrm{VaR}_\alpha(X)\big\} \subseteq \big\{\mathrm{VaR}_{U(\cdot)}(X) = \mathrm{VaR}_\alpha(X)\big\}.$$

Overall we have derived

$$\big\{\mathrm{VaR}_{U(\cdot)}(X) \geq \mathrm{VaR}_\alpha(X)\big\} = \{U \geq \alpha\} \cup \Big(\big\{\mathrm{VaR}_{U(\cdot)}(X) \geq \mathrm{VaR}_\alpha(X)\big\} \cap \{U < \alpha\}\Big)$$

where $\mathrm{VaR}_{U(\omega)}(X) = \mathrm{VaR}_\alpha(X)$ for all $\omega \in \{\mathrm{VaR}_{U(\cdot)}(X) \geq \mathrm{VaR}_\alpha(X)\} \cap \{U < \alpha\}$. From this and Lemma 2.3 follows

$$\begin{aligned}
\int_\alpha^1 \mathrm{VaR}_u(X)\,\mathrm{d}u &= \mathrm{E}\big(\mathrm{VaR}_{U(\cdot)}(X)1_{\{U\geq\alpha\}}\big)\\
&= \mathrm{E}\big(\mathrm{VaR}_{U(\cdot)}(X)(1_{\{\mathrm{VaR}_{U(\cdot)}(X)\geq \mathrm{VaR}_\alpha(X)\}}\\
&\quad - 1_{\{\mathrm{VaR}_{U(\cdot)}(X)\geq \mathrm{VaR}_\alpha(X)\}\cap\{U<\alpha\}})\big)\\
&= \mathrm{E}\big(\mathrm{VaR}_{U(\cdot)}(X)1_{\{\mathrm{VaR}_{U(\cdot)}(X)\geq \mathrm{VaR}_\alpha(X)\}}\big)\\
&\quad - \mathrm{VaR}_\alpha(X)\,\mathrm{E}(1_{\{\mathrm{VaR}_{U(\cdot)}(X)\geq \mathrm{VaR}_\alpha(X)\}\cap\{U<\alpha\}})\\
&= \mathrm{E}(X1_{\{X\geq \mathrm{VaR}_\alpha(X)\}}) - \mathrm{VaR}_\alpha(X)\,\mathrm{E}(1_{\{\mathrm{VaR}_{U(\cdot)}(X)\geq \mathrm{VaR}_\alpha(X)\}\setminus\{U\geq\alpha\}})\\
&= \mathrm{E}(X1_{\{X> \mathrm{VaR}_\alpha(X)\}}) + \mathrm{E}(X1_{\{X= \mathrm{VaR}_\alpha(X)\}})\\
&\quad - \mathrm{VaR}_\alpha(X)\,\mathrm{E}(1_{\{\mathrm{VaR}_{U(\cdot)}(X)\geq \mathrm{VaR}_\alpha(X)\}}) + \mathrm{VaR}_\alpha(X)\,\mathrm{E}(1_{\{U\geq\alpha\}})\\
&= \mathrm{E}(X1_{\{X> \mathrm{VaR}_\alpha(X)\}})\\
&\quad + \mathrm{VaR}_\alpha(X)\big(\mathbf{P}\big(X = \mathrm{VaR}_\alpha(X)\big) - \mathbf{P}\big(X \geq \mathrm{VaR}_\alpha(X)\big) + 1 - \alpha\big)\\
&= \mathrm{E}(X1_{\{X> \mathrm{VaR}_\alpha(X)\}}) + \mathrm{VaR}_\alpha(X)\big(\mathbf{P}\big(X \leq \mathrm{VaR}_\alpha(X)\big) - \alpha\big)\\
&= \mathrm{E}(X1_{X,\mathrm{VaR}_\alpha(X),\alpha}),
\end{aligned}$$

where in the last step we used that Lemma 2.4, in the special case

$$\mathbf{P}\big(X = \mathrm{VaR}_\alpha(X)\big) = 0,$$

implies the equation $\mathbf{P}(X \leq \mathrm{VaR}_\alpha(X)) - \alpha = 0$. □

Proposition 2.2 *Assume* $\alpha \in [0, 1[$. *For*

$$\lambda_\alpha = \frac{1 - \mathbf{P}(X \leq \mathrm{VaR}_\alpha(X))}{1-\alpha}$$

there holds $\lambda_\alpha \in [0, 1]$ *and*

$$\mathrm{ES}_\alpha(X) = \lambda_\alpha\,\mathrm{TailVaR}_\alpha(X) + (1-\lambda_\alpha)\,\mathrm{VaR}_\alpha(X).$$

In particular, the tail value at risk and expected shortfall coincide for continuous distributions.

Proof $\lambda_\alpha \in [0, 1]$ follows directly from Lemma 2.4. We calculate

$$\begin{aligned}(1-\alpha)\,\mathrm{ES}_\alpha(X) &= \mathrm{E}(X 1_{X,\mathrm{VaR}_\alpha(X),\alpha}) \\ &= \mathrm{E}(X 1_{\{X>\mathrm{VaR}_\alpha(X)\}}) + \mathrm{VaR}_\alpha(X)\big(\mathbf{P}\big(X \le \mathrm{VaR}_\alpha(X)\big) - \alpha\big) \\ &= \mathbf{P}\big(X > \mathrm{VaR}_\alpha(X)\big)\,\mathrm{TailVaR}_\alpha(X) \\ &\quad + \mathrm{VaR}_\alpha(X)\big(1-\alpha-\big(1-\mathbf{P}\big(X \le \mathrm{VaR}_\alpha(X)\big)\big)\big) \\ &= (1-\alpha)\lambda_\alpha\,\mathrm{TailVaR}_\alpha(X) - (1-\alpha)\,\mathrm{VaR}_\alpha(X)(1-\lambda_\alpha).\end{aligned}$$

If X is continuous, then by Lemma 2.4, $\lambda_\alpha = 1$, so that $\mathrm{ES}_\alpha(X) = \mathrm{TailVaR}_\alpha(X)$ follows. □

In general, the expected shortfall has better mathematical properties than the tail value at risk (see Sect. 2.3). The following presentation of the expected shortfall serves as motivation in Sect. 2.4.4. Additionally it allows a simple proof of the important approximation result given in Proposition 2.4.

Definition 2.5 Let $\mathbf{P}, \mathbf{Q}$ be measures on the σ-algebra $\mathscr{A}$. Then $\mathbf{Q}$ is *absolutely continuous* with respect to $\mathbf{P}$ if $\mathbf{P}(A) = 0$ implies $\mathbf{Q}(A) = 0$ for all $A \in \mathscr{A}$. In this case we write $\mathbf{Q} \ll \mathbf{P}$.

Proposition 2.3 *Assume* $\mathscr{M}(\Omega, \mathbb{R}) \subseteq L^1(\Omega, \mathbb{R})$ *and*

$$\mathscr{W}_\alpha = \left\{\mathbf{Q}\colon \mathbf{Q} \text{ is a probability measure with } \mathbf{Q} \ll \mathbf{P} \text{ and } \frac{\mathrm{d}\mathbf{Q}}{\mathrm{d}\mathbf{P}} \le \frac{1}{1-\alpha}\right\}.$$

Then for $X \in \mathscr{M}(\Omega, \mathbb{R})$ *there holds*

$$\mathrm{ES}_\alpha(X) = \sup_{\mathbf{Q} \in \mathscr{W}_\alpha} \big\{\mathrm{E}_\mathbf{Q}(X)\big\}.$$

Proof X is integrable with respect to $\mathbf{Q}$, because X is integrable with respect to $\mathbf{P}$, $\mathbf{Q} \ll \mathbf{P}$, and $\frac{\mathrm{d}\mathbf{Q}}{\mathrm{d}\mathbf{P}}$ is bounded. The special choice of $\mathbf{Q}$ defined by

$$\frac{\mathrm{d}\mathbf{Q}}{\mathrm{d}\mathbf{P}} = \frac{1}{1-\alpha} 1_{X,\mathrm{VaR}_\alpha(X),\alpha}$$

(see Lemma 2.5) satisfies both the conditions $\mathbf{Q} \ll \mathbf{P}$ and $\frac{\mathrm{d}\mathbf{Q}}{\mathrm{d}\mathbf{P}} \le (1-\alpha)^{-1}$. Since

$$\mathrm{ES}_\alpha(X) = \frac{1}{1-\alpha}\mathrm{E}(X\, 1_{X,\mathrm{VaR}_\alpha(X),\alpha}) = \mathrm{E}_\mathbf{Q}(X)$$

holds (Lemma 2.5(iii)), there follows

$$\mathrm{ES}_\alpha(X) \le \sup_{\mathbf{R} \in \mathscr{W}_\alpha} \big\{\mathrm{E}_\mathbf{R}(X)\big\}.$$

Now let $\mathbf{R}$ be another probability measure that satisfies the requirements $\mathbf{R} \ll \mathbf{P}$ and $\frac{d\mathbf{R}}{d\mathbf{P}} \leq (1-\alpha)^{-1}$. We must show $\mathrm{E}_{\mathbf{R}}(X) \leq \mathrm{E}_{\mathbf{Q}}(X)$. The set

$$A = \{\omega\colon 1_{X,\mathrm{VaR}_\alpha(X),\alpha}(\omega) > 0\}$$

satisfies $\mathrm{E}_{\mathbf{Q}}(1_A) = 1$. By the construction of $1_{X,\mathrm{VaR}_\alpha(X),\alpha}$ we also have $X(\omega) \leq \inf_{\tilde{\omega}\in A} X(\tilde{\omega})$ for all $\omega \in \Omega \setminus A$. From this follows the inequality

$$\begin{aligned}\mathrm{E}_{\mathbf{R}}(X) &= \mathrm{E}_{\mathbf{P}}\left(\frac{d\mathbf{R}}{d\mathbf{P}} X\, 1_A\right) + \mathrm{E}_{\mathbf{P}}\left(\frac{d\mathbf{R}}{d\mathbf{P}} X\, 1_{\Omega\setminus A}\right) \\ &\leq \mathrm{E}_{\mathbf{P}}\left(\frac{d\mathbf{R}}{d\mathbf{P}} X\, 1_A\right) + \inf_{\tilde{\omega}\in A} X(\tilde{\omega})\mathbf{R}(\Omega\setminus A).\end{aligned}$$

From

$$\mathrm{E}_{\mathbf{P}}\left(\frac{d\mathbf{Q}}{d\mathbf{P}} 1_A\right) = \mathrm{E}_{\mathbf{P}}\left(\frac{d\mathbf{Q}}{d\mathbf{P}}\right) = \mathrm{E}_{\mathbf{Q}}(1) = 1$$

follows

$$\mathrm{E}_{\mathbf{P}}\left(\left(\frac{d\mathbf{Q}}{d\mathbf{P}} - \frac{d\mathbf{R}}{d\mathbf{P}}\right) 1_A\right) = 1 - \mathbf{R}(A) = \mathbf{R}(\Omega\setminus A).$$

Since for all $\omega \in \{X > \inf_{\tilde{\omega}\in A} X(\tilde{\omega})\} \subseteq A$ it holds that

$$\frac{d\mathbf{Q}}{d\mathbf{P}} = \frac{1}{1-\alpha} \geq \frac{d\mathbf{R}}{d\mathbf{P}},$$

on this set we have

$$X\left(\frac{d\mathbf{Q}}{d\mathbf{P}} - \frac{d\mathbf{R}}{d\mathbf{P}}\right) \geq \inf_{\tilde{\omega}\in A} X(\tilde{\omega})\left(\frac{d\mathbf{Q}}{d\mathbf{P}} - \frac{d\mathbf{R}}{d\mathbf{P}}\right).$$

This inequality is fulfilled trivially on $\{X = \inf_{\tilde{\omega}\in A} X(\tilde{\omega})\}$, so that because $A \subseteq \{X \geq \inf_{\tilde{\omega}\in A} X(\tilde{\omega})\}$ it holds on A. We thus deduce

$$\begin{aligned}\mathrm{E}_{\mathbf{R}}(X) &\leq \mathrm{E}_{\mathbf{P}}\left(X\frac{d\mathbf{R}}{d\mathbf{P}} 1_A\right) + \inf_{\tilde{\omega}\in A} X(\tilde{\omega})\,\mathrm{E}_{\mathbf{P}}\left(\left(\frac{d\mathbf{Q}}{d\mathbf{P}} - \frac{d\mathbf{R}}{d\mathbf{P}}\right) 1_A\right) \\ &\leq \mathrm{E}_{\mathbf{P}}\left(X\frac{d\mathbf{R}}{d\mathbf{P}} 1_A\right) + \mathrm{E}_{\mathbf{P}}\left(X\left(\frac{d\mathbf{Q}}{d\mathbf{P}} - \frac{d\mathbf{R}}{d\mathbf{P}}\right) 1_A\right) \\ &= \mathrm{E}_{\mathbf{P}}\left(X\frac{d\mathbf{Q}}{d\mathbf{P}} 1_A\right) = \mathrm{E}_{\mathbf{Q}}(X).\end{aligned}$$

□

Proposition 2.4 *Assume Y an integrable positive function and $\{X_k\}_{k\in\mathbb{N}}$ a sequence of random variables with $|X_k| \leq Y$ almost surely, that converge pointwise almost surely to the random variable X. Then it holds that* $\mathrm{ES}_\alpha(X_n) \to \mathrm{ES}_\alpha(X)$.

Proof Let $\varepsilon > 0$ and $\mathbf{Q} \in \mathscr{W}_\alpha$ with $\mathrm{E}_\mathbf{Q}(X) \geq \mathrm{ES}_\alpha(X) - \varepsilon$. Since for each $\mathbf{R} \in \mathscr{W}_\alpha$ the inequality $0 \leq \frac{\mathrm{d}\mathbf{R}}{\mathrm{d}\mathbf{P}} \leq \frac{1}{1-\alpha}$ holds, the sequence $\{\frac{\mathrm{d}\mathbf{Q}}{\mathrm{d}\mathbf{P}} X_k\}_{k \in \mathbb{N}}$ is dominated by the integrable random variables $\frac{1}{1-\alpha} Y$. In addition, $\frac{\mathrm{d}\mathbf{Q}}{\mathrm{d}\mathbf{P}} X_k$ converges almost everywhere to $\frac{\mathrm{d}\mathbf{Q}}{\mathrm{d}\mathbf{P}} X$. Lebesgue's Theorem thus implies $\mathrm{E}_\mathbf{Q}(X_k) \to \mathrm{E}_\mathbf{Q}(X)$. Since $\varepsilon > 0$ was arbitrary, according to Proposition 2.3 this implies $\liminf_{k\to\infty} \mathrm{ES}_\alpha(X_k) \geq \mathrm{ES}_\alpha(X)$.

There exists a subsequence $\{X_{k_j}\}_{j \in \mathbb{N}}$ with

$$\lim_{j\to\infty} \mathrm{ES}_\alpha(X_{k_j}) = \limsup_{k\to\infty} \mathrm{ES}_\alpha(X_k).$$

Assume $\mathbf{Q}_{k_j} \in \mathscr{W}_\alpha$ with

$$\left|\mathrm{ES}_\alpha(X_{k_j}) - \mathrm{E}_{\mathbf{Q}_{k_j}}(X_{k_j})\right| \leq \frac{1}{j}.$$

Since for each j the Radon-Nikodym derivative $\frac{\mathrm{d}\mathbf{Q}_{k_j}}{\mathrm{d}\mathbf{P}}$ is measurable and we have $0 \leq \frac{\mathrm{d}\mathbf{Q}_{k_j}}{\mathrm{d}\mathbf{P}} \leq \frac{1}{1-\alpha}$, the function $f = \limsup_{j\to\infty} \frac{\mathrm{d}\mathbf{Q}_{k_j}}{\mathrm{d}\mathbf{P}}$ is also measurable with $0 \leq f \leq \frac{1}{1-\alpha}$. The measure defined by $\frac{\mathrm{d}\tilde{\mathbf{Q}}}{\mathrm{d}\mathbf{P}} = f$ is clearly in $\mathscr{W}_\alpha$, so that we have $\mathrm{ES}_\alpha(X) \geq \mathrm{E}_{\tilde{\mathbf{Q}}}(X)$. Since the X_{n_k} converge almost everywhere to X, there holds $\limsup_{j\to\infty} \frac{\mathrm{d}\mathbf{Q}_{k_j}}{\mathrm{d}\mathbf{P}} X_{k_j} = fX$. Because $|\frac{\mathrm{d}\mathbf{Q}_{n_k}}{\mathrm{d}\mathbf{P}} X_{n_k}| \leq \frac{1}{1-\alpha} Y$ we can apply Fatou's Lemma and obtain

$$\begin{aligned}
\mathrm{E}_{\tilde{\mathbf{Q}}}(X) &= \mathrm{E}_\mathbf{P}(fX) \\
&= \mathrm{E}_\mathbf{P}\left(\limsup_{j\to\infty} \frac{\mathrm{d}\mathbf{Q}_{k_j}}{\mathrm{d}\mathbf{P}} X_{k_j}\right) \\
&\geq \limsup_{j\to\infty}\left(\mathrm{E}_\mathbf{P}\left(\frac{\mathrm{d}\mathbf{Q}_{k_j}}{\mathrm{d}\mathbf{P}} X_{k_j}\right)\right) \\
&\geq \limsup_{j\to\infty}\left(\mathrm{ES}_\alpha(X_{k_j}) - \frac{1}{j}\right) \\
&= \limsup_{k\to\infty}\left(\mathrm{ES}_\alpha(X_k)\right).
\end{aligned}$$

Thus it also holds that $\mathrm{ES}_\alpha(X) \geq \limsup_{k\to\infty} \mathrm{ES}_\alpha(X_k)$. □

Proposition 2.4 suggests preferring the expected shortfall over the tail value at risk. This is because for large enough n it is not possible to distinguish X_n and X. Therefore the values of the corresponding risk measures should also be practically indistinguishable. This not the case for the tail value at risk, but Proposition 2.4 shows the expected shortfall does have this property which is needed for the interpretation.

Lemma 2.6 *Let $X\colon \Omega \to \mathbb{R}$ be a normally distributed random variable with mean m and standard deviation s. Assume $f\colon X(\Omega) \to \mathbb{R}$ is a strongly monotone continuous map. If $\Phi_{0,1}$ denotes the distribution function and $\varphi_{0,1} = \frac{\mathrm{d}}{\mathrm{d}x}\Phi_{0,1}$ the density of the standard normal distribution, then there holds*

$$\mathrm{ES}_\alpha(f \circ X) = \int_{{\Phi_{0,1}}^{-1}(\alpha)}^{\infty} f(m + sx)\varphi_{0,1}(x)\,\mathrm{d}x = \mathrm{TailVaR}_\alpha(f \circ X).$$

Proof From Proposition 2.1 it follows that

$$\mathrm{ES}_\alpha(f \circ X) = \frac{1}{1-\alpha}\int_\alpha^1 \mathrm{VaR}_p(f \circ X)\,\mathrm{d}p = \frac{1}{1-\alpha}\int_\alpha^1 f\big(m + s{\Phi_{0,1}}^{-1}(p)\big)\,\mathrm{d}p.$$

By making the substitution $p = \Phi_{0,1}(x)$ we obtain $\mathrm{d}p = \varphi_{0,1}(x)\,\mathrm{d}x$ and thus

$$\mathrm{ES}_\alpha(f \circ X) = \frac{1}{1-\alpha}\int_{{\Phi_{0,1}}^{-1}(\alpha)}^{\infty} f(m + sx)\varphi_{0,1}(x)\,\mathrm{d}x.$$

The distribution function is continuous so we have $\mathrm{ES}_\alpha(f \circ X) = \mathrm{TailVaR}_\alpha(f \circ X)$. □

In the two important special cases of normally distributed and log-normally distributed random variables, the integral above can be evaluated explicitly.

Proposition 2.5 *Let $X\colon \Omega \to \mathbb{R}$ be a normally distributed random variable with mean μ and standard deviation σ. If $\Phi_{0,1}$ denotes the distribution function and $\varphi_{0,1} = \frac{\mathrm{d}}{\mathrm{d}x}\Phi_{0,1}$ the density of the standard normal distribution, then*

$$\mathrm{ES}_\alpha(X) = \mu + \sigma\frac{\varphi_{0,1}({\Phi_{0,1}}^{-1}(\alpha))}{1-\alpha} = \mathrm{TailVaR}_\alpha(X).$$

Proof In this case we have in Lemma 2.6 that $f(x) = x$, so that the integral simplifies to

$$\mathrm{ES}_\alpha(X) = \mu + \frac{\sigma}{1-\alpha}\int_{{\Phi_{0,1}}^{-1}(\alpha)}^{\infty} x\varphi_{0,1}(x)\,\mathrm{d}x.$$

Using the fact $\varphi_{0,1}'(x) = -x\varphi_{0,1}(x)$ we obtain

$$\mathrm{ES}_\alpha(X) = \mu - \frac{\sigma}{1-\alpha}\big[\varphi_{0,1}(p)\big]_{{\Phi_{0,1}}^{-1}(\alpha)}^{\infty} = \mu + \frac{\sigma}{1-\alpha}\varphi_{0,1}\big({\Phi_{0,1}}^{-1}(\alpha)\big).$$ □

Proposition 2.6 *Let $X\colon \Omega \to \mathbb{R}$ be a log-normally distributed random variable, i.e., $\ln X \sim N(m, s^2)$. If $\Phi_{0,1}$ denotes the distribution function and $\varphi_{0,1} = \frac{\mathrm{d}}{\mathrm{d}x}\Phi_{0,1}$ the density of the standard normal distribution, then*

$$\mathrm{ES}_\alpha(X) = \frac{\exp(m + \frac{s^2}{2})}{1-\alpha}\Phi_{0,1}\big(s - {\Phi_{0,1}}^{-1}(\alpha)\big).$$

Proof In this case we have in Lemma 2.6 that $f(x) = \exp(x)$. Thus the integral simplifies to

$$\begin{aligned}
\mathrm{ES}_\alpha(X) &= \frac{1}{1-\alpha}\frac{1}{\sqrt{2\pi}}\int_{\Phi_{0,1}{}^{-1}(\alpha)}^{\infty} \exp(m+sx)\exp\left(-\frac{1}{2}x^2\right)\mathrm{d}x \\
&= \frac{1}{1-\alpha}\frac{1}{\sqrt{2\pi}}\int_{\Phi_{0,1}{}^{-1}(\alpha)}^{\infty} \exp\left(m+\frac{s^2}{2}-\frac{1}{2}(x-s)^2\right)\mathrm{d}x \\
&= \frac{\exp(m+\frac{s^2}{2})}{1-\alpha}\frac{1}{\sqrt{2\pi}}\int_{\Phi_{0,1}{}^{-1}(\alpha)-s}^{\infty} \exp\left(-\frac{1}{2}y^2\right)\mathrm{d}y \\
&= \frac{\exp(m+\frac{s^2}{2})}{1-\alpha}\left(1-\Phi_{0,1}\left(\Phi_{0,1}{}^{-1}(\alpha)-s\right)\right) \\
&= \frac{\exp(m+\frac{s^2}{2})}{1-\alpha}\Phi_{0,1}\left(s-\Phi_{0,1}{}^{-1}(\alpha)\right).
\end{aligned}$$

In this calculation we used, in the last equality, the symmetry of the standard normal distribution. □

2.2.4 Spectral Measures

The expected shortfall can be directly generalized to take into account individual risk aversion. Instead of averaging over all $\mathrm{VaR}_z(X)$ for $z \geq \alpha$ with a uniform weight, one can employ a more general weighting function ϕ.

Definition 2.6 Let $(A, \mathscr{A}, \mu)$ be a probability space with σ-Algebra $\mathscr{A}$ and probability measure μ. Then an integrable map $\phi\colon A \to \mathbb{R}$ is called a *weight function*, if ϕ has the following properties:

(i) $\phi(\alpha) \geq 0$ for almost every $\alpha \in A$,
(ii) $\int_A \phi(\alpha)\,\mathrm{d}\mu(\alpha) = 1$.

Definition 2.7 Let $\phi \in L^1([0, 1])$ be a weight function. The risk measure

$$M_\phi(X) = \int_0^1 \mathrm{VaR}_p(X)\phi(p)\,\mathrm{d}p$$

is called the *spectral measure* of ϕ.

With a spectral measure the risk measure is also weighted by its dependence on the unlikeliness of the occurrence of a loss. The concept of a spectral measure thus permits the representation of an individual profile of risk aversion. Obviously ES_α is an example of a spectral measure. The measure VaR can be thought of as a limit value of spectral measures, since we have $\mathrm{VaR}_\alpha(X) = \int_0^1 \mathrm{VaR}_p(X)\delta_\alpha(p)\,\mathrm{d}p$, where δ_α denotes the Dirac distribution.

2.3 Choosing a Good Risk Measure

2.3.1 Risk Measures and the Intuition of Risk

An important requirement for a good risk measure is that it describe as well as possible the risk intuition of the user. A risk measure that the user may think he understands straight away does not necessarily fulfill this requirement. We would like to illustrate this point more precisely. The following set of axioms, due to Artzner et al. [1], describes properties that correspond to our intuitive notion of risk.

Definition 2.8 A risk measure ρ is called *coherent*, if it has the following properties:

Translation invariance: $\rho(X + \alpha) = \rho(X) + \alpha$ for every $X \in \mathcal{M}(\Omega, \mathbb{R})$ and any constant α.
Positive Homogeneity: $\rho(\alpha X) = \alpha\rho(X)$ for every $X \in \mathcal{M}(\Omega, \mathbb{R})$ and any positive constant α.
Monotony: $X_1 \geq X_2$ almost everywhere $\Rightarrow \rho(X_1) \geq \rho(X_2)$ for any $X_1, X_2 \in \mathcal{M}(\Omega, \mathbb{R})$.[2]
Subadditivity: $\rho(X_1 + X_2) \leq \rho(X_1) + \rho(X_2)$ for any $X_1, X_2 \in \mathcal{M}(\Omega, \mathbb{R})$.

To see how well these axioms really express our intuition of risk, we should consider what each of these four conditions says.

Translation invariance says that certain losses must be completely covered by capital but do not influence the remaining risk: A loss of which one is certain is not a risk because it is entirely predictable. From translation invariance there follows in addition that $\rho(X - \rho(X)) = 0$. The risk capital $\rho(X)$ is thus exactly the amount of money that must be retained so as to absorb the risk at the safety level defined by ρ. In this sense risk measures that satisfy translation invariance are *acceptable* [1].

Positive homogeneity is an invariance under scaling: It is inessential whether one measures risks in cents or euros. If positive homogeneity does not hold then the arbitrarily chosen unit of currency has an influence on the amount of capital, which naturally should not be true.

Monotony means that a portfolio that shows higher losses than another portfolio in all possible situations, must mean more capital is at risk. As an example consider two identical portfolios, where however one of the portfolios features a retroactively paid discount for the premiums depending on the losses.

---

[2] In the original article of Artzner et.al. [1] they refer to gain $Y = -X$, thus monotony is defined differently there.

Subadditivity says that there are diversification effects resulting from the combination of risky portfolios. Subadditivity is especially intuitive for an insurer since the business model of insurance rests on the effect of diversification.[3]

Remark 2.3 Sometimes both positive homogeneity and subadditivity are viewed critically.

- It is tempting to interpret homogeneity in the sense that multiplying the insured sums in a portfolio by a factor brings with it a corresponding multiplication of the risk. This is plausible for small portfolios. However for large portfolios there are additional liquidity risks as large losses incur large payments. This would contradict positive homogeneity.

  In this argument there is the implicit assumption that scaling the insured sums of all contracts by a factor λ would also scale the loss function X by the same factor. However, this is not the case as the loss distribution of the scaled portfolio be directly affected by the additional liquidity risk and thus differ from λX.
- Regarding subadditivity consider the merger of two corporations with loss distributions X_1 and X_2. The merged corporation may have a risk capital larger than $\rho(X_1) + \rho(X_2)$ as there may be internal power struggles which increase the risk.

  Again, the solution to this conundrum is that the loss distribution X of the combined corporation satisfies $X > X_1 + X_2$, because X also has to account for these power struggles.

While we have seen that these criticisms of positive homogeneity and subadditivity are rather a criticism of a naive combination of loss distributions, risk measures that are just translation invariant, monotone and convex have been proposed as an alternative to coherent risk measures: *Convex risk measures* are defined by the requirement that for each $\alpha \in [0, 1]$ and for any two loss distributions $X_1, X_2 \in \mathscr{M}(\Omega, \mathbb{R})$ the inequality

$$\rho\big(\alpha X_1 + (1-\alpha)X_2\big) \leq \alpha\rho(X_1) + (1-\alpha)\rho(X_2)$$

holds. It is clear that convexity is a weaker condition, and follows from subadditivity and positive homogeneity.

Coherent risk measures satisfy intuitive expectations in many situations. While we do not have realistic examples, there could however be areas where what is expected of a risk measure contradicts coherence. This has to be judged case-by-case. If a risk measure does not fulfill the requirements of coherence, then it has

[3]There is a subtle distinction between pooling and diversification, whereby it is argued that the insurance business depends predominantly on pooling. The distinction is based on the idea that the pooling effect can only be achieved with costs (brokers must find insurance clients) whereas diversification is in principle free (a diversified stock portfolio costs exactly the same as an undiversified one with the same market prices). In our context in which we are only concerned with risk effects this distinction is, however, secondary.

to be judged to what extent this is the result of the situation being described, and whether this property is desirable or negligible.

The following technical theorem allows the construction of new coherent risk measures on the basis of existing coherent risk measures. We shall use it later in the proof of Theorem 2.4, in which a descriptive construction of coherent measures will be given.

Theorem 2.1 *Let $(A, \mathscr{A}, \mu)$ be a probability space with σ-Algebra $\mathscr{A}$ and probability measure μ. Let $\{\rho_\alpha\}_{\alpha \in A}$ be a family of risk measures and $\mathscr{M}$ a vector space of real-valued random variables X, such that $\rho_\alpha(X)$ are μ-almost everywhere defined and μ-integrable. If all ρ_α are translation invariant, positively homogeneous, monotone resp. subadditive, then the risk measure $\rho\colon \mathscr{M} \to \mathbb{R}$, $X \mapsto \rho(X) = \int_A \rho_\alpha(X)\,\mathrm{d}\mu(\alpha)$ also has the corresponding property.*

Proof Consider $c \in \mathbb{R}$ and arbitrary random variables X and Y.

Translation invariance:

$$\begin{aligned}\rho(X+c) &= \int_A \rho_\alpha(X+c)\,\mathrm{d}\mu(\alpha) = \int_A \bigl(\rho_\alpha(X)+c\bigr)\,\mathrm{d}\mu(\alpha)\\ &= \int_A \rho_\alpha(X)\,\mathrm{d}\mu(\alpha) + c\int_A \mathrm{d}\mu(\alpha) = \rho(X)+c,\end{aligned}$$

since μ is a probability measure.

Positive homogeneity: For $c \geq 0$ we have

$$\rho(cX) = \int_A \rho_\alpha(cX)\,\mathrm{d}\mu(\alpha) = \int_A c\rho_\alpha(X)\,\mathrm{d}\mu(\alpha) = c\rho(X).$$

Monotony: Suppose $X \geq Y$ almost everywhere. Then it follows from $\rho_\alpha(X) \geq \rho_\alpha(Y)$ that

$$\rho(X) = \int_A \rho_\alpha(X)\,\mathrm{d}\mu(\alpha) \geq \int_A \rho_\alpha(Y)\,\mathrm{d}\mu(\alpha) = \rho(Y).$$

Subadditivity:

$$\rho(X+Y) = \int_A \rho_\alpha(X+Y)\,\mathrm{d}\mu(\alpha) \leq \int_A \bigl(\rho_\alpha(X)+\rho_\alpha(Y)\bigr)\,\mathrm{d}\mu(\alpha) = \rho(X)+\rho(Y).$$

□

In general the risk measure VaR_α, which perhaps does appear at first glance to be the most plausible model for risk measurement, does not satisfy the important axiom of subadditivity. The value at risk is therefore not coherent, and so does not describe our intuition of risk to a desirable extent.

Example 2.2 Let the discrete distribution X be given by

$$\begin{cases} \mathbf{P}(X=-1)=0.96 \\ \mathbf{P}(X=10)=0.04. \end{cases}$$

We can interpret $-X$ as the profit distribution for an insurance contract. The premium is set as 1. There is a loss with a probability of 4 %, and the payment in the event of a loss is always equal to 11. In this simple example costs and investment return are ignored. The business is profitable since $E(-X)=0.56$. We are interested in the risk measure $\mathrm{VaR}_{95\,\%}$. Since losses occur only with the probability $4\,\% < 1-95\,\%$, we have

$$\mathrm{VaR}_{95\,\%}(X)=-1.$$

With our low confidence level there is thus no positive risk.

Consider now a second distribution $Y \sim X$ that is independent of X. The total distribution $X+Y$ is then completely described by

$$\begin{cases} \mathbf{P}(X+Y=-2)=0.96^2=0.9216 \\ \mathbf{P}(X+Y=9)=2\times 0.96\times 0.04=0.0768 \\ \mathbf{P}(X+Y=20)=0.04^2=0.0016. \end{cases}$$

Obviously

$$\mathrm{VaR}_{95\,\%}(X+Y)=9>2(-1)=\mathrm{VaR}_{95\,\%}(X)+\mathrm{VaR}_{95\,\%}(Y).$$

If one were to use the value at risk as the risk measure for this example, one would have to deduce that diversification increases the risk instead of reducing it.

There are nonetheless special cases in which the value at risk is a coherent risk measure (see Theorem 2.2). First we need to do a little preparation.

Remark 2.4 We will recall some properties of Euclidean space here that we will need for the formulation and proof of the following Lemma 2.7. We consider $\mathbb{R}^n$ with a scalar product $\langle\,,\rangle\colon \mathbb{R}^n\times\mathbb{R}^n\to\mathbb{R}$. The pair $(\mathbb{R}^n,\langle\,,\rangle)$ is called *Euclidean space* and is the basis for elementary geometry. A linear map $O\colon \mathbb{R}^n\to\mathbb{R}^n$, $u\mapsto Ou$ is said to be *orthogonal* (or *an isometry*), if $\langle Ox, Oy\rangle=\langle x, y\rangle$ holds for all $x, y\in\mathbb{R}^n$. In particular, O is invertible. The *transposed map* $O^\top$ is defined by the property $\langle Ox, y\rangle=\langle x, O^\top y\rangle$ for all $x, y\in\mathbb{R}^n$. Then $O^\top O=\mathrm{id}_{\mathbb{R}^n}$, which follows from

$$\langle O^\top Ox, y\rangle=\langle Ox, Oy\rangle=\langle x, y\rangle \quad \forall x, y\in\mathbb{R}^n.$$

$O^\top$ is itself also an orthogonal map, since

$$\langle O^\top Ox, O^\top Oy\rangle=\langle x, y\rangle=\langle Ox, Oy\rangle$$

for all $x, y\in\mathbb{R}^n$, and O is invertible.

Lemma 2.7 *Let $X\colon \Omega \to \mathbb{R}^n$ be a random variable and $\phi_X\colon \mathbb{R}^n \to \mathbb{R}$, $u \mapsto E(\mathrm{e}^{\mathrm{i}\langle u,X\rangle})$ its characteristic function. The following assertions are equivalent:*

(i) *For each orthogonal linear map $O\colon \mathbb{R}^n \to \mathbb{R}^n$ one has $OX \sim X$.*
(ii) *There is a function $\psi_X\colon \mathbb{R}^+ \to \mathbb{R}$ with $\phi_X(u) = \psi_X(\|u\|^2)$.*
(iii) *For each $a \in \mathbb{R}^n$ we have $\langle a, X\rangle \sim \|a\|X_1$, where X_1 is the first component of the vector X.*

Proof "(i)⇒(ii)": For each orthogonal linear Map O and each $u \in \mathbb{R}^n$ we have

$$\phi_X(u) = \phi_{OX}(u) = \mathrm{E}\big(\mathrm{e}^{\mathrm{i}\langle u,OX\rangle}\big) = \mathrm{E}\big(\mathrm{e}^{\mathrm{i}\langle O^\top u,X\rangle}\big) = \phi_X\big(O^\top u\big)$$

The characteristic function $\phi_X(\cdot)$ is therefore invariant under orthogonal transformations and the property (ii) follows.

"(ii)⇒(iii)": Assume $a \in \mathbb{R}^n$. Then we get for each $t \in \mathbb{R}$

$$\phi_{\langle a,X\rangle}(t) = E\big(\mathrm{e}^{\mathrm{i}t\langle a,X\rangle}\big) = \mathrm{E}\big(\mathrm{e}^{\mathrm{i}\langle ta,X\rangle}\big) = \phi_X(ta) = \psi_X\big(t^2\|a\|^2\big).$$

On the other hand, we have

$$\phi_{\|a\|X_1}(t) = \mathrm{E}\big(\mathrm{e}^{\mathrm{i}t\|a\|X_1}\big) = \mathrm{E}\big(\mathrm{e}^{\mathrm{i}\langle t\|a\|\mathbf{e}_1,X\rangle}\big) = \phi_X\big(t\|a\|\mathbf{e}_1\big) = \psi_X\big(t^2\|a\|^2\big),$$

and the property (iii) follows from the uniqueness of the characteristic function.

"(iii)⇒(i)": By the uniqueness of the characteristic function it is enough to show that the characteristic function of X is invariant under orthogonal transformations O. We have

$$\begin{aligned}\phi_{OX}(u) &= \mathrm{E}\big(\mathrm{e}^{\mathrm{i}\langle u,OX\rangle}\big) = \mathrm{E}\big(\mathrm{e}^{\mathrm{i}\langle O^\top u,X\rangle}\big) = \phi_{\langle O^\top u,X\rangle}(1) = \phi_{\|O^\top u\|X_1}(1)\\ &= \phi_{\|u\|X_1}(1) = \phi_{\langle u,X\rangle}(1) = \mathrm{E}\big(\mathrm{e}^{\mathrm{i}\langle u,X\rangle}\big) = \phi_X(u).\end{aligned}$$

□

Lemma 2.8 *The risk measure* VaR_α *is translation invariant, positive homogeneous and monotone.*

Proof Assume $a \in \mathbb{R}$ and that X, Y are arbitrary random variables.

Translation invariance: Obviously we have $F_{X+a}(x) = \mathbf{P}(X + a \le x) = \mathbf{P}(X \le x - a) = F_X(x - a)$. It follows that

$$\begin{aligned}\mathrm{VaR}_\alpha(X + a) &= \inf\{x : F_{X+a}(x) \ge \alpha\} = \inf\{x : F_X(x - a) \ge \alpha\}\\ &= \inf\{x + a : F_X(x) \ge \alpha\} = a + \inf\{x : F_X(x) \ge \alpha\}\\ &= \mathrm{VaR}_\alpha(X) + a.\end{aligned}$$

Positive Homogeneity: For $a = 0$ positive homogeneity holds trivially. If $a > 0$, then $F_{aX}(x) = \mathbf{P}(aX \le x) = \mathbf{P}(X \le \frac{x}{a}) = F_X(\frac{x}{a})$. Thus it follows that

$$\mathrm{VaR}_\alpha(aX) = \inf\{x : F_{aX}(x) \ge \alpha\} = \inf\left\{x : F_X\left(\frac{x}{a}\right) \ge \alpha\right\}$$

$$= \inf\{ax : F_X(x) \geq \alpha\} = a \inf\{x : F_X(x) \geq \alpha\}$$
$$= a \operatorname{VaR}_\alpha(X).$$

Monotony: Suppose $X \geq Y$ almost everywhere. Then $F_X(x) = \mathbf{P}(X \leq x) \leq \mathbf{P}(Y \leq x) = F_Y(x)$ and so $\{x : F_X(x) \geq \alpha\} \subseteq \{x : F_Y(x) \geq \alpha\}$. We deduce that

$$\operatorname{VaR}_\alpha(X) = \inf\{x : F_X(x) \geq \alpha\} \geq \inf\{x : F_Y(x) \geq \alpha\} = \operatorname{VaR}_\alpha(Y). \qquad \square$$

Theorem 2.2 *When restricted to a vector space of normally distributed random variables the risk measure* $\operatorname{VaR}_\alpha$ *is coherent for any* $\alpha \in]\frac{1}{2}, 1[$.

Proof By Lemma 2.8 we only need to show subadditivity. Suppose $X, Y : \Omega \to \mathbb{R}$ are arbitrary normal random variables from the vector space. Because of the vector space property any linear combination of X and Y is normally distributed, so that the vector (X, Y) is multivariately normally distributed. Therefore there exist a two-dimensional random vector $Z = (Z_1, Z_2)$ which is standard normally distributed, a linear map $A : \mathbb{R}^2 \to \mathbb{R}^2$ and a vector $b = (b_1, b_2) \in \mathbb{R}^2$, so that $(X, Y)^\top = AZ + b$ holds. Since $\phi_Z(u) = \mathrm{e}^{-\|u\|^2/2}$ by Lemma 2.7 for each vector $a \in \mathbb{R}^2$ the relation $\langle a, Z\rangle \sim \|a\| Z_1$ holds. We have

$$X - b_1 = \langle A^\top \mathbf{e}_1, Z\rangle \sim \|A^\top \mathbf{e}_1\| Z_1,$$
$$Y - b_2 = \langle A^\top \mathbf{e}_2, Z\rangle \sim \|A^\top \mathbf{e}_2\| Z_1,$$
$$X + Y - b_1 - b_2 = \langle A^\top \mathbf{e}_1 + A^\top \mathbf{e}_2, Z\rangle \sim \|A^\top \mathbf{e}_1 + A^\top \mathbf{e}_2\| Z_1.$$

Thus by translation invariance and positive homogeneity of $\operatorname{VaR}_\alpha$ we have

$$\operatorname{VaR}_\alpha(X) = \|A^\top \mathbf{e}_1\| \operatorname{VaR}_\alpha(Z_1) + b_1,$$
$$\operatorname{VaR}_\alpha(Y) = \|A^\top \mathbf{e}_2\| \operatorname{VaR}_\alpha(Z_1) + b_2,$$
$$\operatorname{VaR}_\alpha(X + Y) = \|A^\top \mathbf{e}_1 + A^\top \mathbf{e}_2\| \operatorname{VaR}_\alpha(Z_1) + b_1 + b_2.$$

The subadditivity now follows from

$$\|A^\top \mathbf{e}_1 + A^\top \mathbf{e}_2\| \leq \|A^\top \mathbf{e}_1\| + \|A^\top \mathbf{e}_2\|$$

and $\operatorname{VaR}_\alpha(Z_1) \geq 0$ for $\alpha \geq \frac{1}{2}$, since Z_1 has a standard normal distribution. $\square$

Remark 2.5 A random variable that satisfies one of the equivalent conditions of Lemma 2.7 is called *spherical*. An affine transform of a spherical random variable is called *elliptical*. In the proof of Theorem 2.2 we only used the property of the normal distribution that multi-normal distributions are elliptic. The theorem can thus be generalized to distributions that can be written as linear combinations of components of elliptic distributions. For a precise formulation see [6, Theorem 6.8].

Theorem 2.3 *The expected shortfall* ES_α *is coherent.*

Proof Let $\mathrm{d}p$ be the Lebesgue measure. Then translation invariance, positive homogeneity and monotony follow directly from Theorem 2.1 and Lemma 2.8 with $\rho_p = \mathrm{VaR}_p$ and $(A, \mathscr{A}, \mu) = ([\alpha, 1], \mathscr{B}, \frac{1}{1-\alpha}\,\mathrm{d}p)$.

It remains to show subadditivity. For arbitrary random variables X, Y we obtain using Lemma 2.5(iii)

$$\begin{aligned}
&(1-\alpha)\big(\mathrm{ES}_\alpha(X) + \mathrm{ES}_\alpha(Y) - \mathrm{ES}_\alpha(X+Y)\big) \\
&\quad = \mathrm{E}\big(X 1_{X,\mathrm{VaR}_\alpha(X),\alpha} + Y 1_{Y,\mathrm{VaR}_\alpha(Y),\alpha} - (X+Y) 1_{(X+Y),\mathrm{VaR}_\alpha(X+Y),\alpha}\big) \\
&\quad = \mathrm{E}\big(X(1_{X,\mathrm{VaR}_\alpha(X),\alpha} - 1_{(X+Y),\mathrm{VaR}_\alpha(X+Y),\alpha})\big) \\
&\qquad + \mathrm{E}\big(Y(1_{Y,\mathrm{VaR}_\alpha(Y),\alpha} - 1_{(X+Y),\mathrm{VaR}_\alpha(X+Y),\alpha})\big).
\end{aligned}$$

We now consider the expression $\mathrm{E}(X(1_{X,\mathrm{VaR}_\alpha(X),\alpha} - 1_{(X+Y),\mathrm{VaR}_\alpha(X+Y),\alpha}))$. By the construction of $1_{X,x,\alpha}$ we have for $X(\omega) < x$ the equation $1_{X,x,\alpha}(\omega) = 0$ and for $X(\omega) > x$ the equation $1_{X,x,\alpha}(\omega) = 1$. Since by Lemma 2.5(i) the inequality

$$0 \le 1_{X+Y,\mathrm{VaR}_\alpha(X+Y),\alpha} \le 1$$

holds, we obtain

$$1_{X,\mathrm{VaR}_\alpha(X),\alpha} - 1_{(X+Y),\mathrm{VaR}_\alpha(X+Y),\alpha} \begin{cases} \le 0, & \text{if } X(\omega) < \mathrm{VaR}_\alpha(X) \\ \ge 0, & \text{if } X(\omega) > \mathrm{VaR}_\alpha(X). \end{cases}$$

From this we see in both cases (and trivially also for $X = \mathrm{VaR}_\alpha(X)$) the inequality

$$\begin{aligned}
&X(1_{X,\mathrm{VaR}_\alpha(X),\alpha} - 1_{(X+Y),\mathrm{VaR}_\alpha(X+Y),\alpha}) \\
&\quad \ge \mathrm{VaR}_\alpha(X)(1_{X,\mathrm{VaR}_\alpha(X),\alpha} - 1_{(X+Y),\mathrm{VaR}_\alpha(X+Y),\alpha}).
\end{aligned}$$

Lemma 2.5(ii) now implies

$$\begin{aligned}
&\mathrm{E}\big(X(1_{X,\mathrm{VaR}_\alpha(X),\alpha} - 1_{(X+Y),\mathrm{VaR}_\alpha(X+Y),\alpha})\big) \\
&\quad \ge \mathrm{VaR}_\alpha(X)\,\mathrm{E}\big((1_{X,\mathrm{VaR}_\alpha(X),\alpha} - 1_{(X+Y),\mathrm{VaR}_\alpha(X+Y),\alpha})\big) \\
&\quad = \mathrm{VaR}_\alpha(X)\big((1-\alpha) - (1-\alpha)\big) = 0.
\end{aligned}$$

The same argument also implies

$$\mathrm{E}\big(Y(1_{Y,\mathrm{VaR}_\alpha(Y),\alpha} - 1_{(X+Y),\mathrm{VaR}_\alpha(X+Y),\alpha})\big) \ge 0.$$

Therefore, summing up, we obtain

$$(1-\alpha)\big(\mathrm{ES}_\alpha(X) + \mathrm{ES}_\alpha(Y) - \mathrm{ES}_\alpha(X+Y)\big) \ge 0 + 0 = 0. \qquad \square$$

Theorem 2.4 *A spectral measure M_ϕ is coherent, if the weight function ϕ is (almost everywhere) monotone increasing.*

Proof Since ϕ is monotone increasing, we can define a measure on $([0,1],\mathscr{B})$ by $\phi(p) =: \nu([0,p])$. By Fubini's Theorem it follows that

$$\begin{aligned}
M_\phi(X) &= \int_0^1 \mathrm{VaR}_p(X)\phi(p)\,\mathrm{d}p = \int_0^1 \mathrm{VaR}_p(X)\Big(\int_0^p \mathrm{d}\nu(\alpha)\Big)\,\mathrm{d}p \\
&= \int_0^1 \Big(\int_0^1 1_{[0,p]}(\alpha)\,\mathrm{VaR}_p(X)\,\mathrm{d}\nu(\alpha)\Big)\,\mathrm{d}p \\
&= \int_0^1 \Big(\int_0^1 1_{[\alpha,1]}(p)\,\mathrm{VaR}_p(X)\,\mathrm{d}\nu(\alpha)\Big)\,\mathrm{d}p \\
&= \int_0^1 \Big(\int_0^1 1_{[\alpha,1]}(p)\,\mathrm{VaR}_p(X)\,\mathrm{d}p\Big)\,\mathrm{d}\nu(\alpha) = \int_0^1 \Big(\int_\alpha^1 \mathrm{VaR}_p(X)\,\mathrm{d}p\Big)\,\mathrm{d}\nu(\alpha) \\
&= \int_0^1 (1-\alpha)\,\mathrm{ES}_\alpha(X)\,\mathrm{d}\nu(\alpha),
\end{aligned}$$

where we have used the identity $1_{[0,p]}(\alpha) = 1_{[\alpha,1]}(p)$ for $\alpha, p \in [0,1]$. The assertion now follows from Theorem 2.1 with $\mathrm{d}\mu(\alpha) = (1-\alpha)\,\mathrm{d}\nu(\alpha)$, since

$$\begin{aligned}
\int_0^1 \mathrm{d}\mu(\alpha) &= \int_0^1 (1-\alpha)\,\mathrm{d}\nu(\alpha) = \int_0^1 \Big(\int_\alpha^1 \mathrm{d}p\Big)\,\mathrm{d}\nu(\alpha) \\
&= \int_0^1 \Big(\int_0^1 1_{[\alpha,1]}(p)\,\mathrm{d}p\Big)\,\mathrm{d}\nu(\alpha) = \int_0^1 \Big(\int_0^1 1_{[\alpha,1]}(p)\,\mathrm{d}\nu(\alpha)\Big)\,\mathrm{d}p \\
&= \int_0^1 \Big(\int_0^1 1_{[0,p]}(\alpha)\,\mathrm{d}\nu(\alpha)\Big)\,\mathrm{d}p = \int_0^1 \nu\big([0,p]\big)\,\mathrm{d}p = \int_0^1 \phi(p)\,\mathrm{d}p \\
&= 1.
\end{aligned}$$

□

A spectral measure is thus coherent exactly when the individual risk aversion puts higher weights on the higher losses.

2.3.2 Practical Considerations

Some risk measures such as VaR_α or $\mathrm{TailVaR}_\alpha$ are defined with a confidence level α. This confidence level permits a first intuitive impression of the desired safety level. It is necessary however to exercise a certain caution, since the safety level depends both on the risk measure and on the time horizon being considered. For example, we have seen above that a tail value at risk for a confidence level α always offers a greater safety level than a value at risk with the same confidence level α. In addition, given a confidence level α, it is clear that the safety level that can be attained increases with the time horizon under consideration.

A further important requirement is the practicability of a risk measure.

- If one knows the class of the distribution, then the problem of determining the risk reduces to estimating the parameters of the given distribution. But even if the distributions of the individual component risks are known, the aggregation to an overall distribution already brings considerable numerical problems, even in the simplest case where the random variables are independent. Therefore in practice one usually calculates the overall distribution using Monte Carlo simulations.
- Variance-reducing techniques can be employed to bring down the number of scenarios that are needed. Furthermore an approximate portfolio evaluation can reduce the numerical burden.
- If we assume that the risk distribution is determined by Monte Carlo simulations, then VaR_α and spectral measures can be calculated with similar effort. If the risk has to be more carefully studied then spectral measures have advantages, since they are defined through an integration and thus may be more stable. In Sect. 5.2.4 we will give more detail about this property using the example of a particularly intuitive allocation scheme for risk capital.

The result of the risk measure $\rho: \mathscr{M}(\Omega, \mathbb{R}) \to \mathbb{R}$ is itself no random variable, but like the expected value is a deterministic value. In Monte Carlo Simulations these deterministic quantities are approximated by an estimator, i.e., a random variable $R_k^{\rho,X}$ on the basis of k independent realizations of X. Here "approximation" means that for a given small bound $\varepsilon > 0$ and a given "meta-confidence level" $\tilde{\alpha}$ we have the inequality

$$\mathbf{P}\big(\big|\rho(X) - R_k^{\rho,X}\big| > \varepsilon\big) < 1 - \tilde{\alpha}. \tag{2.3}$$

The theoretical background is provided by the Weak Law of Large Numbers.

Example 2.3 Let $\rho = \mathrm{VaR}_\alpha$. In order to estimate $\mathrm{VaR}_\alpha(X)$ in a numerically stable way, we must choose a large enough number k of scenarios such that sufficiently many scenarios show a loss higher than $\mathrm{VaR}_\alpha(X)$. For example, to get more than 100 scenarios with higher losses we choose $k \in \mathbb{N}$ so large that $(1-\alpha)k > 100$. We shall use the notation

$$\mathrm{MAX}_m\big(\{a_1, \dots, a_k\}\big)$$

for the m-highest value in the set $\{a_1, \dots, a_k\}$. Now we can set

$$R_k^{\mathrm{VaR}_\alpha,X}(X_1, \dots, X_k) = \mathrm{MAX}_{\lfloor(1-\alpha)k+1\rfloor}\big(\{X_1, \dots, X_k\}\big),$$

where $\lfloor a \rfloor$ is the integer part of the real number a. For given $\varepsilon, \tilde{\alpha}$ we now choose k so large that the inequality (2.3) holds. That such a choice is possible follows intuitively from the definition of value at risk and the law of large numbers. In practice, one would not offer a proof but simply choose k so large that for successive Monte Carlo simulations the values of $R_k^{\rho,X}$ hardly differ from one another.

The number of simulations is often pragmatically determined by the available computing capacity and acceptable run times. This can lead to results which are actually not stable. In particular, if X is a heavy-tailed distribution (e.g., a Pareto distribution) then more than 100,000 simulations can easily be necessary to obtain stable results for $\mathrm{VaR}_{99.5\,\%}(X)$.

One can easily see that the estimate $R_k^{\mathrm{VaR}_\alpha,X}(X_1,\dots,X_k)$ coincides with the value at risk of the empirical distribution function F_k of the sample values; this is because

$$F_k\big(R_k^{\mathrm{VaR}_\alpha,X}\big)=1-\frac{\lfloor(1-\alpha)k\rfloor}{k}\in[\alpha,\alpha+1/k[.$$

By the Glivenko-Cantelli Theorem the empirical distribution functions F_k converge uniformly to the distribution F of X, so that here the value at risk of the empirical distribution can be used as an approximation to the value at risk of the theoretical distribution.

Example 2.4 Let $\rho=\mathrm{ES}_\alpha$. We set

$$R_k^{\mathrm{ES}_\alpha,X}(X_1,\dots,X_k)=\frac{\sum_{m=1}^{\lfloor(1-\alpha)k\rfloor}\mathrm{MAX}_m(\{X_1,\dots,X_k\})}{\lfloor(1-\alpha)k\rfloor}$$

and otherwise proceed as in the Example 2.3.

The theoretical background is provided by the following version of the Law of Large Numbers [9].

Theorem 2.5 *For a sequence $(X_k)_{k\in\mathbb{N}}$ of integrable i.i.d. random quantities on the probability space $(\Omega,\mathbf{P})$ the equation*

$$\lim_{k\to\infty}\frac{\sum_{m=1}^{\lfloor(1-\alpha)k\rfloor}\mathrm{MAX}_m(X_1,\dots,X_k)}{\lfloor(1-\alpha)k\rfloor}=\mathrm{ES}_\alpha(X_1)$$

holds almost surely, where $\lfloor\ \rfloor$ denotes the integer part.

Proof Let F be the distribution of X_1. Then

$$y\mapsto\mathrm{VaR}_y(X_1)=\inf\{x:F(x)\ge y\}$$

is integrable, since by Lemma 2.1

$$\int_0^1|\mathrm{VaR}_y(X_1)|\,\mathrm{d}y=\int_0^1|\mathrm{VaR}_y(X_1)|\,\mathrm{d}F\big(\mathrm{VaR}_y(X_1)\big)=\int_{-\infty}^{\infty}|x|\,\mathrm{d}F(x)<\infty.$$

We put $U_i:=F(X_i)$, $i=1,\dots,k$. Since $\mathbf{P}(\mathrm{VaR}_{U_i}(X_i)=X_i)=1$ by Lemma 2.2, the X_i are identically distributed and $t\mapsto\mathrm{VaR}_t(X)$ monotone increasing, we have

$$\mathrm{MAX}_m(X_1,\dots,X_k)=\mathrm{MAX}_m\big(\mathrm{VaR}_{F_{X_1}}(X_1),\dots,\mathrm{VaR}_{F_{X_k}}(X_k)\big)$$

$$= \mathrm{MAX}_m\big(\mathrm{VaR}_{F_{X_1}}(X_1), \ldots, \mathrm{VaR}_{F_{X_k}}(X_1)\big)$$
$$= \mathrm{VaR}_{\mathrm{MAX}_m(F(X_1),\ldots,F(X_k))}(X_1)$$
$$= \mathrm{VaR}_{\mathrm{MAX}_m(U_1,\ldots,U_k)}(X_1) \quad \text{a.s.}$$

Therefore it is enough to show

$$\lim_{k\to\infty} \frac{\sum_{m=1}^{\lfloor(1-\alpha)k\rfloor} \mathrm{VaR}_{\mathrm{MAX}_m(U_1,\ldots,U_k)}(X_1)}{\lfloor(1-\alpha)k\rfloor} = \frac{1}{1-\alpha}\int_\alpha^1 \mathrm{VaR}_y(X_1)\,\mathrm{d}y \qquad \text{a.s.}$$

We shall show more generally that for each integrable function $g\colon]0,1[\to \mathbb{R}$ the relationship

$$\lim_{k\to\infty} \frac{\sum_{m=1}^{\lfloor(1-\alpha)k\rfloor} g(\mathrm{MAX}_m(U_1,\ldots,U_k))}{\lfloor(1-\alpha)k\rfloor} = \frac{1}{1-\alpha}\int_\alpha^1 g(x)\,\mathrm{d}x \qquad \text{a.s.}$$

holds. For this we define piecewise constant maps

$$g_k\colon]0,1[\times\Omega \to \mathbb{R},$$

$k \in \mathbb{N}$, by

$$g_k(t) := g\big(\mathrm{MAX}_{\lfloor(1-t)k\rfloor+1}(U_1,\ldots,U_k)\big).$$

It follows that

$$\int_{(\lfloor\alpha k\rfloor+1)/k}^1 g_k(t)\,\mathrm{d}t = \sum_{m=1}^{\lfloor(1-\alpha)k\rfloor} g\big(\mathrm{MAX}_m(U_1,\ldots,U_k)\big).$$

With the notation

$$J_k(t) = \begin{cases} 0 & \text{for } 0 \le t \le \frac{\lfloor\alpha k\rfloor+1}{k}, \\ \frac{k}{\lfloor(1-\alpha)k\rfloor} & \text{for } \frac{\lfloor\alpha k\rfloor+1}{k} < t \le 1, \end{cases}$$

it is thus sufficient to show

$$\lim_{k\to\infty}\int_0^1 g_k(t)J_k(t)\,\mathrm{d}t = \frac{1}{1-\alpha}\int_\alpha^1 g(t)\,\mathrm{d}t \quad \text{a.s.} \tag{2.4}$$

Let λ be the Lebesgue measure on $]0,1[$. We next show that with probability 1 with respect to $(\Omega, \mathbf{P})$ we have

$$\lim_{k\to\infty} \lambda\big(\big\{t : \big|g_k(t) - g(t)\big| \ge \delta\big\}\big) = 0 \quad \forall\delta > 0. \tag{2.5}$$

For $\varepsilon > 0$ we can find by Lusin's Theorem a Borel set $B \subseteq]0,1[$ and a continuous function $\tilde{g}\colon]0,1[\to \mathbb{R}$, so that $g = \tilde{g}$ on $]0,1[\setminus B$ and $\lambda(B) \le \varepsilon$. We now put

$$\tilde{g}_k(t) := \tilde{g}\big(\mathrm{MAX}_{\lfloor (1-t)k\rfloor+1}(U_1,\dots,U_k)\big),$$
$$B_k := \big\{t\colon \mathrm{MAX}_{\lfloor (1-t)k\rfloor+1}(U_1,\dots,U_k)\in B\big\}.$$

$\tilde{g}_k$ is correspondingly piecewise constant, and we have $\{t\colon \tilde{g}_k(t)\neq g_k(t)\}\subseteq B_k$. Since the U_i are identically distributed and independent,

$$\lambda(B_k)=\frac{1}{k}\sum_{i=1}^{k}1_B(U_i)$$

converges a.s., according to the Strong Law of Large Numbers, to

$$\mathrm{E}\big(1_B(U_1)\big)=\mathbf{P}(U_1\in B)=\lambda(B)\le\varepsilon,$$

so, in particular, $\limsup_k \lambda(B_k)\le\varepsilon$ a.s. Since $\mathrm{MAX}_{\lfloor (1-t)k\rfloor+1}(U_1,\dots,U_k)$, as the $\frac{\lfloor tk\rfloor+1}{k}$-quantile of the empirical distribution of the samples $(U_1,\dots,U_k)$, converges to the t-quantile of the uniform distribution and $\tilde{g}$ is continuous, it also holds that

$$\lim_{k\to\infty}\tilde{g}_k(t)=\tilde{g}\quad\text{a.s.}$$

All in all we conclude that

$$\begin{aligned}\limsup_k \lambda\big(\big\{t:\big|g_k(t)-g(t)\big|\ge\delta\big\}\big) &\le \limsup_k \lambda\big(\big\{t:\big|\tilde{g}(t)-g(t)\big|\ge\delta\big\}\big)\\ &\quad+\limsup_k \lambda\big(\big\{t:\big|g_k(t)-\tilde{g}_k(t)\big|\ge\delta\big\}\big)\\ &\quad+\limsup_k \lambda\big(\big\{t:\big|\tilde{g}_k(t)-\tilde{g}(t)\big|\ge\delta\big\}\big)\\ &\le \lambda(B)+\limsup_k \lambda(B_k)\\ &\quad+\limsup_k \lambda\big(\big\{t:\big|\tilde{g}_k(t)-\tilde{g}(t)\big|\ge\delta\big\}\big)\\ &\le 2\varepsilon.\end{aligned}$$

With this the relation (2.5) has been shown.

Since we also have

$$\lim_{k\to\infty}\int_0^1|g_k|\,\mathrm{d}\lambda=\lim_{k\to\infty}\frac{1}{k}\sum_{i=1}^{k}\big|g(U_i)\big|=\int_0^1|g|\,\mathrm{d}\lambda,$$

we can for almost every $\omega\in\Omega$ apply the Theorem of Vitali relative to $(]0,1[,\lambda)$ to obtain

$$\lim_{k\to\infty}\int_0^1|g_k-g|\,\mathrm{d}\lambda=0\quad\text{a.s.}$$

Since the sequence J_k, $k\in\mathbb{N}$, is bounded and converges to $\frac{1}{1-\alpha}\mathbf{1}_{(\alpha,1)}$, we finally obtain the desired convergence (2.4). □

2.4 Dynamic Risk Measures

The risk measures that we have so far studied are as a rule intended for a time horizon of one year. On the other hand, insurance contracts and the liabilities associated to them are often at risk for many years. This temporal asymmetry raises the following questions:

- How should a risk measure reflect new information that becomes available with the passage of time?
- How should a risk measure react to the changes in the risk profile over a multi-year time horizon?
- How should one take account of temporal dependencies?

Temporal dependencies can be induced by external trends that are relevant to the development of losses. An example from life insurance is the improvement in life expectancy that results from medical progress. The nature of the loss to be insured can also change with time. For example, older persons have a greater probability of dying than younger ones, and the associated volatility is correspondingly greater. Therefore life insurances have a risk profile that changes with time. This can have consequences for the risk capital needed.

Example 2.5 A company takes over at the time $t = 0$ the obligations of a competitor for a purchase price of V_0. The inventory is exhausted in n years. The company expects that in year t it should have reserves V_t (deterministically calculated at time 0). Furthermore suppose the insurance benefit L_t in year t should follow a standard normal distribution with expectation μ_t and standard deviation σ_t.

The cash flow at time t is then given by the equation

$$\mathrm{Cf}_t = (1 + s_t)V_{t-1} - V_t - L_t,$$

where we have denoted the risk-free interest rate (assumed deterministic) by s_t. With the notation

$$v_t = \prod_{\tau=1}^{t} (1 + s_\tau)^{-1}$$

for the discount factor, the present value of the cash flow is given by

$$W_1 = \sum_{t=1}^{n} v_t \mathrm{Cf}_t = V_0 - \sum_{t=1}^{n} v_t L_t.$$

Obviously W_1 is also normally distributed, and we have

$$\mathrm{E}(W_1) = V_0 - \sum_{t=1}^{n} v_t \mu_t.$$

In this the index $_1$ is for the start of the first time period (see Fig. 2.4).

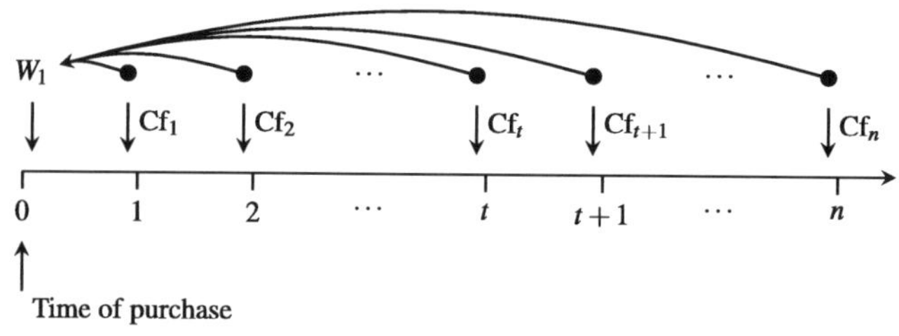

Fig. 2.4 The value W_1 of the portfolio purchased

The random vector $(L_1,\ldots,L_n)^\top$ has the covariance matrix

$$\operatorname{cov}\big((L_1,\ldots,L_n)^\top\big)_{ij} = \operatorname{corr}_{ij}\sigma_i\sigma_j,$$

where we have set $\operatorname{corr}(L_s,L_t) = \operatorname{corr}_{st}$. From

$$\operatorname{cov}\left(\sum_{t=1}^n v_t L_t\right) = (v_1,\ldots,v_t)\operatorname{cov}\big((L_1,\ldots,L_n)^\top\big)(v_1,\ldots,v_t)^\top$$

we obtain

$$\sigma(W_1) = \sqrt{\sum_{i,j=1}^n \operatorname{corr}_{ij} v_i\sigma_i v_j\sigma_j}$$

and with Proposition 2.5

$$\mathrm{ES}_\alpha(W_1) = -V_0 + \sum_{t=1}^n v_t\mu_t + \frac{\varphi_{0,1}(\Phi_{0,1}^{-1}(\alpha))}{1-\alpha}\sqrt{\sum_{i,j=1}^n \operatorname{corr}_{ij} v_i\sigma_i v_j\sigma_j},$$

where the time horizon stretches over n periods. The time dependence of the insurance benefit L_t increases the risk and thus the required risk capital, since for $\operatorname{corr}_{ij} > 0$ the inequality

$$\sum_{i,j=1}^n \operatorname{corr}_{ij} v_i\sigma_i v_j\sigma_j - \sum_{i,j=1}^n (v_i\sigma_i)^2 = 2\sum_{i<j} \operatorname{corr}_{ij} v_i\sigma_i v_j\sigma_j > 0$$

holds.

This example shows that it is not possible to describe the risk over several periods purely based on the corresponding one-period risks. If we want to capture correctly multi-year risks we must consider multi-period risk measures. For risk management purposes it is thus of interest to describe the changes in risks with the passage of time. We are therefore interested not only in $\mathrm{ES}_\alpha(W_1)$ but also in $\mathrm{ES}_\alpha(W_t)$, where $\mathrm{ES}_\alpha(W_t)$is determined for the time horizon $n-t+1$. At time $t=0$, $\mathrm{ES}_\alpha(W_t)$ is a random variable, since the evolution of losses for the first $t-1$ periods is still unknown.

In Sect. 2.4.2 we will present a framework for the description of these dynamic aspects. As preparation we need some fundamental facts about filtrations.

2.4.1 Filtrations

In this section we introduce terminology that allows us to describe the interaction of time and known information.

Definition 2.9 Let $\mathbb{T} = \{0, \dots, n\}$ and $\mathscr{F}_0 = \{\emptyset, \Omega\}$. A *filtration* $(\mathscr{F}_t)_{t \in \mathbb{T}}$ is a set of σ-algebras on Ω with

$$\mathscr{F}_0 \subseteq \mathscr{F}_1 \subseteq \cdots \subseteq \mathscr{F}_n.$$

Remark 2.6 The condition that $\mathscr{F}_0$ is the trivial σ-algebra, is often not required. We do require it here to express the fact that the initial values can be considered known. On technical grounds, it is frequently required that $\mathscr{F}_0$ be the σ-algebra generated by the null sets with respect to a given measure. This extra property can always be achieved by completing the σ-algebra, and we don't need it for what follows.

Remark 2.7 Definition 2.9 can also be generalized to infinite sets $\mathbb{T} \subseteq \mathbb{N}$ and continuous index sets $\mathbb{T} \subseteq \mathbb{R}$. For our purposes, however, finitely many discrete time steps are sufficient.

Definition 2.10 Let $(\mathscr{F}_t)_{t \in \mathbb{T}}$ be a filtration on the set Ω. An *adapted stochastic process with values in* $\mathbb{R}^k$ is a map

$$X : \Omega \times \mathbb{T} \to \mathbb{R}^k, \quad (\omega, t) \mapsto X_t(\omega),$$

where for each $t \in \mathbb{T}$ the map $X_t : \Omega \to \mathbb{R}^k$ is measurable with respect to $\mathscr{F}_t$.

This definition can be interpreted as follows: Each $\omega \in \Omega$ describes a possible history of the process under consideration. At time t the part of the history that corresponds to the past $\{0, \dots, t\}$ is known. The adaptedness condition means that the value $X_t(\omega)$ at time t is known with certainty. This interpretation is particularly striking for the product filtration (Sect. 2.4.3), see Corollary 2.2.

Example 2.6 (Finitely generated filtration) The σ-algebras that occur in practical modeling are mostly finitely generated, because computers are used. Let $\mathscr{P}_1 = \{A_1, \dots, A_{m_1}\}$ be a finite *partition* of Ω. In other words, suppose $A_j \subset \Omega$, $A_i \cap A_j = \emptyset$ for $i \neq j$ and $\bigcup_{i=1}^{m_1} A_i = \Omega$. Let $\mathscr{F}_1$ be the σ-algebra generated by $\mathscr{P}_1$. Now we can construct inductively a filtration as successive refinements of $\mathscr{P}_1$. For each $A \in \mathscr{P}_t$ we choose a partition $\mathscr{P}(A) = \{B_1(A), \dots, B_{k(A)}(A)\}$ and set

$\mathscr{P}_{t+1} = \bigcup_{A \in \mathscr{P}_t} \mathscr{P}(A)$. This family of subsets of Ω is then again a partition of Ω and thus defines a σ-algebra $\mathscr{F}_{t+1}$. A concrete example is given by

$$\begin{bmatrix} \mathscr{P}_0 = \{\{1,2,3,4,5\}\} \\ \mathscr{P}_1 = \{\{1,2\},\{3,4,5\}\} \\ \mathscr{P}_2 = \{\{1,2\},\{3\},\{4,5\}\} \end{bmatrix} \to \begin{bmatrix} \mathscr{F}_0 = \{\emptyset, \{1,2,3,4,5\}\} \\ \mathscr{F}_1 = \{\emptyset, \{1,2\},\{3,4,5\},\{1,2,3,4,5\}\} \\ \mathscr{F}_2 = \{\emptyset, \{3\},\{1,2\},\{4,5\},\{1,2,3\}, \\ \{3,4,5\},\{1,2,4,5\},\{1,2,3,4,5\}\} \end{bmatrix}.$$

From

$$X_t(\omega) = \begin{cases} t & \text{for } \omega = 1 \\ t & \text{for } \omega = 2 \\ 3t + 4t(t-1) & \text{for } \omega = 3 \\ 3t & \text{for } \omega = 4 \\ 3t & \text{for } \omega = 5 \end{cases} \quad \text{and} \quad Y(\omega) = \begin{cases} t & \text{for } \omega = 1 \\ t & \text{for } \omega = 2 \\ 3t + 4t^2 & \text{for } \omega = 3 \\ 3t & \text{for } \omega = 4 \\ 3t & \text{for } \omega = 5 \end{cases}$$

we see X_t is an adapted stochastic process, though Y_t is not: Obviously Y_0 and X_0 are constant and so $\mathscr{F}_0$-measurable. We have $X_1(\omega) = 1 \Leftrightarrow \omega \in \{1,2\} \in \mathscr{F}_1$, and $X_1(\omega) = 3 \Leftrightarrow \omega \in \{3,4,5\} \in \mathscr{F}_1$. Therefore X_1 is $\mathscr{F}_1$-measurable. On the other hand, $Y_1(\omega) = 7 \Leftrightarrow \omega \in \{3\} \notin \mathscr{F}_1$, so that Y_1 is not $\mathscr{F}_1$-measurable. Finally we have $X_2(\omega) = 2 \Leftrightarrow \omega \in \{1,2\} \in \mathscr{F}_2$, $X_2(\omega) = 14 \Leftrightarrow \omega \in \{3\} \in \mathscr{F}_2$, and $X_2(\omega) = 6 \Leftrightarrow \omega \in \{4,5\} \in \mathscr{F}_2$, so that X_2 is also $\mathscr{F}_2$-measurable.

Suppose X is the final result of a long-term investment. At the time of the investment the company expected the result $\mathrm{E}(X)$. In the course of time, however, this assessment will change because of economic uncertainty. This updating process can be described in terms of conditional expectations with respect to a filtration.

Definition 2.11 Let $(\Omega, \mathscr{A}, \mathbf{P})$ be a probability space, $\tilde{\mathscr{A}}$ a sub-σ-algebra of $\mathscr{A}$ and X an $\mathscr{A}$-measurable random variable. The *conditional expectation* of X with respect to $\tilde{\mathscr{A}}$ is the (a.s. uniquely determined) $\tilde{\mathscr{A}}$-measurable random variable $\mathrm{E}(X \mid \tilde{\mathscr{A}})$ with the property that for any bounded $\tilde{\mathscr{A}}$-measurable random variable Z there holds

$$\mathrm{E}(XZ) = \mathrm{E}\big(\mathrm{E}(X \mid \tilde{\mathscr{A}})Z\big).$$

For a proof that this is well defined we cite Theorem 23.4 in [5].

Lemma 2.9 *Let $(\mathscr{F}_t)_{t \in \mathbb{T}}$ be a filtration on Ω and X a $\mathscr{F}_n$-measurable random variable. Then*

$$(t, \omega) \mapsto X_t(\omega) := \mathrm{E}(X \mid \mathscr{F}_t)_{|\omega}$$

is an adapted stochastic process.

In addition, X_t satisfies the martingale property

$$\mathrm{E}(X_{t+1} \mid \mathscr{F}_t) = X_t$$

for all $t \in \{0, \ldots, n-1\}$.

Proof The first assertion follows directly from the $\mathscr{F}_t$-measurability of $\mathrm{E}(X \mid \mathscr{F}_t)$.

For the second assertion, let Z be a $\mathscr{F}_t$-measurable function. Then Z is also $\mathscr{F}_{t+1}$-measurable, and

$$\mathrm{E}(X_{t+1} Z) = \mathrm{E}\big(\mathrm{E}(X \mid \mathscr{F}_{t+1})\, Z\big) = \mathrm{E}(X Z) = \mathrm{E}\big(\mathrm{E}(X \mid \mathscr{F}_t)\, Z\big) = \mathrm{E}(X_t\, Z).$$

The claim follows from the $\mathscr{F}_t$-measurability of X_t and the uniqueness of the conditional expectation. □

By the martingale property one can interpret $\mathrm{E}(X \mid \mathscr{F}_t)$ as the best estimate for the result X at time t. $\mathrm{E}(X \mid \mathscr{F}_t)$ is itself a random variable that reflects the uncertainty between the times 0 and t.

2.4.1.1 Product Filtrations

For concrete applications to cash flows, as a rule, filtrations are constructed based on successive new pieces of information. To describe this construction we need the following notation:

Definition 2.12 Let $\mathscr{A}_1, \ldots, \mathscr{A}_k$ be σ-algebras on the sets $\Omega_1, \ldots, \Omega_k$. The *product-$\sigma$-algebra* on the product set $\Omega_1 \times \Omega_2 \times \cdots \times \Omega_k$ is then given by

$$\bigotimes_{t=1}^{k} \mathscr{A}_t = \mathscr{A}_1 \otimes \cdots \otimes \mathscr{A}_k = \sigma\big(A_1 \times \cdots \times A_k \mid A_t \in \mathscr{A}_t \text{ for } t \in \{1, \ldots .k\}\big),$$

where we have used the convention

$$A_1 \times \cdots \times A_{t-1} \times \emptyset \times A_{t+1} \times \cdots \times A_k = \emptyset.$$

In general, we have $\mathscr{A}_1 \otimes \mathscr{A}_2 \neq \{A_1 \times A_2 \mid A_1 \in \mathscr{A}_1, A_2 \in \mathscr{A}_2\}$. We have equality, however, if $\mathscr{A}_2$ is the trivial σ-algebra.

Lemma 2.10 *Let Ω_1 and Ω_2 be sets and let $\mathscr{A}_1$ be a σ-algebra on Ω_1. Then*

$$A \in \mathscr{A}_1 \otimes \{\emptyset, \Omega_2\} \quad \Leftrightarrow \quad A = A_1 \times \Omega_2,$$

with $A_1 \in \mathscr{A}_1$.

Proof This follows immediately from the fact that the sets of the form

$$\{A_1 \times \Omega_2 \colon A_1 \in \mathscr{A}_1\}$$

make up a σ-algebra. □

Lemma 2.11 (Associativity) *Assume $\mathscr{A}_1, \ldots, \mathscr{A}_{j+k}$ to be σ-algebras. Then*

$$\left(\bigotimes_{t=1}^{j} \mathscr{A}_t\right) \otimes \left(\bigotimes_{t=j+1}^{j+k} \mathscr{A}_t\right) = \bigotimes_{t=1}^{j+k} \mathscr{A}_t.$$

Proof See [3, page 142f]. □

A σ-algebra models what events are in principle possible. To describe economic dynamics we assume that, as a matter of principle, the same events are possible in each time period. For example, in each time period t the event E_t can occur that the stock price of a corporation increases by more than 10 %. On the other hand, the probability of the events depends on the dynamics, and is thus in general different for each period of time: If the corporation at the end of period $t-1$ brings a new very promising product to market, then the probability of the event E_t will be higher than the probability of E_{t-1}. In this approach we assign each time period (considered in isolation) the same σ-algebra $\mathscr{A}$ but not necessarily the same probability measure. The following example illustrates this idea:

Example 2.7 (AR(1)-Process) We consider a stock index S_t, whose dynamics are governed by the equation

$$S_t = \alpha S_{t-1} + s\omega_t, \quad t \in \{1, \ldots, n\},$$

where $\alpha, s > 0$ are constants and $\omega_1, \omega_2, \ldots$ are drawn independently from a standard normal distribution. For each fixed period t, our σ-algebra is exactly the Borel algebra $\mathscr{B}(\mathbb{R})$ on $\mathbb{R}$. To describe the whole dynamics we need to draw n times, and clearly the associated σ-algebra is exactly the Borel algebra

$$\mathscr{B}\left(\mathbb{R}^n\right) = \bigotimes_{t=1}^{n} \mathscr{B}(\mathbb{R})$$

on the set $\Omega = \mathbb{R}^n$. To describe the dynamics up to the period t, it would seem natural to choose as the σ-algebra the Borel algebra $\mathscr{B}(\mathbb{R}^t)$. However, this has the disadvantage that the set on which the σ-algebra is defined changes with each time step. If we multiply $\mathscr{B}(\mathbb{R}^t)$ together with $n-t$ copies of the trivial σ-algebra $\mathscr{A}_0 = \{\emptyset, \mathbb{R}\}$, the resulting σ-algebra is defined on the whole space $\Omega = \mathbb{R}^n$. It has moreover the same measurability structure as the σ-algebra $\mathscr{B}(\mathbb{R}^t)$. This is because, since the maps $f(\omega_1, \ldots, \omega_n)$ measurable with respect to the σ-algebra $\mathscr{B}(\mathbb{R}^t) \times \bigotimes_{s=t+1}^{n} \mathscr{A}_0$ do not depend on $(\omega_{t+1}, \ldots, \omega_n)$, there is a bijection

$$\Psi(g)(\omega_1, \ldots, \omega_n) = g(\omega_1, \ldots, \omega_t)$$

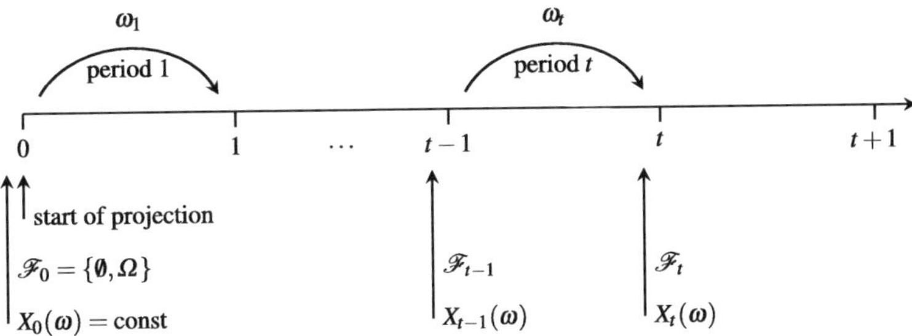

Fig. 2.5 Illustration for Definition 2.13. The random events occurring in the period t are described by the drawing $\omega_t \in \Omega_t$. If X_t is an adapted stochastic process then at time t the values $X_0(\omega), \dots, X_t(\omega)$ are known, since they depend only on $(\omega_1, \dots, \omega_t)$ (see Corollary 2.2 further below)

of the space of $\mathscr{B}(\mathbb{R}^t)$-measurable maps onto the space of $\mathscr{B}(\mathbb{R}^t) \times \bigotimes_{s=t+1}^n \mathscr{A}_0$-measurable maps. With this the natural filtration for our process is

$$\mathscr{F}_t = \bigotimes_{s=1}^{t} \mathscr{B}(\mathbb{R}) \otimes \bigotimes_{s=t+1}^{n} \mathscr{A}_0.$$

The process describing the dynamics

$$S\colon \mathbb{R}^n \times \{1, \dots, n\} \to \mathbb{R},$$
$$(\omega_1, \dots, \omega_n, t) \mapsto S_t(\omega_1, \dots, \omega_n)$$

is an adapted stochastic process for this filtration. Note that S_t does not depend on $\omega_{t+1}, \dots, \omega_n$. This expresses that the future is not known.

Remark also that for $\alpha \notin \{0, 1\}$ neither the S_t for $t \in \{1, \dots, n\}$ nor the increments $S_{t+1} - S_t$ for $t \in \{1, \dots, n-1\}$ are independently distributed. The distribution of the processes S therefore shows a non-trivial dependence structure.

The construction described in Example 2.7 can be generalized in the following way:

Definition 2.13 Assume $\mathbb{T} = \{0, \dots, n\}$ and for $t \in \mathbb{T} \setminus \{0\}$ that $\mathscr{A}_t$ is a σ-algebra on the set Ω_t. The *product filtration* on the Cartesian product $\Omega = \prod_{t=1}^n \Omega_t$ is given by

$$\mathscr{F}_t = \begin{cases} \{\emptyset, \Omega\} & \text{if } t = 0, \\ \bigotimes_{s=1}^t \mathscr{A}_s \otimes \bigotimes_{s=t+1}^n \{\emptyset, \Omega_s\} & \text{otherwise.} \end{cases} \tag{2.6}$$

The σ-algebra $\mathscr{F}_t$ can be understood as the restriction of the σ-Algebra $\mathscr{F}_n$ to the time interval from 0 to t (see the Fig. 2.5). In practical applications the $\mathscr{A}_s$ are almost always the same. The somewhat greater generality of Definition 2.13 does not lead to any additional difficulties.

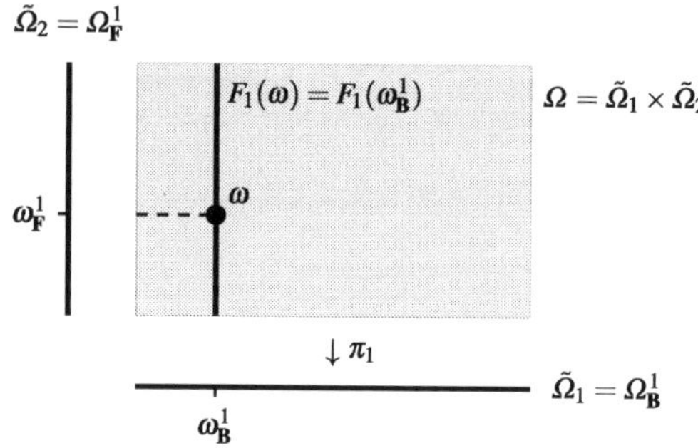

Fig. 2.6 Illustration of the product structure in Definition 2.14 in a two-dimensional example

Definition 2.14 Let $\Omega = \prod_{t=1}^n \Omega_t$ and

$$\pi_t : \Omega \to \prod_{s=1}^{t} \Omega_s, \quad \omega \mapsto \pi_t(\omega) = (\omega_1, \dots, \omega_t)$$

be the projection on the first t factors. For $\omega \in \Omega$ and $t \in \{1, \dots, n\}$ the *t-fiber through* ω is given by

$$F_t(\omega) = \pi_t^{-1}\big(\pi_t(\omega)\big),$$

and we put $F_0(\omega) = \Omega$. For $w \in \pi_t(\Omega)$ the *t-fiber over* w is the set $F_t(w) = \pi_t^{-1}(w)$.

We write $\Omega^t_{\mathbf{B}} = \prod_{s=1}^t \Omega_s$ and $\Omega^t_{\mathbf{F}} = \prod_{s=t+1}^n \Omega_s$.[4] Furthermore we use the notation $\pi_t(\omega) = \omega^t_{\mathbf{B}} \in \Omega^t_{\mathbf{B}}$ and define $\omega^t_{\mathbf{F}} \in \Omega^t_{\mathbf{F}}$ by $\omega = (\omega^t_{\mathbf{B}}, \omega^t_{\mathbf{F}})$.

Definition 2.14 is illustrated in Fig. 2.6 and in Fig. 2.7.

Corollary 2.1 *Let* $\Omega = \prod_{t=1}^n \Omega_t$. *For each* $\omega \in \Omega$ *we thus have*

$$\{\omega\} = F_n(\omega) \subseteq F_{n-1}(\omega) \subseteq \cdots \subseteq F_1(\omega) \subseteq F_0(\omega) = \Omega.$$

The relationships in Definition 2.14 are consistent in the sense that the equation $F_t(\omega) = F_t(\pi_t(\omega))$ holds for every $\omega \in \Omega$.

The following lemma, resp. the subsequent corollary, shows that a stochastic process adapted to a product filtration depends at time t only on the uncertainty up to the time t, and not on future imponderables. This conforms to the experience that the present depends on events in the past, but is not influenced by events in the future.

Lemma 2.12 *A map* $g : \Omega \to \mathbb{R}$ *is exactly then* $\mathscr{F}_t$*-measurable when it is* $\mathscr{F}_n$*-measurable and constant on the fiber* $F_t(\omega)$.

[4] The index $\mathbf{B}$ stands for "Basis" and the index $\mathbf{F}$ for "Fiber".

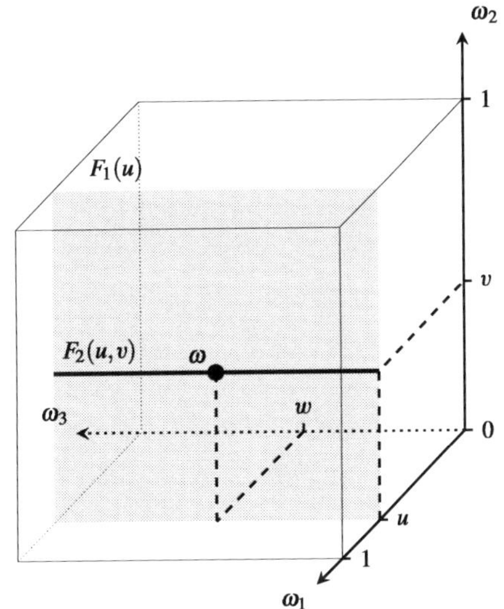

Fig. 2.7 Illustration for Definition 2.14. We consider the space $\Omega =]0,1[^3$, where **P** is the Lebesgue measure; $\omega = (u, v, w)$ is the random value that has to be realized. At time $t = 1$ at the beginning of the period $t+1$, $\omega_1 = u$ is known. The fiber $F_1(u)$ describes the probability space for the remaining uncertainty. At time $t = 2$, (u, v) is known and the fiber $F_2(u, v) \subseteq F_1(u)$ describes the remaining uncertainty. At time $t = 3$, ω is known. Since there is no further uncertainty, the fiber $F_3(\omega)$ is reduced to the point ω

Proof "⇒": Let g be $\mathscr{F}_t$-measurable. $\mathscr{F}_t \subseteq \mathscr{F}_n$, so g is also $\mathscr{F}_n$-measurable. From Lemma 2.10 follows that $\mathscr{F}_t$ is made up of the $\mathscr{F}_n$-measurable subsets of the form $A = \tilde{A} \times \Omega^t_{\mathbf{F}}$. We now assume that g is not constant on the fibers. Because $F_t(\omega) = \{\omega^t_{\mathbf{B}}\} \times \Omega^t_{\mathbf{F}}$, there exist $x, y \in \Omega^t_{\mathbf{F}}$ with $g(\omega^t_{\mathbf{B}}, x) \neq g(\omega^t_{\mathbf{B}}, y)$. Thus there are open intervals B_x, B_y with $g(\omega^t_{\mathbf{B}}, x) \in B_x$, $g(\omega^t_{\mathbf{B}}, y) \in B_y$ and $B_x \cap B_Y = \emptyset$. It follows that $g^{-1}(B_x) \cap g^{-1}(B_y) = \emptyset$. Since $(\omega^t_{\mathbf{B}}, x) \in g^{-1}(B_x) \setminus g^{-1}(B_y)$, we deduce that $g^{-1}(B_y) \cap F_t(\omega) \neq F_t(\omega)$. Thus we have a contradiction to $g^{-1}(B_y) \in \mathscr{F}_t$.

"⇐": Since g is constant on the fibers $F_t(\omega)$, for each Borel set $B \subset \mathbb{R}$ there exists a set $A \subset \Omega^t_{\mathbf{B}}$ with $g^{-1}(B) = A \times \Omega^t_{\mathbf{F}}$. Since the set $g^{-1}(B)$ is measurable with respect to $\mathscr{F}_n$, it is also $\mathscr{F}_t$-measurable by Lemma 2.10. □

Corollary 2.2 *Let $(\mathscr{F}_t)_{t\in\mathbb{T}}$ be a product filtration on $\Omega = \prod_{t=1}^n \Omega_t$ and $(\omega, t) \mapsto X_t(\omega)$ an adapted stochastic process. Then $X_t(\omega)$ depends only on the first t components $\omega_1, \dots, \omega_t$.*

Proof This follows because $X_t(\omega)$ is constant on $F_t(\omega) = \{\omega^t_{\mathbf{B}}\} \times \prod_{s=t+1}^n \Omega_s$. □

Since $\mathscr{F}_t$-measurable functions g only depend on ω through $\omega^t_{\mathbf{B}}$, we shall write $g(\omega^t_{\mathbf{B}})$ in some places instead of $g(\omega)$.

Definition 2.15 For $t \in \{1, \dots, n\}$, let (Ω_t, μ_t) be a measure space and $(\mathscr{F}_t)_{t\in\mathbb{T}}$ the product filtration on $\Omega = \prod_{t=1}^n \Omega_t$. Suppose the probability measure **P** on $(\Omega, \mathscr{F}_n)$ is absolutely continuous with respect to the product measure $\mu = \bigotimes_{t=1}^n \mu_t$. Then $(\Omega, (\mathscr{F}_t)_{t\in\mathbb{T}}, \mathbf{P})$ is a *filtered product economy*.

We write $\mu^t_{\mathbf{B}} = \bigotimes_{s=1}^t \mu_s$ and $\mu^t_{\mathbf{F}} = \bigotimes_{s=t+1}^n \mu_s$.

Remark 2.8 Since for each $A \subseteq F_t(\omega)$ there is a unique set $\tilde{A} \subseteq \prod_{s=t+1}^{n} \Omega_s$ with $A = (\omega_1, \dots, \omega_t) \times \tilde{A}$, $\mu_{\mathbf{F}}^t$ induces, in a canonical way, through

$$\mu_{\mathbf{F}}^{t,F_t(\omega)}(A) = \mu_{\mathbf{F}}^t(\tilde{A})$$

a measure on $F_t(\omega)$. To simplify the notation, we will also write $\mu_{\mathbf{F}}^t$ for this measure on the fiber.

In a filtered product economy, n discrete time periods are modeled, so that, during each period t, additional economic uncertainty arises that is described by $\omega_t \in \Omega_t$. The $\omega = (\omega_1, \dots, \omega_n)$ thus describes the cumulative uncertainty over all periods. At the start of period $t+1$, $\omega_{\mathbf{B}}^t$ is a known quantity while $F_t(\omega_{\mathbf{B}}^t)$ describes the remaining risk. An economic product economy does not describe the setting of prices, but describes only the random events that can influence fixing prices. The price dynamics of an investment are described by adapted stochastic processes on economic product economies (see Example 2.7).

Remark 2.9 The measures μ_t on $(\Omega_t, \mathscr{A}_t)$ model the sources of randomness in the individual time periods t. By choosing the product measure $\mu = \bigotimes_{t=1}^{n} \mu_t$ on $\Omega = \prod_{t=1}^{n} \Omega_t$ we are supposing the independence of the sources of randomness. Each period thus brings, independently of all other periods, a random influence ω_t to the cumulative uncertainty $\omega = (\omega_1, \dots, \omega_n)$. This is, however, just a mathematical trick to simplify the modeling, and the product measure μ has, as a rule, no direct economic interpretation. In particular, the independence of the sources of randomness does not mean that an economic price dynamics on $(\Omega, (\mathscr{F}_t)_{t \in \mathbb{T}}, \mu)$ has independent increments, as Example 2.7 shows.

As we are merely assuming that $\mathbf{P}$ is absolutely continuous relative to μ, we have defined a modeling framework that is sufficiently general for practical applications.

Example 2.8 Consider two projection periods $t \in \{1, 2\}$ and a loss function that is exponentially distributed in each period. We further assume that the economic cycle has the effect that at the end of period 1 the expected loss for period 2 is 10 % higher than the losses incurred in period 1. Let (X_1, X_2) be the loss process over the two periods. Then the corresponding distribution function $\mathbf{P}(X_1 < y_1, X_2 < y_2)$ has the density

$$p(y_1, y_2) = a \exp(-a y_1) \frac{10}{11 y_1} \exp\left(-\frac{10 y_2}{11 y_1}\right) 1_{(0,\infty)\times(0,\infty)}(y_1, y_2)$$

with respect to the two-dimensional Lebesgue measure, where $\mathrm{E}(X_1) = 1/a$. In each simulation, one draws a random number x_1 from the exponential distribution with rate a, and then a random number x_2 from an exponential distribution with rate $(1.1\, x_1)^{-1}$.

Thus, from the standpoint of practical modeling, using a probability measure $\mathbf{P}$ which is absolutely continuous relative to μ is sufficiently general, and leads to significant technical simplifications (see e.g., Lemma 2.13 and Proposition 2.7).

Lemma 2.13 *Let $(\Omega, (\mathscr{F}_t)_{t\in\mathbb{T}}, \mathbf{P})$ be a filtered product economy. On the fiber $F_t(\omega)$ there is a probability measure on the σ-algebra induced by $\mathscr{F}_n$, given by*

$$\mathbf{P}_{\omega_\mathbf{B}^t}(A) = \frac{\int 1_A\, p(\omega_\mathbf{B}^t, \omega_\mathbf{F}^t)\, \mathrm{d}\mu_\mathbf{F}^t}{\int p(\omega_\mathbf{B}^t, \omega_\mathbf{F}^t)\, \mathrm{d}\mu_\mathbf{F}^t},$$

where p is the density of $\mathbf{P}$ relative to μ.

Proof Clearly $\mathbf{P}_{\omega_\mathbf{B}^t}$ is a measure on $F_t(\omega) = \{\omega_1\} \times \cdots \times \{\omega_t\} \times \Omega_{t+1} \times \cdots \times \Omega_n$. The assertion thus follows from

$$\mathbf{P}_{\omega_\mathbf{B}^t}\big(F_t(\omega)\big) = \frac{\int 1_{F_t(\omega)}\, p(\omega_\mathbf{B}^t, \omega_\mathbf{F}^t)\, \mathrm{d}\mu_\mathbf{F}^t}{\int p(\omega_\mathbf{B}^t, \omega_\mathbf{F}^t)\, \mathrm{d}\mu_\mathbf{F}^t} = \frac{\int p(\omega_\mathbf{B}^t, \omega_\mathbf{F}^t)\, \mathrm{d}\mu_\mathbf{F}^t}{\int p(\omega_\mathbf{B}^t, \omega_\mathbf{F}^t)\, \mathrm{d}\mu_\mathbf{F}^t} = 1. \qquad \square$$

We have already seen that an observer at time t knows the portion $\omega_\mathbf{B}^t$ of his history ω, and that the remaining uncertainty can be described by $F_t(\omega_\mathbf{B}^t)$. The probability measure $\mathbf{P}_{\omega_\mathbf{B}^t}$ is of use to the observer in determining the probabilities of future events.

Proposition 2.7 *Let $(\Omega, (\mathscr{F}_t)_{t\in\mathbb{T}}, \mathbf{P})$ be a filtered product economy and assume $\mathbf{P} = p\,\mu$. Then we have $\mathbf{P}$-almost everywhere*

$$\mathrm{E}(g \mid \mathscr{F}_t)_{|\omega_\mathbf{B}^t} = \frac{\int_{\Omega_\mathbf{F}^t} g(\omega) p(\omega)\, \mathrm{d}\mu_\mathbf{F}^t}{\int_{\Omega_\mathbf{F}^t} p(\omega)\, \mathrm{d}\mu_\mathbf{F}^t}.$$

Proof By Definition 2.11 the conditional expectation satisfies

$$\int_\Omega g(\omega) Z\big(\omega_\mathbf{B}^t\big) p(\omega)\, \mathrm{d}\mu = \int_\Omega \mathrm{E}(g \mid \mathscr{F}_t)\big(\omega_\mathbf{B}^t\big) Z\big(\omega_\mathbf{B}^t\big) p(\omega)\, \mathrm{d}\mu$$

for any integrable $\mathscr{F}_t$-measurable function $\omega_\mathbf{B}^t \mapsto Z(\omega_\mathbf{B}^t)$, and is uniquely determined by this condition. The functions

$$\tilde{g}\big(\omega_\mathbf{B}^t\big) = \int_{\Omega_\mathbf{F}^t} g(\omega) p(\omega)\, \mathrm{d}\mu_\mathbf{F}^t$$

$$\tilde{p}\big(\omega_\mathbf{B}^t\big) = \int_{\Omega_\mathbf{F}^t} p(\omega)\, \mathrm{d}\mu_\mathbf{F}^t$$

are clearly $\mathscr{F}_t$-measurable. Thus their quotient is also $\mathscr{F}_t$-measurable. We make the calculation

$$\begin{aligned}\int_\Omega g(\omega) Z\big(\omega_\mathbf{B}^t\big) p(\omega)\, \mathrm{d}\mu &= \int_{\Omega_\mathbf{B}^t} \tilde{g}\big(\omega_\mathbf{B}^t\big) Z\big(\omega_\mathbf{B}^t\big)\, \mathrm{d}\mu_\mathbf{B}^t \\ &= \int_{\Omega_\mathbf{B}^t} \frac{\tilde{g}(\omega_\mathbf{B}^t)}{\tilde{p}(\omega_\mathbf{B}^t)} \tilde{p}\big(\omega_\mathbf{B}^t\big)\, Z\big(\omega_\mathbf{B}^t\big)\, \mathrm{d}\mu_\mathbf{B}^t\end{aligned}$$

$$= \int_{\Omega} \frac{\tilde{g}(\omega_{\mathbf{B}}^t)}{\tilde{p}(\omega_{\mathbf{B}}^t)} Z(\omega_{\mathbf{B}}^t) p(\omega)\, \mathrm{d}\mu.$$

The assertion thus follows immediately from the definition of conditional expectation. □

2.4.2 General Dynamic Risk Measures

Definition 2.16 Let $(\Omega, \mathbf{P})$ be a probability space and $(\mathscr{F}_t)_{t\in\{0,\dots,n\}}$ a filtration on Ω. For each t, let $\mathscr{M}_t(\Omega, \mathbb{R})$ be a vector space of $\mathscr{F}_t$-measurable functions. A *dynamic risk measure* is a family $(\rho_t)_{t\in\{0,\dots,n\}}$ of maps

$$\rho_t : \mathscr{M}_n(\Omega, \mathbb{R}) \to \mathscr{M}_t(\Omega, \mathbb{R}),$$

so that we have:

(i) $X_1 \geq X_2$ a.s. $\Rightarrow \rho_t(X_1) \geq \rho_t(X_2)$ a.s. (Monotony);
(ii) For $K \in \mathscr{M}_t(\Omega, \mathbb{R})$ there holds $\rho_t(X + K) = \rho_t(X) + K$ a.s. (Translation invariance).

Definition 2.17 A dynamic risk measure $(\rho_t)_{t\in\{0,\dots,n\}}$ is called *coherent*, if:

(i) $\rho_t(KX) = K\rho_t(X)$ a.s. for all $K \in \mathscr{M}_t(\Omega, \mathbb{R})$ with $K \geq 0$ a.s., and $KX \in \mathscr{M}_n(\Omega, \mathbb{R})$ (Homogeneity);
(ii) $\rho_t(X_1 + X_2) \leq \rho_t(X_1) + \rho_t(X_2)$ a.s. (Subadditivity).

The definitions are analogous to Definition 2.8. The map ρ_t is the risk measure determined at time t, which has the time horizon $]t, n[$. Since ρ_t is based on the evolution of risk up to the time t, ρ_t is not real-valued, but rather an $\mathscr{M}_t(\Omega, \mathbb{R})$-valued map. On the same grounds, K is taken as an element of $\mathscr{M}_t(\Omega, \mathbb{R})$.

Remark 2.10 In the literature it is often taken that $\mathscr{M}_t(\Omega, \mathbb{R}) = L^\infty(\Omega, \mathscr{F}_t)$ (see e.g., [8]).

In Sect. 2.4.3, we will examine dynamic risk measures on filtered product economics, and thus explicitly employ the product structure. This will lead us to dynamic risk measures with properties relevant to actual practice. In Sect. 2.4.4 we describe an alternative approach for general filtrations, that is favored in the mathematical literature. We shall nonetheless see that this alternative approach is problematic from a practical point of view. Section 2.4.4 can therefore be skipped by readers who are primarily interested in applications.

2.4.3 Dynamic Risk Measures on Filtered Product Economies

Let $(\Omega, (\mathscr{F}_t)_{t\in\mathbb{T}}, \mathbf{P})$ be a filtered product economy and $\rho\colon \mathscr{M}(\Omega,\mathbb{R}) \to \mathbb{R}$ a risk measure. If ρ is "sufficiently generic" then it can be transferred "pointwise", in a natural way, to the fibers $F_t(\omega) \subseteq \Omega$, where the probability measure $\mathbf{P}_{\omega^t_{\mathbf{B}}}$ is used instead of $\mathbf{P}$. This produces, for each fiber $F_t(\omega)$, a risk measure $\rho_t(\omega)\colon \mathscr{M}(F_t(\omega),\mathbb{R}) \to \mathbb{R}$. Since the t-fiber through $\tilde{\omega} \in F_t(\omega)$ is just $F_t(\tilde{\omega}) = F_t(\omega)$, one would also expect that $\rho_t(\omega) = \rho_t(\tilde{\omega})$. The conjecture suggests itself that $(\omega, X) \mapsto \rho_t(\omega)(X(\omega^t_{\mathbf{B}}, \cdot))$ defines a dynamic risk measure. In Theorem 2.6 this general construction will be carried out for the value at risk, and in Theorem 2.7 for the expected shortfall.

Definition 2.18 Let $(\Omega, (\mathscr{F}_t)_{t\in\mathbb{T}}, \mathbf{P})$ be a filtered product economy, $\alpha \in \,]0,1[$, and $\mathscr{M}_t(\Omega,\mathbb{R})$ the space of $\mathscr{F}_t$-measurable functions. The family of maps parametrized by $t \in \mathbb{T}$

$$\mathrm{VaR}_{\alpha,t}\colon \mathscr{M}_n(\Omega,\mathbb{R}) \to \mathscr{M}_t(\Omega,\mathbb{R}), \quad X \mapsto \mathrm{VaR}_{\alpha,t}(X)$$

with

$$\mathrm{VaR}_{\alpha,t}(X)_{|\omega^t_{\mathbf{B}}} = \inf\bigl\{x \in \mathbb{R}\colon \mathbf{P}_{\omega^t_{\mathbf{B}}}\bigl(X\bigl(\omega^t_{\mathbf{B}}, \cdot\bigr) \le x\bigr) \ge \alpha\bigr\}$$

is called *dynamic value at risk*.

Theorem 2.6 *The dynamic value at risk is a dynamic risk measure.*

Proof For each $\omega \in \Omega$, $\mathrm{VaR}_{\alpha,t}$ is just the ordinary value at risk for the probability space $(F_t(\omega), \mathbf{P}_{\omega^t_{\mathbf{B}}})$. Because the inequality $X > Y$ carries over in a trivial way to subsets, the monotone property of the value at risk carries over pointwise to $\mathrm{VaR}_{\alpha,t}$ for each $\omega^t_{\mathbf{B}} \in \Omega^t_{\mathbf{B}}$.

Since $\mathscr{F}_t$-measurable functions are constant on the sets

$$\bigl\{\omega^t_{\mathbf{B}}\bigr\} \times \Omega^t_{\mathbf{F}} \subseteq \Omega,$$

the same pointwise argument also gives translation invariance.

It remains to show the $\mathscr{F}_t$-measurability of $\mathrm{VaR}_{\alpha,t}(X)$. First we assume that X is bounded below. Then there is an increasing sequence $\{X_k\}_{k\in\mathbb{N}}$ of simple functions with $\limsup_{k\to\infty} X_k = \lim_{k\to\infty} X_k = X$. Since X_k is measurable and takes on only finitely many values, the map $\omega^t_{\mathbf{B}} \mapsto \mathrm{VaR}_{\alpha,t}(X_k)_{|\omega^t_{\mathbf{B}}}$ is likewise measurable. By the monotony of $\mathrm{VaR}_{\alpha,t}$, $\{\mathrm{VaR}_{\alpha,t}(X_k)\}_{k\in\mathbb{N}}$ is an increasing sequence of measurable simple functions, which is why

$$\lim_{k\to\infty} \mathrm{VaR}_{\alpha,t}(X_k)$$

is measurable. On the grounds that $X_k \le X$ and of monotony, we have

$$\lim_{k\to\infty} \mathrm{VaR}_{\alpha,t}(X_k) \le \mathrm{VaR}_{\alpha,t}(X).$$

Assume there is an $\varepsilon > 0$ with $\lim_{k\to\infty} \mathrm{VaR}_{\alpha,t}(X_k) < \mathrm{VaR}_{\alpha,t}(X) - \varepsilon$. Since X_k is an increasing sequence, for all k it is true that $\mathrm{VaR}_{\alpha,t}(X_k) < \mathrm{VaR}_{\alpha,t}(X) - \varepsilon$. This implies

$$\mathbf{P}_{\omega^t_{\mathbf{B}}}\left(X_k\left(\omega^t_{\mathbf{B}}, \cdot\right) \le \mathrm{VaR}_{\alpha,t}(X) - \varepsilon/2\right) \ge \alpha$$

for all k. From X_k's convergence to X follows

$$\mathbf{P}_{\omega^t_{\mathbf{B}}}\left(X\left(\omega^t_{\mathbf{B}}, \cdot\right) \le \mathrm{VaR}_{\alpha,t}(X) - \varepsilon/2\right) \ge \alpha$$

contradicting the definition of $\mathrm{VaR}_{\alpha,t}$.

If X is not bounded below, let $\tilde{X}_k = \max(X, -k)$. Then $\{\hat{X}_k\}_{k\in\mathbb{N}}$ is a decreasing sequence of measurable functions bounded below that converges pointwise to X. Therefore $\lim_{k\to\infty} \mathrm{VaR}_{\alpha,t}(\tilde{X}_k)$ is measurable, and we have

$$\lim_{k\to\infty} \mathrm{VaR}_{\alpha,t}(\tilde{X}_k) \ge \mathrm{VaR}_{\alpha,t}(X).$$

If there were an $\varepsilon > 0$ with $\liminf_{k\to\infty} \mathrm{VaR}_{\alpha,t}(\tilde{X}_k) > \mathrm{VaR}_{\alpha,t}(X) + \varepsilon$, then we would have

$$\mathbf{P}_{\omega^t_{\mathbf{B}}}\left(\tilde{X}_k\left(\omega^t_{\mathbf{B}}, \cdot\right) \le \mathrm{VaR}_{\alpha,t}(X)_{|\omega^t_{\mathbf{B}}} + \varepsilon/2\right) < \alpha$$

for all k, and by the convergence of $\tilde{X}_k$

$$\mathbf{P}_{\omega^t_{\mathbf{B}}}\left(X\left(\omega^t_{\mathbf{B}}, \cdot\right) \le \mathrm{VaR}_{\alpha,t}(X)_{|\omega^t_{\mathbf{B}}} + \varepsilon/2\right) < \alpha$$

in contradiction to the definition of $\mathrm{VaR}_{\alpha,t}$. □

Definition 2.19 Let $(\Omega, (\mathscr{F}_t)_{t\in\mathbb{T}}, \mathbf{P})$ be a filtered product economy, $\alpha \in]0, 1[$ and $\mathscr{M}_t(\Omega, \mathbb{R})$ the space of integrable, $\mathscr{F}_t$-measurable functions. The family of maps parametrized by $t \in \mathbb{T}$

$$\mathrm{ES}_{\alpha,t} \colon \mathscr{M}_n(\Omega, \mathbb{R}) \to \mathscr{M}_t(F_t, \mathbb{R}), \quad X \mapsto \mathrm{ES}_{\alpha,t}(X)$$

with

$$\mathrm{ES}_{\alpha,t}(X)_{|\omega^t_{\mathbf{B}}} = \frac{1}{1-\alpha} \mathrm{E}_{\mathbf{P}_{\omega^t_{\mathbf{B}}}}\left(X\left(\omega^t_{\mathbf{B}}, \cdot\right) 1_{X(\omega^t_{\mathbf{B}}, \cdot), \mathrm{VaR}_{\alpha,t}(X)_{|\omega^t_{\mathbf{B}}}, \alpha}\right)$$

is called *dynamic expected shortfall*, where $1_{X,x,\alpha}$ is defined as in Lemma 2.5.

Theorem 2.7 *The dynamic expected shortfall is a coherent, dynamic risk measure.*

Proof For each $\omega \in \Omega$, $\mathrm{ES}_{\alpha,t}$ is the ordinary expected shortfall on the probability space $(F_t(\omega), \mathbf{P}_{\omega^t_{\mathbf{B}}})$. Therefore monotony and subadditivity carry over directly to $\mathrm{ES}_{\alpha,t}(X)$. Since $\mathscr{F}_t$-measurable functions on the sets

$$\left\{\omega^t_{\mathbf{B}}\right\} \times \Omega^t_{\mathbf{F}} \subseteq \Omega$$

are constant, the same pointwise argument delivers translation invariance and homogeneity.

To show the $\mathscr{F}_t$-measurability of $\mathrm{ES}_{\alpha,t}(X)$, we first assume that X is bounded below. Then there is an increasing sequence $\{X_k\}_{k\in\mathbb{N}}$ of simple functions for which $\lim_{k\to\infty} X_k = X$ almost everywhere. Since X_k and $\mathrm{VaR}_{\alpha,t}(X_k)$ are measurable and take on only finitely many values, for each $w \in \pi_t(\omega)$ the map

$$\omega^t_{\mathbf{B}} \mapsto 1_{X_k(\omega^t_{\mathbf{B}},w),\mathrm{VaR}_{\alpha,t}(X_k),\alpha},$$

and thus also

$$\omega^t_{\mathbf{B}} \mapsto \mathrm{ES}_{\alpha,t}(X_k)_{|\omega^t_{\mathbf{B}}},$$

are $\mathscr{F}_t$-measurable. Due to the monotony of $\mathrm{ES}_{\alpha,t}$, $\{\mathrm{ES}_{\alpha,t}(X_k)\}_{k\in\mathbb{N}}$ is an increasing sequence of measurable, simple functions, from which we see

$$\limsup_{k\to\infty} \mathrm{ES}_{\alpha,t}(X_k) = \limsup_{k\to\infty} \frac{\int_{F_t(\omega^t_{\mathbf{B}})} 1_{X_k(\omega^t_{\mathbf{B}},\,\cdot\,),\mathrm{VaR}_{\alpha,t}(X_k),\alpha}\, p(\omega^t_{\mathbf{B}},\,\cdot\,)\,\mathrm{d}\mu^t_{\mathbf{F}}}{(1-\alpha)\int_{F_t(\omega^t_{\mathbf{B}})} p(\omega^t_{\mathbf{B}},\,\cdot\,)\,\mathrm{d}\mu^t_{\mathbf{F}}} = \mathrm{ES}_{\alpha,t}(X)$$

is measurable.

If X is not bounded from below, we can consider the sequence

$$\tilde{X}_k = \big\{\max(X,-k)\big\}_{k\in\mathbb{N}}.$$

We have just seen that $\mathrm{ES}_{\alpha,t}(\tilde{X}_k)$ is measurable for each k. Therefore also measurable is the function

$$\begin{aligned}\liminf_{k\to\infty} \mathrm{ES}_{\alpha,t}(\tilde{X}_k) &= \liminf_{k\to\infty} \frac{\int_{F_t(\omega^t_{\mathbf{B}})} 1_{X_k(\omega^t_{\mathbf{B}},\,\cdot\,),\mathrm{VaR}_{\alpha,t}(\max(X,-k)),\alpha}\, p(\omega^t_{\mathbf{B}},\,\cdot\,)\,\mathrm{d}\mu^t_{\mathbf{F}}}{(1-\alpha)\int_{F_t(\omega^t_{\mathbf{B}})} p(\omega^t_{\mathbf{B}},\,\cdot\,)\,\mathrm{d}\mu^t_{\mathbf{F}}}\\ &= \mathrm{ES}_{\alpha,t}(X).\end{aligned}$$

□

It is conceivable, that a badly chosen risk measure can, over time, lead to contradictory estimates of risk. In the following definition we therefore formalize a minimum requirement upon dynamic risk measures, that should not be violated by risk measures used in practice.

Definition 2.20 Let (Ω,μ) be a measure space, $B \subset \Omega$ and $f\colon \Omega \to \bar{\mathbb{R}}$ a map. The *essential supremum* of f over B is defined by

$$\operatorname{ess\,sup}_B(f) = \inf\big\{a \in \mathbb{R} \colon \mu\big(\big\{x\colon f(x) > a\big\} \cap B\big) = 0\big\} \in \bar{\mathbb{R}}$$

and the *essential infimum* of f over B by

$$\operatorname{ess\,inf}_B(f) = \sup\big\{a \in \mathbb{R} \colon \mu\big(\big\{x\colon f(x) < a\big\} \cap B\big) = 0\big\} \in \bar{\mathbb{R}}.$$

Definition 2.21 Let $(\mathscr{F}_t)_{t\in\mathbb{T}}$ be a product filtration on $\Omega = \prod_{s=1}^{n} \Omega_s$. A dynamic risk measure $(\rho_t)_{t\in\mathbb{T}}$ is *time-consistent*, if for each random variable X, almost every $\omega \in \Omega$ and each $\mathscr{F}_{t+1}$-measurable subset $B \subseteq F_t(\omega)$ with $\mathbf{P}_{\omega^t_{\mathbf{B}}}(B) > 0$ and

$$\operatorname{ess\,inf}_B\big(\rho_{t+1}(X)\big) > \rho_t(X)_{|\omega},$$

there is an $\mathscr{F}_{t+1}$ -measurable subset $C \subseteq F_t(\omega)$ with $\mathbf{P}_{\omega^t_{\mathbf{B}}}(C) > 0$ and

$$\operatorname{ess\,sup}_C\big(\rho_{t+1}(X)\big) < \rho_t(X)_{|\omega}.$$

For an illustration of Definition 2.21 see Fig. 2.8.

Definition 2.21 expresses the idea that the capital requirement cannot increase in time for each possible future. Let ρ_t be a time-consistent, dynamic risk measure and $\rho_t(X)_{|\omega} = K$. If at time t the probability that $\rho_{t+1}(X)_{|\omega} > K$ is greater than 0, so more capital must be provided, then the probability that $\rho_{t+1}(X)_{|\omega} < K$, so some capital can be freed, is also greater than 0.

Definition 2.22 Let $(\mathscr{F}_t)_{t\in\mathbb{T}}$ be a product filtration on $\Omega = \prod_{s=1}^{n} \Omega_s$. A dynamic risk measure ρ_t is *weakly time-consistent*, if for almost every ω and each pair (X, t) there holds

$$\operatorname{ess\,inf}_{F_t(\omega)}\big(\rho_{t+1}(X)\big) \leq \rho_t(X)_{|\omega}.$$

Remark 2.11 In practice it is impossible to distinguish between time-consistency and weak time-consistency. This is because if ρ_t is weakly time-consistent and X is a given random variable, then an arbitrarily small change in X is enough to make X time-consistent.

Corollary 2.3 *Time-consistency implies weak time-consistency.*

Make the assumption that an enterprise uses, in setting capital levels, a dynamic risk measure ρ_t that violates weak time-consistency at some time τ. Suppose further that at time τ the business puts up sufficient capital $\rho_\tau(X)$. With probability 1 it will have to provide more capital one period later, although in the interim no cash has flowed. Worse risk management is hardly imaginable. On these grounds one will certainly want to require of any risk measure to be used in practice that it is time-consistent.

Remark 2.12 In the literature the notion "time-consistent" is often used, to express another condition in which two random variables are compared with each other (see Definition 2.24). We will, however, see in Sect. 2.4.4, that the condition in Definition 2.24 is less useful for practical risk management than requiring time-consistency.

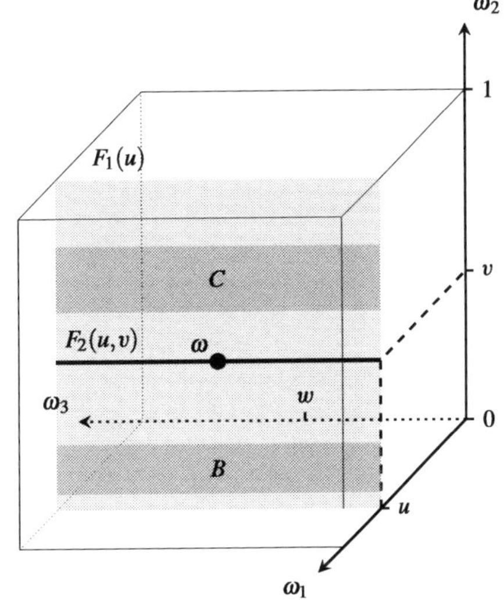

Fig. 2.8 Illustration of Definition 2.21. We consider the space $\Omega =]0, 1[^3$, where **P** is the Lebesgue measure, and write $\omega = (u, v, w)$. The quantity $\rho_1(u)$ is constant on $F_1(u)$ and lies between the values of the risk measure ρ_2 on B and C

Remark 2.13 Remark that Definition 2.21 and Definition 2.22 are not time-symmetric. There are good reasons for this, since the risk should lessen as the time horizon shrinks, as in less time fewer losses can occur.

Example 2.9 Let $\Omega =]0, 1[\times]0, 1[$, and **P** be Lebesgue measure on Ω and $\alpha \in]0, 1[$. Assume the random variable X is defined by

$$X(\omega) = \begin{cases} 2 & \text{if } \omega_1 < (1-\alpha)/2, \\ 1 & \text{otherwise.} \end{cases}$$

Then we have $\mathrm{VaR}_{\alpha,0}(X)_{|\omega} = 1$ for all $\omega \in \Omega$ but

$$\mathrm{VaR}_{\alpha,1}(X)_{|\tilde{\omega}} = \begin{cases} 2 & \text{if } \tilde{\omega}_1 < (1-\alpha)/2, \\ 1 & \text{otherwise.} \end{cases}$$

This shows the dynamic value at risk is not time-consistent.

Theorem 2.8 *Let $(\Omega, (\mathscr{F}_t)_{t\in\mathbb{T}}, \mathbf{P})$ be a filtered product economy. Then the dynamic value at risk on $(\Omega, (\mathscr{F}_t)_{t\in\mathbb{T}}, \mathbf{P})$ is weakly time-consistent.*

Proof Let $\omega \in \Omega$. For $t \in \{0, \ldots, n-1\}$ we put

$$G = \left\{\tilde{\omega} \in F_t(\omega) \colon X(\tilde{\omega}) \geq \mathrm{VaR}_{\alpha,t+1}(X)_{|\tilde{\omega}_{\mathbf{B}}^{t+1}}\right\}.$$

Let $\tilde{\omega} \in F_t(\omega)$. Lemma 2.1 implies

$$\mathbf{P}_{\tilde{\omega}_{\mathbf{B}}^{t+1}}\left(X\left(\tilde{\omega}_{\mathbf{B}}^{t+1}, \cdot\right) \leq \mathrm{VaR}_{\alpha,t+1}(X)_{|\tilde{\omega}_{\mathbf{B}}^{t+1}}\right) = \alpha,$$

since $\mathrm{VaR}_{\alpha,t+1}(X)_{|\tilde{\omega}_{\mathbf{B}}^{t+1}}$ is exactly the value at risk for the random variable $X(\tilde{\omega}_{\mathbf{B}}^{t+1},\cdot)$ on the probability space $(F_{t+1}(\tilde{\omega}),\mathbf{P}_{\tilde{\omega}_{\mathbf{B}}^{t+1}})$. It follows that

$$1-\alpha \le \mathbf{P}_{\tilde{\omega}_{\mathbf{B}}^{t+1}}\big(X\big(\tilde{\omega}_{\mathbf{B}}^{t+1},\cdot\big) \ge \mathrm{VaR}_{\alpha,t+1}(X)\big)$$
$$= \frac{\int_{\Omega_{\mathbf{F}}^{t+1}} 1_G\, p(\tilde{\omega}_{\mathbf{B}}^{t+1},\cdot)\,\mathrm{d}\mu_{\mathbf{F}}^{t+1}}{\int_{\Omega_{\mathbf{F}}^{t+1}} p(\tilde{\omega}_{\mathbf{B}}^{t+1},\cdot)\,\mathrm{d}\mu_{\mathbf{F}}^{t+1}}$$

for each $\tilde{\omega} \in F_t(\omega)$. Thus we obtain

$$\begin{aligned}
\mathbf{P}_{\omega_{\mathbf{B}}^t}(G)\int_{F_t(\omega)} p\big(\omega_{\mathbf{B}}^t,\cdot\big)\,\mathrm{d}\mu_{\mathbf{F}}^t &= \int_{F_t(\omega)} 1_G\, p\big(\omega_{\mathbf{B}}^t,\cdot\big)\,\mathrm{d}\mu_{\mathbf{F}}^t \\
&= \int_{\Omega_{t+1}}\int_{\Omega_{\mathbf{F}}^{t+1}} 1_G\, p\big(\omega_{\mathbf{B}}^t,\cdot\big)\,\mathrm{d}\mu_{\mathbf{F}}^{t+1}\,\mathrm{d}\mu_{t+1} \\
&\ge \int_{\Omega_{t+1}} (1-\alpha)\int_{\Omega_{\mathbf{F}}^{t+1}} p\big(\omega_{\mathbf{B}}^t,\cdot\big)\,\mathrm{d}\mu_{\mathbf{F}}^{t+1}\,\mathrm{d}\mu_{t+1} \\
&= (1-\alpha)\int_{F_t(\omega)} p\big(\omega_{\mathbf{B}}^t,\cdot\big)\,\mathrm{d}\mu_{\mathbf{F}}^t.
\end{aligned}$$

Therefore it holds that $\mathbf{P}_{\omega_{\mathbf{B}}^t}(G) \ge 1-\alpha$ and $X(\tilde{\omega}) \ge \operatorname{ess\,inf}_{F_{t+1}(\tilde{\omega})}(\mathrm{VaR}_{\alpha,t+1}(X))$ for almost every $\tilde{\omega} \in G$. This implies $\mathrm{VaR}_{\alpha,t}(X)_{|\omega} \ge \operatorname{ess\,inf}_{F_t(\omega)}(\mathrm{VaR}_{\alpha,t+1}(X))$. □

Theorem 2.9 *Let $(\Omega,(\mathscr{F}_t)_{t\in\mathbb{T}},\mathbf{P})$ be a filtered product economy, then the dynamic expected shortfall on $(\Omega,(\mathscr{F}_t)_{t\in\mathbb{T}},\mathbf{P})$ is time-consistent.*

Proof Assume $\omega \in \Omega$ and $u = \pi_t(\omega)$. We suppose that there are a random variable X and an $\mathscr{F}_{t+1}$-measurable subset $B \subseteq F_t(\omega)$ with $\mathbf{P}_u(B) > 0$ and

$$\operatorname{ess\,inf}_B\big(\mathrm{ES}_{\alpha,t+1}(X)\big) > \mathrm{ES}_{\alpha,t}(X)_{|\omega}.$$

If there is no $\mathscr{F}_{t+1}$-measurable set $C \subseteq F_t(\omega)$ with $\mathbf{P}_u(C) > 0$ and

$$\operatorname{ess\,sup}_C\big(\mathrm{ES}_{\alpha,t+1}(X)\big) < \mathrm{ES}_{\alpha,t}(X)_{|\omega},$$

then for almost every $v \in \Omega_{t+1}$

$$\mathrm{ES}_{\alpha,t+1}(X)_{|(u,v)} \ge \mathrm{ES}_{\alpha,t}(X)_{|u}. \tag{2.7}$$

It suffices therefore to show that the inequality (2.7) leads to a contradiction.

We assume that (2.7) holds and put

$$G = \big\{\tilde{\omega} \in F_t(\omega)\colon X(\tilde{\omega}) > \mathrm{VaR}_{\alpha,t+1}(X)_{|\pi_{t+1}(\tilde{\omega})}\big\},$$
$$H = \big\{\tilde{\omega} \in F_t(\omega)\colon X(\tilde{\omega}) = \mathrm{VaR}_{\alpha,t+1}(X)_{|\pi_{t+1}(\tilde{\omega})}\big\}$$

as well as

$$\beta\colon \Omega_{t+1} \to [0,1],$$
$$v \mapsto \beta(v) = \begin{cases} \frac{1-\alpha-\mathbf{P}_{(u,v)}(G\cap F_{t+1}(\omega))}{\mathbf{P}_{(u,v)}(H\cap F_{t+1}(\omega))} & \text{if } \mathbf{P}_{(u,v)}(H\cap F_{t+1}(\omega)) > 0, \\ 0 & \text{otherwise.} \end{cases}$$

Note that because of the $(t+1)$-fiberwise definition of G and H we have in general $G \neq \{X > \mathrm{VaR}_{\alpha,t}(X)\}$ and $H \neq \{X = \mathrm{VaR}_{\alpha,t}(X)\}$.

By Lemma 2.5, there holds for all $v \in \Omega_{t+1}$

$$\begin{aligned} &(1-\alpha)\,\mathrm{ES}_{\alpha,t+1}(X)_{|(u,v)} \\ &= \int_{F_{t+1}(u,v)} X\left(1_G + \beta(v)\,1_H\right) \mathrm{d}\mathbf{P}_{(u,v)} \\ &= \frac{\int_{F_{t+1}(u,v)} X\,1_G\,p\,\mathrm{d}\mu_{\mathbf{F}}^{t+1}}{\int_{F_{t+1}(u,v)} p\,\mathrm{d}\mu_{\mathbf{F}}^{t+1}} + \frac{\int_{F_{t+1}(u,v)} X\,\beta(v)\,1_H\,p\,\mathrm{d}\mu_{\mathbf{F}}^{t+1}}{\int_{F_{t+1}(u,v)} p\,\mathrm{d}\mu_{\mathbf{F}}^{t+1}}. \end{aligned}$$

Since the inequality (2.7) has been assumed, on the set $B \subset F_t(\omega)$ strict inequality actually holds, and B is not a $\mathbf{P}_u$-null set, we obtain by integration over v the inequality

$$\begin{aligned} &(1-\alpha)\,\mathrm{ES}_{\alpha,t}(X)_{|u} \int_{F_t(u)} p\,\mathrm{d}\mu_{\mathbf{F}}^t \\ &< \int_{F_t(u)} X\,1_G\,p\,\mathrm{d}\mu_{\mathbf{F}}^t \\ &\quad + \int_{\Omega_{t+1}} \beta \int_{F_{t+1}(u,\cdot)} X\,1_H\,p\,\mathrm{d}\mu_{\mathbf{F}}^{t+1}\,\mathrm{d}\mu_{t+1} \\ &= \left(\int_G X\,\mathrm{d}\mathbf{P}_u + \int_H \beta X\,\mathrm{d}\mathbf{P}_u\right) \int_{F_t(u)} p\,\mathrm{d}\mu_{\mathbf{F}}^t. \end{aligned} \tag{2.8}$$

From

$$\beta(v) \int_{F_{t+1}(u,v)} 1_H\,p\,\mathrm{d}\mu_{\mathbf{F}}^{t+1} = (1-\alpha) \int_{F_{t+1}(u,v)} p\,\mathrm{d}\mu_{\mathbf{F}}^{t+1} - \int_{F_{t+1}(u,v)} 1_G\,p\,\mathrm{d}\mu_{\mathbf{F}}^{t+1}$$

follows

$$\begin{aligned} \int_{\Omega_{t+1}} \beta \int_{F_{t+1}(u,\cdot)} 1_H\,p\,\mathrm{d}\mu_{\mathbf{F}}^{t+1}\,\mathrm{d}\mu_{t+1} &= (1-\alpha) \int_{\Omega_{t+1}} \int_{F_{t+1}(u,\cdot)} p\,\mathrm{d}\mu_{\mathbf{F}}^{t+1}\,\mathrm{d}\mu_{t+1} \\ &\quad - \int_{\Omega_{t+1}} \int_{F_{t+1}(u,\cdot)} 1_G\,p\,\mathrm{d}\mu_{\mathbf{F}}^{t+1}\,\mathrm{d}\mu_{t+1} \\ &= (1-\alpha) \int_{F_t(u)} p\,\mathrm{d}\mu_{\mathbf{F}}^t - \int_{F_t(u)} 1_G\,p\,\mathrm{d}\mu_{\mathbf{F}}^t \end{aligned}$$

and from this $\int_G \mathrm{d}\mathbf{P}_u + \int_H \beta \, \mathrm{d}\mathbf{P}_u = 1 - \alpha$. Let

$$V = \{\tilde{\omega} \in F_t(\omega) \colon X(\tilde{\omega}) > \mathrm{VaR}_{\alpha,t}(X)_{|u}\},$$

$$c = \begin{cases} 1 - \frac{\int_{H \cap V} (1-\beta) \, \mathrm{d}\mathbf{P}_u}{\int_{H \setminus V} \beta \, \mathrm{d}\mathbf{P}_u} & \text{if } \int_{H \setminus V} \beta \, \mathrm{d}\mathbf{P}_u > 0 \\ 0 & \text{otherwise.} \end{cases}$$

Then

$$\begin{aligned} 1 - \alpha &= \int_G \mathrm{d}\mathbf{P}_u + \int_H \beta \, \mathrm{d}\mathbf{P}_u \\ &= \int_G \mathrm{d}\mathbf{P}_u + c \int_{H \setminus V} \beta \, \mathrm{d}\mathbf{P}_u + \int_{H \cap V} \mathrm{d}\mathbf{P}_u \\ &= \int_{(G \cup H) \cap V} \mathrm{d}\mathbf{P}_u + \int_{G \setminus V} \mathrm{d}\mathbf{P}_u + c \int_{H \setminus V} \beta \, \mathrm{d}\mathbf{P}_u. \end{aligned} \tag{2.9}$$

We now show that, in addition,

$$\int_H \beta X \, \mathrm{d}\mathbf{P}_u \leq c \int_{H \setminus V} \beta X \, \mathrm{d}\mathbf{P}_u + \int_{H \cap V} X \, \mathrm{d}\mathbf{P}_u. \tag{2.10}$$

From $\inf_{\tilde{\omega} \in H \cap V} X(\tilde{\omega}) \geq \sup_{\tilde{\omega} \in H \setminus V} X(\tilde{\omega})$ and $0 \leq \beta \leq 1$ follows

$$\begin{aligned} \int_{H \cap V} X \, \mathrm{d}\mathbf{P}_u &\geq \int_{H \cap V} \Big(\inf_{\tilde{\omega} \in H \cap V} X(\tilde{\omega})(1 - \beta) + \beta X \Big) \, \mathrm{d}\mathbf{P}_u \\ &\geq \sup_{\tilde{\omega} \in H \setminus V} X(\tilde{\omega}) \left(\int_{H \cap V} (1 - \beta) \, \mathrm{d}\mathbf{P}_u \right) + \int_{H \cap V} \beta X \, \mathrm{d}\mathbf{P}_u \end{aligned}$$

and thus

$$\begin{aligned} \int_{H \cap V} \beta X \, \mathrm{d}\mathbf{P}_u \int_{H \setminus V} \beta \, \mathrm{d}\mathbf{P}_u &\leq \int_{H \cap V} X \, \mathrm{d}\mathbf{P}_u \int_{H \setminus V} \beta \, \mathrm{d}\mathbf{P}_u \\ &\quad - \sup_{\tilde{\omega} \in H \setminus V} X(\tilde{\omega}) \left(\int_{H \cap V} (1 - \beta) \, \mathrm{d}\mathbf{P}_u \right) \int_{H \setminus V} \beta \, \mathrm{d}\mathbf{P}_u \\ &\leq \int_{H \cap V} X \, \mathrm{d}\mathbf{P}_u \int_{H \setminus V} \beta \, \mathrm{d}\mathbf{P}_u \\ &\quad - \int_{H \setminus V} \beta X \, \mathrm{d}\mathbf{P}_u \int_{H \cap V} (1 - \beta) \, \mathrm{d}\mathbf{P}_u. \end{aligned}$$

We conclude

$$\begin{aligned}
\int_H \beta X \, \mathrm{d}\mathbf{P}_u \int_{H\setminus V} \beta \, \mathrm{d}\mathbf{P}_u &= \int_{H\setminus V} \beta X \, \mathrm{d}\mathbf{P}_u \int_{H\setminus V} \beta \, \mathrm{d}\mathbf{P}_u + \int_{H\cap V} \beta X \, \mathrm{d}\mathbf{P}_u \int_{H\setminus V} \beta \, \mathrm{d}\mathbf{P}_u \\
&\leq \int_{H\setminus V} \beta X \, \mathrm{d}\mathbf{P}_u \Big(\int_{H\setminus V} \beta \, \mathrm{d}\mathbf{P}_u - \Big(\int_{H\cap V} (1-\beta) \, \mathrm{d}\mathbf{P}_u \Big) \Big) \\
&\quad + \int_{H\cap V} X, \mathrm{d}\mathbf{P}_u \int_{H\setminus V} \beta \, \mathrm{d}\mathbf{P}_u \\
&= \int_{H\setminus V} \beta \, \mathrm{d}\mathbf{P}_u \Big(c \int_{H\setminus V} \beta X \, \mathrm{d}\mathbf{P}_u + \int_{H\cap V} X \, \mathrm{d}\mathbf{P}_u \Big),
\end{aligned}$$

which implies the inequality (2.10).

Since H and G are disjoint, it follows from the inequalities (2.8) and (2.10) that

$$\begin{aligned}
(1-\alpha) \, \mathrm{ES}_{\alpha,t}(X)_{|u} &< \int_{G\cup(H\cap V)} X \, \mathrm{d}\mathbf{P}_u + c \int_{H\setminus V} \beta \, X \, \mathrm{d}\mathbf{P}_u \\
&= \int_{(G\cup H)\cap V} X \, \mathrm{d}\mathbf{P}_u + \int_{G\setminus V} X \, \mathrm{d}\mathbf{P}_u + c \int_{H\setminus V} \beta \, X \, \mathrm{d}\mathbf{P}_u \\
&\leq \int_{(G\cup H)\cap V} X \, \mathrm{d}\mathbf{P}_u + \inf_{\tilde{\omega}\in V} X(\tilde{\omega}) \Big(\int_{G\setminus V} \mathrm{d}\mathbf{P}_u + c \int_{H\setminus V} \beta \, \mathrm{d}\mathbf{P}_u \Big) \\
&\overset{(*)}{=} \int_{(G\cup H)\cap V} X \, \mathrm{d}\mathbf{P}_u + \inf_{\tilde{\omega}\in V} X(\tilde{\omega}) \Big(1 - \alpha - \int_{(G\cup H)\cap V} \mathrm{d}\mathbf{P}_u \Big) \\
&\overset{(**)}{\leq} \int_V X \, \mathrm{d}\mathbf{P}_u + \mathrm{VaR}_{\alpha,t}(X)_{|u} \Big(1 - \alpha - \int_V \mathrm{d}\mathbf{P}_u \Big) \\
&\overset{(**)}{=} (1-\alpha) \, \mathrm{ES}_{\alpha,t}(X)_{|u},
\end{aligned}$$

where we have additionally used in $(*)$ the equation (2.9) and in $(**)$ the definition of the set V. This is a contradiction, so that our assumption that

$$\mathrm{ES}_{\alpha,t+1}(X)_{|(u,v)} \geq \mathrm{ES}_{\alpha,t}(X)_{|u} \quad \text{for almost all } v \in \Omega_{t+1}$$

must have been false. □

This demonstrates that both the dynamic value at risk and the dynamic expected shortfall are good candidates for multi-period risk management.

2.4.4 A Class of Dynamic Risk Measures on General Filtrations

In the pertinent literature, another class of dynamic risk measures is studied that are defined in an elegant way on general filtrations. Proposition 2.9 gives the method of constructing this class. Example 2.10 provides, in our view, a reason why this construction, in spite of its elegance, is hardly suitable for practical risk management.

For Proposition 2.9 we need a modification, for probability spaces, of the (pointwise) supremum over a set of functions.

Definition 2.23 Let $(\Omega, \mathscr{A}, \mu)$ be a measure space, and $\mathscr{S} \subset \mathscr{B}(\Omega, \mathbb{R})$ a subset of measurable functions. Then the *essential supremum*

$$\operatorname{ess\,sup}(\mathscr{S}) \in \mathscr{B}(\Omega, \bar{\mathbb{R}})$$

is defined by the following properties:

(i) $\operatorname{ess\,sup}(\mathscr{S}) \geq f$ holds a.s. for any $f \in \mathscr{S}$.
(ii) If $g \in \mathscr{B}(\Omega, \mathbb{R})$ satisfies the inequality $g \geq f$ a.s for each $f \in \mathscr{S}$, then it follows that $g \geq \operatorname{ess\,sup}(\mathscr{S})$ a.s.

Remark that $\operatorname{ess\,sup}(\mathscr{S})$ can take the value ∞ on a set with genuinely positive measure.

Remark 2.14 Observe that the equation $\operatorname{ess\,sup}\{f\} = f$ holds for any f. Hence for singleton sets of functions Definition 2.23 does not reduce to Definition 2.20.

Lemma 2.14 *Let* $(\Omega, \mathscr{A}, \mathbf{P})$ *be a probability space and* $\mathscr{S} \subset \mathscr{B}(\Omega, \mathbb{R})$, $\mathscr{S} \neq \emptyset$. *Then there exists* $\operatorname{ess\,sup}(\mathscr{S}) \in \mathscr{B}(\Omega, \bar{\mathbb{R}})$.

Proof Let

$$\tilde{\mathscr{S}} = \left\{\omega \mapsto \max_{g \in S}\{g(\omega)\} : S \subseteq \mathscr{S} \text{ is a finite set}\right\}$$

be the set of pointwise maxima of all finite subsets of $\mathscr{S}$ and

$$\theta(x) = \begin{cases} -\frac{\pi}{2} & \text{when } x = -\infty, \\ \arctan(x) & \text{when } -\infty < x < \infty, \\ \frac{\pi}{2} & \text{when } x = \infty. \end{cases}$$

Since $\mathbf{P}$ is a probability measure, and for any measurable function g the composition $\theta \circ g$ is bounded and measurable, there exists $\int_\Omega |\theta \circ g| \, \mathrm{d}\mathbf{P}$. From this we see $\theta \circ g$ is integrable, and

$$\alpha = \sup\left\{\int_\Omega \theta \circ g \, \mathrm{d}\mathbf{P} : g \in \tilde{\mathscr{S}}\right\} \in \bar{\mathbb{R}}$$

is well-defined. Let $\{g_k\}_{k \in \mathbb{N}} \subseteq \tilde{\mathscr{S}}$ be a sequence with

$$\lim_{k \to \infty} \int_\Omega \theta \circ g_k \, \mathrm{d}\mathbf{P} = \alpha.$$

Since for any k we can replace the function g_k with $\max\{g_1, \dots, g_k\}$, we can assume without any loss of generality that $g_{k+1} \geq g_k$ for each $k \in \mathbb{N}$. Since the sequence $\{g_k\}_{k \in \mathbb{N}}$ is increasing, we have for $f(\omega) = \sup_{k \in \mathbb{N}} g_k(\omega)$ and for each $\omega \in \Omega$

$$f(\omega) = \lim_{k \in \mathbb{N}} g_k(\omega) \in \bar{\mathbb{R}}.$$

f is also measurable as the supremum of measurable functions, because of which $\int_\Omega \theta \circ f \, \mathrm{d}\mathbf{P}$ is well defined. Since θ is bounded, it follows from the theorem of dominated convergence and the continuity of θ that

$$\alpha = \lim_{k \to \infty} \int_\Omega \theta \circ g_k \, \mathrm{d}\mathbf{P} = \int_\Omega \lim_{k \to \infty} \theta \circ g_k \, \mathrm{d}\mathbf{P} = \int_\Omega \theta \circ \lim_{k \to \infty} g_k \, \mathrm{d}\mathbf{P} = \int_\Omega \theta \circ f \, \mathrm{d}\mathbf{P}.$$

Let $g \in \mathscr{S}$. Since $\max(g_k, g) \in \tilde{\mathscr{S}}$ for any $k \in \mathbb{N}$ and

$$\max\{f, g\} = \max\left\{\lim_{k \to \infty} g_k, g\right\} = \lim_{k \to \infty} \max\{g_k, g\},$$

it follows that

$$\int_\Omega \theta \circ \max\{f, g\} \, \mathrm{d}\mathbf{P} = \int_\Omega \lim_{k \to \infty} \theta \circ \max\{g_k, g\} \, \mathrm{d}\mathbf{P} = \lim_{k \to \infty} \int_\Omega \theta \circ \max\{g_k, g\} \, \mathrm{d}\mathbf{P} \leq \alpha.$$

Thus the integral over $\theta \circ f - \theta \circ \max\{f, g\}$ is non-negative. However,

$$\theta \circ \max\{f, g\} - \theta \circ f$$

is also non-negative since θ is monotone. This is only possible if $\theta \circ \max\{f, g\} = \theta \circ f$ a.s. It follows that $f \geq g$ a.s.

Now let $g \in \mathscr{B}(\Omega, \mathbb{R})$ be a measurable function that satisfies the inequality $g \geq h$ a.s for any $h \in \mathscr{S}$. For each $k \in \mathbb{N}$ then it obviously holds that $g \geq g_k$. It follows that $g \geq \sup_{k \in \mathbb{N}} g_k = f$, and so $f = \operatorname{ess\,sup}(\mathscr{S})$. □

The expected shortfall of a random variable X can be expressed as the supremum of the expectations of X relative to a class of probability measures (Proposition 2.3). The following proposition shows that this sort of representation leads in a natural way to coherent risk measures.

Proposition 2.8 *Let* $\mathscr{W}$ *be a subset of the probability measures on a probability space* $(\Omega, \mathscr{A}, \mathbf{P})$ *with the following properties:*

(i) *For each* $X \in \mathscr{M}(\Omega, \mathbb{R})$ *and each* $Q \in \mathscr{W}$ *there exists* $\mathrm{E}_Q(X)$.
(ii) $\mathbf{Q} \ll \mathbf{P}$ *for all* $\mathbf{Q} \in \mathscr{W}$.

Then a coherent risk measure is defined by

$$\rho^{\mathscr{W}}(X) = \sup_{\mathbf{Q} \in \mathscr{W}} \left\{\mathrm{E}_{\mathbf{Q}}(X)\right\}.$$

We will prove Proposition 2.8 after Proposition 2.9 as an elementary corollary.

If one chooses $\mathscr{M}_n(\Omega, \mathbb{R})$ to be the space of almost everywhere measurable functions, then one can show that a large class of coherent risk measures can be represented according to Proposition 2.8 [4, Theorem 3.2]. This motivates also constructing coherent dynamic risk measures using this representation.

Proposition 2.9 *Let $(\Omega, \mathscr{A}, \mathbf{P})$ be a probability space and $(\mathscr{F}_t)_{t\in\{0,\dots,n\}}$ a filtration with $\mathscr{F}_n = \mathscr{A}$. For $t \in \{0, \dots, n\}$ let $\mathscr{M}_t(\Omega, \mathbb{R})$ be the vector space of a.e. bounded, $\mathscr{F}_t$-measurable functions. Let $\mathscr{W}$ be a set of probability measures on Ω.*

Then the family of maps

$$\rho_t^{\mathscr{W}} : \mathscr{M}_n(\Omega, \mathbb{R}) \to \mathscr{M}_t(\Omega, \mathbb{R}), \quad X \mapsto \rho_t^{\mathscr{W}}(X) = \operatorname{ess}\sup_{\mathbf{Q}\in\mathscr{W}} \mathrm{E}_{\mathbf{Q}}(X \mid \mathscr{F}_t)$$

is a coherent, dynamic risk measure.

Proof We show first that $\rho_t^{\mathscr{W}}$ is a dynamic risk measure.

By Lemma 2.14 the map $\rho_t^{\mathscr{W}}(X)$ exists and it is $\mathscr{F}_t$-measurable. Since X is almost everywhere bounded, we have $\rho_t^{\mathscr{W}}(X)_{|\omega} \in \mathbb{R}$ a.s.

Monotony: Suppose $X_1 \geq X_2$ a.s. Then for each $\mathbf{Q} \in \mathscr{W}$ the inequality $\mathrm{E}_{\mathbf{Q}}(X_1) \geq \mathrm{E}_{\mathbf{Q}}(X_2)$ holds. Since the relationship "$\geq$" is preserved in passing to the supremum the monotone property of $\rho_t^{\mathscr{W}}$ follows.

Translation invariance: Let $K \in \mathscr{M}_t(\Omega, \mathbb{R})$ and $\mathbf{Q} \in \mathscr{W}$. Since K is $\mathscr{F}_t$-measurable, we have

$$\mathrm{E}_{\mathbf{Q}}(X + K \mid \mathscr{F}_t) = \mathrm{E}_{\mathbf{Q}}(X \mid \mathscr{F}_t) + K.$$

Passing to the supremum then brings $\rho_t^{\mathscr{W}}(X + K) = \rho_t^{\mathscr{W}}(X) + K$.

Homogeneity: Let $K \in \mathscr{M}_t(\Omega, \mathbb{R})$ with $K \geq 0$ a.s. and $KX \in \mathscr{M}_n(\Omega, \mathbb{R})$. For $\mathbf{Q} \in \mathscr{W}$ it then holds that

$$\mathrm{E}_{\mathbf{Q}}(KX \mid \mathscr{F}_t) = K\, \mathrm{E}_{\mathbf{Q}}(X \mid \mathscr{F}_t)$$

and for this that

$$\rho_t^{\mathscr{W}}(KX) = \sup_{\mathbf{Q}\in\mathscr{W}} \left\{ K\, \mathrm{E}_{\mathbf{Q}}(X) \mid \mathscr{F}_t \right\} = K \sup_{\mathbf{Q}\in\mathscr{W}} \left\{ \mathrm{E}_{\mathbf{Q}}(X \mid \mathscr{F}_t) \right\} = K \rho_t^{\mathscr{W}}(X),$$

where we have used $K \geq 0$ in deriving the second equality.

Subadditivity: For $X_1, X_2 \in \mathscr{M}_n(\Omega, \mathbb{R})$ we have

$$\begin{aligned} \sup_{\mathbf{Q}\in\mathscr{W}} \left\{ \mathrm{E}_{\mathbf{Q}}(X_1 + X_2 \mid \mathscr{F}_t) \right\} &= \sup_{\mathbf{Q}\in\mathscr{W}} \left\{ \mathrm{E}_{\mathbf{Q}}(X_1 \mid \mathscr{F}_t) + \mathrm{E}_{\mathbf{Q}}(X_2 \mid \mathscr{F}_t) \right\} \\ &\leq \sup_{\mathbf{Q}\in\mathscr{W}} \left\{ \mathrm{E}_{\mathbf{Q}}(X_1 \mid \mathscr{F}_t) \right\} + \sup_{\mathbf{Q}\in\mathscr{W}} \left\{ \mathrm{E}_{\mathbf{Q}}(X_2 \mid \mathscr{F}_t) \right\}. \end{aligned}$$

□

Proof of Proposition 2.8 We apply Proposition 2.9 for

$$n = 1, \qquad \mathscr{F}_1 = \mathscr{A}, \qquad \mathscr{F}_0 = \{\emptyset, \Omega\}.$$

Then $\rho_0^{\mathscr{W}}(X) = \operatorname{ess\,sup}_{\mathbf{Q}\in\mathscr{W}} \mathrm{E}_{\mathbf{Q}}(X \mid \mathscr{F}_0) = \sup_{\mathbf{Q}\in\mathscr{W}} \mathrm{E}_{\mathbf{Q}}(X) = \rho^{\mathscr{W}}(X)$ and it immediately follows from the definition of a coherent dynamic risk measure that ρ_0 is a coherent risk measure. □

It would seem mathematically plausible, to extend the expected shortfall to a dynamic risk measure using Proposition 2.9 for a given filtration. In fact, this route has been traveled in the literature (see e.g. [8, Example 25]). Unfortunately, this extension is not well suited for practical applications, since the extended risk measure loses the characteristics of the expected shortfall (see Example 2.10).

Example 2.10 We consider the probability space

$$(\Omega, \mathbf{P}) = \big(]0,1[\times]0,1[, \mathrm{d}\omega\big),$$

where Ω is equipped with the Borel algebra $\mathscr{B}(]0,1[\times]0,1[)$ and $\mathbf{P} = \mathrm{d}\omega = \mathrm{d}\omega_1 \otimes \mathrm{d}\omega_2$ is the Lebesgue measure on $\mathbb{R}^2$. We choose the product filtration

$$\mathscr{F}_0 = \{\emptyset, \Omega\}, \qquad \mathscr{F}_1 = \big\{A\times]0,1[\colon A \in \mathscr{B}(]0,1[)\big\}, \qquad \mathscr{F}_2 = \mathscr{B}\big(]0,1[\times]0,1[\big)$$

For $\alpha \in]0,1[$ let

$$\mathscr{W}_\alpha = \left\{\mathbf{Q}\colon \mathbf{Q} \text{ is a probability measure with } \mathbf{Q} \ll \mathbf{P} \text{ and } \frac{\mathrm{d}\mathbf{Q}}{\mathrm{d}\mathbf{P}} \le \frac{1}{1-\alpha}\right\}.$$

Then by Proposition 2.3 it holds that $\rho_0^{\mathscr{W}_\alpha} = \mathrm{ES}_\alpha$ and $\rho_t^{\mathscr{W}_\alpha}$ is the coherent dynamic risk measure derived from the expected shortfall using Proposition 2.9.

For $\mu \in]0,1[$ let

$$A_\mu = \left\{\omega\colon 0 < \omega_1 < \frac{1}{4},\ 0 < \omega_2 < 2(1-\mu)(1-\alpha)\right\},$$

$$B_\mu = \left\{\omega\colon \frac{3}{4} \le \omega_1 < 1,\ 0 < \omega_2 < 2(1-\mu)(1-\alpha)\right\},$$

$$C_\mu = \{\omega\colon 1 - \mu + \mu\alpha \le \omega_2 < 1\}$$

(see the Fig. 2.9). By construction, we have $A_\mu \cap B_\mu = B_\mu \cap C_\mu = C_\mu \cap A_\mu = \emptyset$ as well as

$$\mathbf{P}(A_\mu) = \mathbf{P}(B_\mu) = \frac{1}{2}(1-\mu)(1-\alpha), \qquad \mathbf{P}(C_\mu) = \mu(1-\alpha),$$

so that $\mathbf{P}(A_\mu \cup B_\mu \cup C_\mu) = 1 - \alpha$. Therefore the measure $\mathbf{Q}_\mu$ defined by the density

$$\frac{\mathrm{d}\mathbf{Q}_\mu}{\mathrm{d}\mathbf{P}} = \frac{1_{A_\mu} + 1_{B_\mu} + 1_{C_\mu}}{1-\alpha}$$

must be a probability measure and satisfies the inequality

$$\frac{d\mathbf{Q}_\mu}{d\mathbf{P}} \leq \frac{1}{1-\alpha}.$$

We have $\mathbf{Q}_\mu \in \mathscr{W}_\alpha$ and thus for all bounded random variables X the inequality

$$\mathrm{E}_{\mathbf{Q}_\mu}(X \mid \mathscr{F}_t) \leq \rho_t^{\mathscr{W}_\alpha}(X).$$

We now look at the random variable

$$X\colon \Omega \to \mathbb{R}, \quad (\omega_1, \omega_2) \mapsto \xi(\omega_2),$$

where ξ depends only on ω_2 and is monotone increasing. For $\omega_1 \in]\frac{1}{4}, \frac{3}{4}[$ it holds by Proposition 2.7 that

$$\begin{aligned}
\mathrm{E}_{\mathbf{Q}_\mu}(X \mid \mathscr{F}_1)_{|\omega_1} &= \frac{\int_0^1 \xi(\omega_2)\frac{d\mathbf{Q}_\mu}{d\mathbf{P}}(\omega_1,\omega_2)\,d\omega_2}{\int_0^1 \frac{d\mathbf{Q}_\mu}{d\mathbf{P}}(\omega_1,\omega_2)\,d\omega_2} \\
&= \frac{\frac{1}{1-\alpha}\int_{1-\mu(1-\alpha)}^1 \xi(\omega_2)\,d\omega_2}{\frac{1}{1-\alpha}\int_{1-\mu(1-\alpha)}^1 d\omega_2} \\
&= \frac{1}{\mu(1-\alpha)}\int_0^{\mu(1-\alpha)} \xi(1-x)\,dx
\end{aligned}$$

and thus

$$\lim_{\mu\to 0} \mathrm{E}_{\mathbf{Q}_\mu}(X \mid \mathscr{F}_1)_{|\omega_1} = \lim_{x\to 0} \xi(1-x) = \sup_{\omega_2 \in]0,1[} \xi(\omega_2) = \sup_{\omega\in\Omega} X(\omega).$$

Now let $\omega_1 \in]0, \frac{1}{4}[\cup]\frac{3}{4}, 1[$ and

$$\tilde{A}_\mu = \frac{1}{4} + A_\mu, \qquad \tilde{B}_\mu = -\frac{1}{4} + B_\mu.$$

We can repeat the same analysis with the risk measure $\tilde{\mathbf{Q}}_\mu$ given by

$$\frac{d\tilde{\mathbf{Q}}_\mu}{d\mathbf{P}} = \frac{1_{\tilde{A}_\mu} + 1_{\tilde{B}_\mu} + 1_{C_\mu}}{1-\alpha}$$

and obtain

$$\lim_{\mu\to 0} \mathrm{E}_{\tilde{\mathbf{Q}}_\mu}(X \mid \mathscr{F}_1)_{|\omega_1} = \sup_{\omega\in\Omega} X(\omega)$$

for all $\omega_1 \in]0, \frac{1}{4}[\cup]\frac{3}{4}, 1[$. Because $\rho_1^{\mathscr{W}_\alpha}(X) \leq \sup_{\omega\in\Omega} X(\omega)$ it follows that $\rho_1^{\mathscr{W}_\alpha}(X) = \sup_{\omega\in\Omega} X(\omega)$ almost surely. This means that the dynamic risk measure

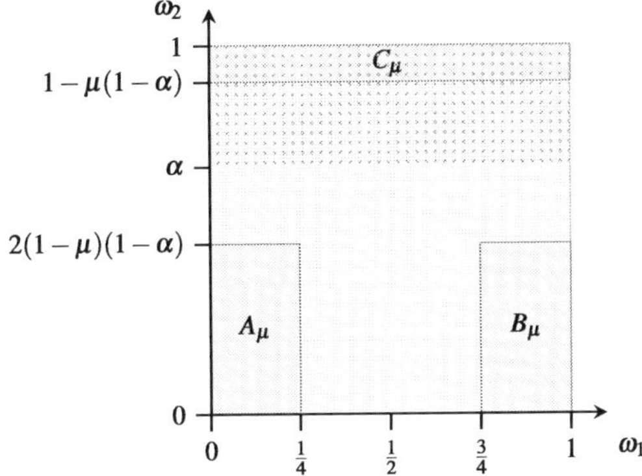

Fig. 2.9 The construction of the measure $\mathbf{Q}_\mu$ in Example 2.10

has completely lost, at time 1, the character of the expected shortfall. This risk measure is naturally totally unsuited for risk management at time $t = 1$. If ξ is not almost certainly constant, then, in addition, there follows the existence of a $c > 0$, so that

$$\rho_0^{\mathscr{W}_\alpha}(X) + c < \rho_1^{\mathscr{W}_\alpha}(X) \quad \text{a.s.}$$

The risk measure $\rho_t^{\mathscr{W}_\alpha}$ is not weakly time-consistent and thus also not time-consistent.

Observe that this example is simple but not at all pathological: the critical feature of our example is that for each $\omega \in \Omega$ we can find a probability measure $\mathbf{Q} \in \mathscr{W}_\alpha$ which, restricted to $F_t(\omega)$, does not vanish only in a small neighborhood of the supremum of X. Since $F_t(\omega)$ is a null set relative to $\mathbf{P}$ we have enough room on $\Omega \setminus F_t(\omega)$ to extend the measure $\mathbf{Q}$ so that we have $\mathbf{Q} \in \mathscr{W}_\alpha$. It follows that $\rho_t^{\mathscr{W}_\alpha}(X) = \sup_{\hat{\omega} \in F_t(\omega)}\{X(\hat{\omega})\}$. This construction can be carried out for nearly any practical example.

In the literature it is often concluded that one must be very cautious in using the expected shortfall in a dynamic context, or that the expected shortfall is simply unsuited to dynamic contexts (see for example [2, Sect. 5.3], [8, Example 25]). The authors limit themselves to a subclass of dynamic risk measures that satisfy additional axiomatically introduced conditions of time-consistence. A typical example of such a condition is the following (see [7, 8]).

Definition 2.24 A dynamic risk measure ρ_t is called *comparison consistent,* if for all random variables $X, Y \in \mathscr{M}(\Omega, \mathbb{R})$ it holds that

$$\rho_{t+1}(X) \geq \rho_{t+1}(Y) \text{ a.s.} \quad \Rightarrow \quad \rho_t(X) \geq \rho_t(Y) \text{ a.s.}$$

Remark 2.15 In [7, 8] the author speaks of "time-consistent" instead of "comparison consistent". We are, however, of the opinion that the consistency condition in Definition 2.24 is not plausible. Since the risk measure at the start of the time period t does not provide a full description of the future cash flow, it could happen that as

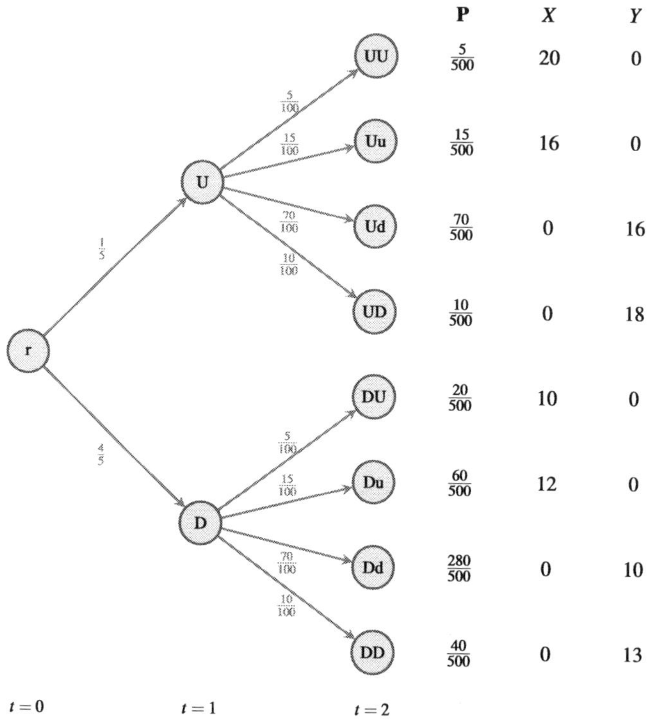

Fig. 2.10 Breakdown of comparison consistency for the expected shortfall

a result of new information at the start of period $t+1$ the risk of the random quantity X is larger than that associated with Y, even if in the earlier time period their relationship was the opposite. Example 2.11 shows how such a gain in information can lead to a violation of comparison consistency.

Therefore we shall reserve "time-consistent" for the variant given in Definition 2.21.

The expected shortfall has properties that have shown themselves useful in the risk management in the single-period case. The extended expected shortfall derived from using Proposition 2.9 fundamentally changes its nature if one breaks up the period into several partial periods. This is not a deficiency of the expected shortfall but the fault of the special construction in the extension. In fact, the generalization given in the Definition 2.19 of the expected shortfall is time-consistent and retains its character with passage of time.

For a dynamic risk measure of the form $\rho_t^{\mathscr{W}}(X) = \operatorname{ess\,sup}_{\mathbf{Q}\in\mathscr{W}} \mathrm{E}_{\mathbf{Q}}(X \mid \mathscr{F}_t)$ the defining set $\mathscr{W}$ does not depend on the time t. It is therefore difficult to describe any gain in information about the risk environment that may happen with time. This seems to be the key problem with this construction and is independent of the choice of the special risk measure "expected shortfall".

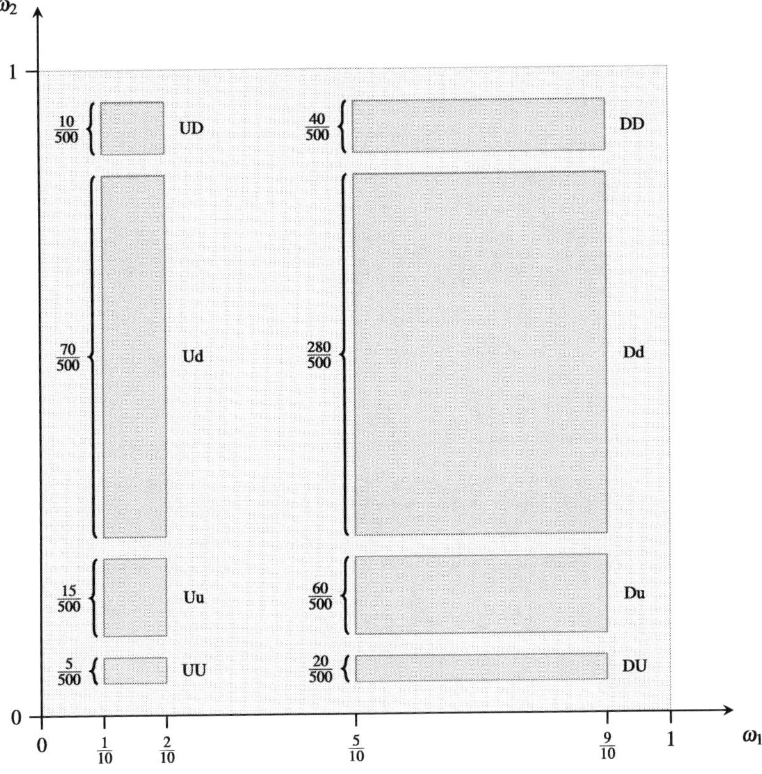

Fig. 2.11 Construction of the discrete Example 2.11 using a product filtration with $\Omega =]0,1[\times]0,1[$ and $\mathbf{P} = \frac{5}{2} \times 1_{\mathrm{UU \cup Uu \cup Ud \cup UD \cup DU \cup Du \cup Dd \cup DD}}$. The filtration is given by $\mathscr{F}_0 = \{\emptyset, \Omega\}$, $\mathscr{F}_1 = \{A \times]0,1[: A \subseteq [0,1[\text{ is Borel-measurable}\}$ and $\mathscr{F}_2 = \mathscr{B}(]0,1[\times]0,1[)$

The following example shows that the expected shortfall fails to be comparison consistent, even though it is time-consistent.

Example 2.11 Figure 2.10 shows an example in which the comparison consistency of the expected shortfall $\mathrm{ES}_{80\,\%,t}$ breaks down. Note that this discrete example can be constructed using a filtered product economy based on a Borel algebra (see Fig. 2.11). From Lemma 2.5 we obtain at time $t = 1$ the following values:

$$\mathrm{ES}_{80\,\%,1}(X)_{|(U)} = \frac{20 \times 5 + 16 \times 15}{20} = \frac{340}{20} = 17,$$

$$\mathrm{ES}_{80\,\%,1}(X)_{|(D)} = \frac{12 \times 15 + 10 \times 5}{20} = \frac{230}{20} = 11.5,$$

$$\mathrm{ES}_{80\,\%,1}(Y)_{|(U)} = \frac{18 \times 10 + 16 \times 10}{20} = \frac{340}{20} = 17,$$

$$\mathrm{ES}_{80\,\%,1}(Y)_{|(D)} = \frac{13\times 10+10\times 10}{20} = \frac{230}{20} = 11.5,$$

so that the values of X and Y at time 1 agree. Comparison consistency would then imply that one also has

$$\mathrm{ES}_{80\,\%,0}(X)_{|(r)} = \mathrm{ES}_{80\,\%,0}(Y)_{|(r)}.$$

However, direct calculation for the time $t=0$ gives

$$\mathrm{ES}_{80\,\%,0}(X)_{|(r)} = \frac{20\times 5+16\times 15+12\times 60+10\times 20}{100} = \frac{1260}{100} = 12.6$$

$$\mathrm{ES}_{80\,\%,0}(Y)_{|(r)} = \frac{18\times 10+16\times 70+13\times 20}{100} = \frac{1560}{100} = 15.6$$

If we set $\tilde{Y} = Y - c$ for $c \in]0,3[$, we obtain the apparently stronger statements,

$$\mathrm{ES}_{80\,\%,0}(\tilde{Y}) > \mathrm{ES}_{80\,\%,0}(X), \quad \text{but } \mathrm{ES}_{80\,\%,1}(\tilde{Y}) < \mathrm{ES}_{80\,\%,1}(X) \text{ a.s.}$$

We interpret this result as saying that portfolio $\tilde{Y}$ wins in capital efficiency over portfolio X at time $t=1$. Comparison consistency is simply lost because in evaluating $\mathrm{ES}_{80\,\%,0}$ both branches, U and D, contribute to the 20 % of highest losses, while at time 1 the choice is of the 20 % largest losses on each of the branches U resp. D. The relevant information gain in node U is thus that the events that bring losses in D cannot occur again. Analogously one gains the information in node D that the loss-bringing events in U cannot occur. So we do not see a consistency violation in the vernacular sense.

References

1. P. Artzner, F. Delbaen, J.-M. Eber, D. Heath, Coherent measures of risk. Math. Finance **9**(3), 203–228 (1999) (p. 31)
2. P. Artzner, F. Delbaen, J.-M. Eber, D. Heath, H. Ku, Coherent multiperiod risk adjusted values and Bellman's principle. Ann. Oper. Res. **152**(1), 5–22 (2007) (p. 69)
3. H. Bauer, *Measure and Integration Theory*. De Gruyter Studies in Mathematics (de Gruyter, Berlin, 2001) (p. 48)
4. F. Delbaen, Coherent risk measures on general probability spaces, in *Advances in Finance and Stochastics, Essays in Honour of Dieter Sondermann* (Springer, New York, 2002), pp. 1–37 (p. 66)
5. J. Jacod, P.E. Protter, *Probability Essentials*, 2nd edn. Universitext (Springer, Berlin, 2003) (p. 46)
6. A. McNeil, R. Frey, P. Embrechts, *Quantitative Risk Management. Concepts, Techniques, Tools* (Princeton University Press, Princeton, 2005) (p. 36)
7. F. Riedel, Dynamic coherent risk measures, in *Stochastic Processes and Their Applications*, vol. 112, (2004), pp. 185–200 (p. 69)
8. F. Riedel, Dynamic risk measures. Lecture Notes of a Mini-Course at Paris IX, Dauphine, December 2007 (pp. 54, 67, 69)
9. W.R. van Zwet, A strong law for linear functions of order statistics. Ann. Probab. **8**(5), 986–990 (1980) (p. 40)

Chapter 3
Dependencies

3.1 Diversification

"Bad luck never comes alone", they say. Happily this is not exactly true, since it is the circumstance that "bad luck" does not always occur in concentration that makes the business model of "insurance" possible at all. Otherwise insurance companies that insure a 'ive would have to set aside the full risk capital for each individual risk, which would naturally not be "affordable". In this connection the effect is sometimes called "equalization in the collective". We consider the *diversification effect* a bit more generally for an arbitrary (thus, in particular, a possibly small) number of risks which can exhibit loss distributions of various types.

Definition 3.1 Let $\rho\colon \mathcal{M}(\Omega, \mathbb{R}) \to \mathbb{R}$, $X \mapsto \rho(X)$ be a risk measure. We consider an aggregate system with several partial risks, which are described by the loss distributions $X_i \in \mathcal{M}(\Omega, \mathbb{R})$ $(i \in \{1, \ldots, m\})$. Then the *diversification effect of the aggregate system* $\{X_1, \ldots, X_m\}$ *with respect to* ρ is given by

$$\sum_{i=1}^{m} \rho(X_i) - \rho\left(\sum_{i=1}^{m} X_i\right).$$

We say that there is a *diversification effect* if this number is positive.

Proposition 3.1 *For coherent risk measures the diversification effect is never negative.*

Proof This follows directly from subadditivity $\rho(\sum_{i=1}^{m} X_i) \le \sum_{i=1}^{m} \rho(X_i)$. □

Remark 3.1 In general, whether the diversification effect is positive depends both on the system of random variables and on the risk measure. For value at risk, for example, there could be a negative diversification effect, for which reason this risk measure is often criticized (see e.g., Example 2.2 and [1]).

M. Kriele, J. Wolf, *Value-Oriented Risk Management of Insurance Companies*,
EAA Series, DOI 10.1007/978-1-4471-6305-3_3, © Springer-Verlag London 2014

If the risk measure ρ is a given, then the diversification effect varies with the dependency structure of the partial risks X_i.

Example 3.1 One can imagine, for example, an insurance company that has two lines of business, hail insurance with the risk "hail" and comprehensive insurance with the risk "car damage". Now let X_1 be the yearly loss on hail insurance, and X_2 the yearly loss on comprehensive car insurance. In this, both X_1 and X_2 depend a priori on both risks. It is certainly not always the case, if it hails, that every insured person suffers car damage, and conversely it is clear that car accidents cannot cause hailstorms. If we denote the total loss from the insurance policies with the company by $X = X_1 + X_2$, then the question arises how we are to calculate these losses, i.e., what is the distribution of X. Since car damage can also arise from big hail stones, and in a hailstorm the road conditions are particularly bad so that one must take into account more accidents, X_1 and X_2 are not independent. To estimate the distribution of X correctly, one should consider both risks at the same time. This would mean extensive data requirements.

It would be more practical if at first one were able to estimate the two risks for themselves, and then consider, in a second step, their dependence. This is in fact possible: let F_X, F_{X_1}, F_{X_2} be the distribution functions of X, X_1, X_2. Then there is a function $C\colon [0,1] \times [0,1] \to [0,1]$, so that

$$F_X(x_1, x_2) = C\big(F_{X_1}(x_1), F_{X_2}(x_2)\big).$$

Thus one may at first find separately the distribution function for hail and car damage, and then in a second step try to find the function C. If one does this, one knows the joint distribution.

The dependence of hail and car damage is determined in this method by the form of C. The function C is called a *copula* (for a formal description see Definition 3.3). We will see, in Sects. 3.2.1.1 and 3.2.3 that working with copulas is no more difficult than working with the ordinary distribution functions that any loss actuary knows well.

Nonetheless, the use of copulas in companies is not very widespread, and other dependency structures (catchword "correlation matrix") are often employed. We will examine correlations more closely in Sect. 3.3. Here let us remark that they are not really more manageable in stochastic risk models and, in addition, they provide much less information than (simple) copulas. This will be clear when we first study the notion of a copula, and then reflect on what correlations really mean.

3.2 Copulas

Here we follow the treatment in [2]. Since the function $C(F_{X_1}(x_1), F_{X_2}(x_2))$, introduced in Sect. 3.1, is a two-dimensional distribution function, C itself must have certain properties. If an event is "impossible" for the i-th risk factor then it cannot occur in the aggregate risk. From this follows the property

$$C(u_1, 0) = C(0, u_2) = 0 \quad \text{for all } u_1, u_2.$$

For the distribution function $(x_1, x_2) \mapsto F_X(x_1, x_2)$ we can observe that $F_{X_1}(x_1) = \lim_{y\to\infty} F_X(x_1, y)$ and $F_{X_2}(x_2) = \lim_{y\to\infty} F_X(y, x_2)$. This is expressed by the property

$$C(1, u_2) = u_2, \qquad C(u_1, 1) = u_1 \quad \text{for all } u_1, u_2.$$

For a general bivariate distribution function F_X we have

$$\begin{aligned} 0 &\le \mathbf{P}(x_{11} < X_1 \le x_{12},\ x_{21} < X_2 \le x_{22}) \\ &= F(x_{12}, x_{22}) - F(x_{11}, x_{22}) - F(x_{12}, x_{21}) + F(x_{11}, x_{21}), \end{aligned}$$

which, with $u_{ij} = F_{X_i}(x_{ij})$, implies the inequality

$$C(u_{12}, u_{22}) - C(u_{11}, u_{22}) - C(u_{12}, u_{21}) + C(u_{11}, u_{21}) \ge 0.$$

To generalize these necessary conditions for C to m dimensions we need a little more terminology. We denote by $\bar{\mathbb{R}} = \{-\infty\} \cup \mathbb{R} \cup \{\infty\}$ the 2-point compactification of $\mathbb{R}$ with the canonically induced order ($-\infty \le a$, $a \le \infty$ for all $a \in \mathbb{R}$) and write

$$\bar{\mathbb{R}}^m = \overbrace{\bar{\mathbb{R}} \times \cdots \times \bar{\mathbb{R}}}^{m \text{ factors}}.$$

For $a, b \in \bar{\mathbb{R}}^m$ with $a_i \le b_i$ for all i, let $[a, b] = [a_1, b_1] \times \cdots \times [a_m, b_m]$.

Definition 3.2 Let $S_1, \dots, S_m$ be subsets of $\bar{\mathbb{R}}$, $a_i = \inf S_i$ and $b_i = \sup S_i$. A *pre-distribution function*[1] is a map

$$F : S_1 \times \cdots \times S_m \to [0, 1], \quad x \mapsto F(x_1, \dots, x_m)$$

with

(i) $F(x_1, \dots, x_{i-1}, a_i, x_{i+1}, \dots, x_m) = 0\ \forall x \in S_1 \times \cdots \times S_m,\ \forall i \in \{1, \dots, m\}$,
(ii) $V_F(]x_1, x_2]) := \sum_{i_1=1}^{2} \cdots \sum_{i_m=1}^{2} (-1)^{i_1+\cdots+i_m} F((x_{i_1})_1, \dots, (x_{i_m})_m) \ge 0$ for all intervals $[x_1, x_2] \subset S_1 \times \cdots \times S_m$.

If $b_k \in S_k$ for all k, then the i-th *marginal distribution* is given by the map

$$F_{(i)} : S_i \to [0, 1], \quad x \mapsto F(b_1, \dots, b_{i-1}, x, b_i, \dots, b_m).$$

[1]This is not standard terminology but is practical in our context.

Clearly any distribution function F_X is a predistribution function, where we have

$$\mathbf{P}\big((x_1)_1 < X_1 \leq (x_2)_1, \ldots, (x_1)_m < X_m \leq (x_2)_m\big) = V_{F_X}\big(]x_1, x_2]\big).$$

If, conversely, $S_i = \bar{\mathbb{R}}$ for all i and $F(\infty, \ldots, \infty) = 1$, then a distribution function is defined by

$$\mathbf{P}\big(x \in]x_1, x_2]\big) = V_F\big(]x_1, x_2]\big).$$

If F is a distribution with marginal distributions $F_{(i)}$, and $C\colon [0,1]^m \to [0,1]$ is a map with

$$F(x_1, \ldots, x_m) = C\big(F_{(1)}(X_1), \ldots, F_{(m)}(X_m)\big),$$

then C is obviously a copula in the sense of the following definition:

Definition 3.3 A predistribution function $C\colon [0,1]^m \to [0,1]$ with marginal distributions

$$u_i \mapsto C_{(i)}(u_i) = u_i$$

for all $i \in \{1, \ldots, m\}$ is called a *copula*.

Proposition 3.2 *Let $C\colon [0,1]^m \to [0,1]$ be a copula and $F_1, \ldots, F_m\colon \bar{\mathbb{R}} \to [0,1]$ be 1-dimensional distribution functions. Then*

$$x \mapsto F(x_1, \ldots, x_m) := C\big(F_1(x_1), \ldots, F_m(x_m)\big)$$

is an m-dimensional distribution with marginal distribution functions $F_1, \ldots, F_m$.

Proof The property $C_{(i)}(u_i) = u_i$ implies

$$F(\infty, \ldots, \infty, x_i, \infty, \ldots, \infty) = C\big(1, \ldots, 1, F_i(x_i), 1, \ldots, 1\big) = F_i(x_i)$$

and thus, in particular, $F(\infty, \ldots, \infty) = C(1, \ldots, 1) = 1$.
For $i \in \{1, \ldots, m\}$ there holds

$$\begin{aligned} &F(x_1, \ldots, x_{i-1}, -\infty, x_{i+1}, \ldots, x_m) \\ &\quad = C\big(F_1(x_1), \ldots, F_{i-1}(x_{i-1}), 0, F(x_{i+1}), \ldots, F_m(x_m)\big) \\ &\quad = 0. \end{aligned}$$

The proof is finished by verifying that

$$\begin{aligned} V_F\big(]x_1, x_2]\big) &= \sum_{i_1=1}^{2} \cdots \sum_{i_m=1}^{2} (-1)^{i_1+\cdots+i_m} F\big((x_{i_1})_1, \ldots, (x_{i_m})_m\big) \\ &= \sum_{i_1=1}^{2} \cdots \sum_{i_m=1}^{2} (-1)^{i_1+\cdots+i_m} C\big(F_1\big((x_{i_1})_1\big), \ldots, F_m\big((x_{i_m})_m\big)\big) \end{aligned}$$

$$= V_C\left(\left]\begin{pmatrix} F_1((x_1)_1) \\ \vdots \\ F_m((x_1)_m) \end{pmatrix}, \begin{pmatrix} F_1((x_2)_1) \\ \vdots \\ F_m((x_2)_m) \end{pmatrix}\right]\right) \geq 0.$$

□

One can, alternatively, think of a copula as a multivariate distribution function whose marginal distributions are all exactly the uniform distribution on [0, 1].

Lemma 3.1 *Let* $F: S_1 \times \cdots \times S_m \to [0,1]$ *be a predistribution function and assume*

$$x_1, x_2 \in S_1 \times \cdots \times S_m.$$

Then

$$\left|F(x_2) - F(x_1)\right| \leq \sum_{i=1}^{m} \left|F_{(i)}\big((x_2)_i\big) - F_{(i)}\big((x_1)_i\big)\right|.$$

Proof By the triangle inequality we deduce

$$\begin{aligned}
\left|F(x_2) - F(x_1)\right| &= \left|F\big((x_2)_1, \ldots, (x_2)_m\big) - F\big((x_1)_1, \ldots, (x_1)_m\big)\right| \\
&\leq \left|F\big((x_2)_1, \ldots, (x_2)_m\big) - F\big((x_1)_1, (x_2)_2, \ldots, (x_2)_m\big)\right| \\
&\quad + \left|F\big((x_1)_1, (x_2)_2, \ldots, (x_2)_m\big)\right. \\
&\quad \left. - F\big((x_1)_1, (x_1)_2, (x_2)_3, \ldots, (x_2)_m\big)\right| \\
&\quad + \cdots \\
&\quad + \left|F\big((x_1)_1, \ldots, (x_1)_{m-1}, (x_2)_m\big) - F\big((x_1)_1, \ldots, (x_1)_m\big)\right|.
\end{aligned}$$

We have to show that for each i the map

$$\begin{aligned}
(t_1, \ldots, t_{i-1}, t_{i+1}, \ldots, t_m) \mapsto &\left|F(t_1, \ldots, t_{i-1}, s_2, t_{i+1}, \ldots, t_m)\right. \\
&\left. - F(t_1, \ldots, t_{i-1}, s_1, t_{i+1}, \ldots, t_m)\right| \\
&=: g_{s_1,s_2}(t_1, \ldots, t_{i-1}, t_{i+1}, \ldots, t_m)
\end{aligned}$$

is monotone increasing with respect to each t_j ($j \in \{1, \ldots, i-1, i+1, \ldots, m\}$). For then it follows, with $b_i = \sup S_i$, that

$$\begin{aligned}
&\left|F\big((x_1)_1, \ldots, (x_1)_{i-1}, (x_2)_i, (x_2)_{i+1}, \ldots, (x_2)_m\big)\right. \\
&\quad \left. - F\big((x_1)_1, \ldots, (x_1)_{i-1}, (x_1)_i, (x_2)_{i+1}, \ldots, (x_2)_m\big)\right| \\
&\leq \left|F\big(b_1, \ldots, b_{i-1}, (x_2)_i, b_{i+1}, \ldots, b_m\big) - F\big(b_1, \ldots, b_{i-1}, (x_1)_i, b_{i+1}, \ldots, b_m\big)\right| \\
&= \left|F_{(i)}\big((x_2)_i\big) - F_{(i)}\big((x_1)_i\big)\right|.
\end{aligned}$$

To show monotony we can assume, with no loss in generality, that $s_2 > s_1$. In this case the difference is positive, and we can drop the absolute value. Further let

$r_2 > r_1, r_1, r_2 \in S_j$, $j \in \{1, \dots, i-1, i+1, \dots, m\}$. We now apply the property (ii) in Definition 3.2 on the set

$$\left] \sum_{k \in \{1,\dots,m\} \setminus \{i,j\}} t_k \mathbf{e}_k + r_1 \mathbf{e}_j + s_1 \mathbf{e}_i, \sum_{k \in \{1,\dots,m\} \setminus \{i,j\}} t_k \mathbf{e}_k + r_2 \mathbf{e}_j + s_2 \mathbf{e}_i \right],$$

where $\mathbf{e}_k$ is the k-th unit vector in $\mathbb{R}^m$. This gives us

$$\begin{aligned} 0 \leq F&\left(\sum_{k \in \{1,\dots,m\} \setminus \{i,j\}} t_k \mathbf{e}_k + r_2 \mathbf{e}_j + s_2 \mathbf{e}_i \right) \\ &- F\left(\sum_{k \in \{1,\dots,m\} \setminus \{i,j\}} t_k \mathbf{e}_k + r_1 \mathbf{e}_j + s_2 \mathbf{e}_i \right) \\ &- F\left(\sum_{k \in \{1,\dots,m\} \setminus \{i,j\}} t_k \mathbf{e}_k + r_2 \mathbf{e}_j + s_1 \mathbf{e}_i \right) \\ &+ F\left(\sum_{k \in \{1,\dots,m\} \setminus \{i,j\}} t_k \mathbf{e}_k + r_1 \mathbf{e}_j + s_1 \mathbf{e}_i \right) \\ = &\, g_{s_1,s_2}(t_1, \dots, t_{j-1}, r_2, t_{j+1}, \dots, t_{i-1}, t_{i+1}, \dots, t_m) \\ &- g_{s_1,s_2}(t_1, \dots, t_{j-1}, r_1, t_{j+1}, \dots, t_{i-1}, t_{i+1}, \dots, t_m), \end{aligned}$$

since the other terms in the sum cancel pairwise. □

Corollary 3.1 *Copulas are Lipschitz continuous*:

$$\left| C(u_2) - C(u_1) \right| \leq \sum_{i=1}^{m} \left| (u_2)_i - (u_1)_i \right|$$

for all $u_1, u_2 \in [0, 1]^m$.

The practicability of copulas shows itself in the following theorem, that was already hinted at in Sect. 3.1 for 2 risk factors:

Theorem 3.1 (Sklar) *Let* F_X *be a multivariate distribution function with marginal distributions* $F_{X_1}, \dots, F_{X_m}$. *Then there is a copula* C *with*

$$F_X(x_1, \dots, x_m) = C\big(F_{X_1}(x_1), \dots, F_{X_m}(x_m)\big)$$

The copula C *is unique, if the marginal distributions* $F_{X_1}, \dots, F_{X_m}$ *are continuous.*

Proof We will show the theorem in the special case that all marginal distributions are continuous. For the (relatively elaborate) generalization to arbitrary marginal distributions we will limit ourselves to sketching the idea of the proof.

Let $F_{X_1}, \ldots, F_{X_m}$ be arbitrary marginal distributions and $x_1, x_2 \in \bar{\mathbb{R}}^m$. If, for all $i \in \{1, \ldots, m\}$, $F_{X_i}((x_1)_i) = F_{X_i}((x_2)_i)$, then by Lemma 3.1 $F_X(x_1) = F_X(x_2)$. Therefore we can define a map

$$\tilde{C}\colon F_{X_1}(\bar{\mathbb{R}}) \times \cdots \times F_{X_m}(\bar{\mathbb{R}}) \to [0,1]$$
$$(u_1, \ldots, u_m) \mapsto \tilde{C}(u_1, \ldots, u_m),$$

that satisfies $F_X(x) = \tilde{C}(F_X(x_1), \ldots, F_X(x_m))$ for all $x \in \bar{\mathbb{R}}^m$.

We now show that $\tilde{C}$ is a predistribution function, which satisfies $\tilde{C}_{(i)}(u_i) = u_i$ for all $u_i \in F_{X_i}(\bar{\mathbb{R}})$.

(i) $\tilde{C}(u_1, \ldots, u_{i-1}, 0, u_i, \ldots, u_m) = 0$: Since F_{X_i} is a distribution function, we have

$$\lim_{x_i \to -\infty} F_{X_i}(x_i) = 0.$$

From this follows that $0 = F_{X_i}(-\infty) \in F_{X_i}(\bar{\mathbb{R}})$. The assertion now follows from

$$F_X(x_1, \ldots, x_{i-1}, -\infty, x_{i+1}, \ldots, x_m) = 0.$$

(ii) $\tilde{C}_{(i)}(u_i) = u_i$ for all $u_i \in F_{X_i}(\bar{\mathbb{R}})$: Since F_{X_j} is a distribution function, we have

$$\lim_{x_j \to \infty} F_{X_j}(x_j) = 1,$$

so that $1 \in F_{X_j}(\bar{\mathbb{R}})$, and the marginal distributions of $\tilde{C}$ exist. Let $x_i \in \bar{\mathbb{R}}$ be chosen so that $F_{X_i}(x_i) = u_i$. The assertion can now be read from

$$\begin{aligned} u_i &= F_{X_i}(x_i) = F_X(\infty, \ldots, \infty, x_i, \infty, \ldots, \infty) \\ &= \tilde{C}\big(F_{X_1}(\infty), \ldots, F_{X_{i-1}}(\infty), F_{X_i}(x_i), F_{X_{i+1}}(\infty), \ldots, F_{X_m}(\infty)\big) \\ &= \tilde{C}(1, \ldots, 1, u_i, 1, \ldots, 1) = \tilde{C}_{(i)}(u_i). \end{aligned}$$

(iii) For any $u_1, u_2 \in F_{X_1}(\bar{\mathbb{R}}) \times \cdots \times F_{X_m}(\bar{\mathbb{R}})$ with

$$(u_1)_i \le (u_2)_i$$

for all $i \in \{1, \ldots, m\}$, we have $V_{\tilde{C}}(]u_1, u_2]) \ge 0$: Since every F_{X_i} is monotone increasing, there are $x_1, x_2 \in \bar{\mathbb{R}}^m$ with $F_{X_i}((x_1)_i) = (u_1)_i$, $F_{X_i}((x_2)_i) = (u_2)_i$ and $(x_1)_i \le (x_2)_i$. Therefore

$$0 \le V_{F_X}\big(]x_1, x_2]\big) = \sum_{i_1=1}^{2} \cdots \sum_{i_m=1}^{2} (-1)^{i_1+\cdots+i_m} F_X\big((x_{i_1})_1, \ldots, (x_{i_m})_m\big)$$

$$= \sum_{i_1=1}^{2} \cdots \sum_{i_m=1}^{2} (-1)^{i_1+\cdots+i_m} \tilde{C}\big(F_{X_1}\big((x_{i_1})_1\big), \ldots, F_{X_m}\big((x_{i_m})_m\big)\big)$$

$$= \sum_{i_1=1}^{2} \cdots \sum_{i_m=1}^{2} (-1)^{i_1+\cdots+i_m} \tilde{C}\big((u_{i_1})_1, \ldots, (u_{i_m})_m\big) = V_{\tilde{C}}\big(]u_1, u_2]\big).$$

If the marginal distributions are continuous, $F_{X_i}(\bar{\mathbb{R}}) = [0, 1]$, so that $\tilde{C}$ is defined on the whole of $[0, 1]^m$, and thus is a copula.

In the general case, $F_{X_i}(\bar{\mathbb{R}})$ is a proper subset of $[0, 1]$. Then it must be shown that $\tilde{C}$ can be extended to a copula that is defined on all $[0, 1]^m$. This is in fact possible, but not uniquely. We will not prove this here but only sketch how the predistribution function $\tilde{C}$ can be extended to a copula C. The simplest copula is the product copula $\hat{C}(u_1, \ldots, u_m) = u_1 u_2 \cdots u_m$. That this really is a copula, follows from the already proven part of the theorem and the fact that for (in particular, for continuous) 1-dimensional distribution functions $F_1, \ldots, F_m$ the product

$$F(x_1, \ldots, x_,) = F_1(x_1) \cdots F_m(x_m)$$

is an m-dimensional distribution function that has the marginal distributions $F_1, \ldots, F_m$. This construction motivates the definition of C as a multilinear interpolation of $\tilde{C}$: First we notice that, because of the Lipschitz continuity of $\tilde{C}$ (in the proof of Lemma 3.1 we did not actually use the property that C is defined on the whole of $[0, 1]$) we can extend $\tilde{C}$ uniquely to be continuous on the closure $\overline{F_{X_1}(\bar{\mathbb{R}})} \times \cdots \times \overline{F_{X_m}(\bar{\mathbb{R}})}$ of $F_{X_1}(\bar{\mathbb{R}}) \times \cdots \times F_{X_m}(\bar{\mathbb{R}})$. For $u \in [0, 1]^m$, let $u^{[1]}$ and $u^{[2]}$ be defined by

$$u_i^{[1]} = \sup\big\{v_i \in \overline{F_{X_i}(\bar{\mathbb{R}})} : v_i \le u_i\big\}, \qquad u_i^{[2]} = \inf\big\{v_i \in \overline{F_{X_i}(\bar{\mathbb{R}})} : v_i \ge u_i\big\}.$$

With

$$\xi_i(u_i) = \begin{cases} \frac{u_i - u_i^{[1]}}{u_i^{[2]} - u_i^{[1]}} & \text{for } u_i^{[1]} < u_i^{[2]} \\ 1 & \text{for } u_i^{[1]} = u_i^{[2]} \end{cases}$$

set

$$C(u) = \sum_{i_1=1}^{2} \cdots \sum_{i_m=1}^{2} \prod_{k=1}^{m} \big(1 - \xi_k(u_k)\big)^{2-i_k} \prod_{k=1}^{m} \big(\xi_k(u_k)\big)^{i_k - 1} \tilde{C}\big(u_1^{[i_1]}, \ldots, u_m^{[i_m]}\big).$$

It is clear by definition that C agrees with $\tilde{C}$ on $F_{X_1}(\bar{\mathbb{R}}) \times \cdots \times F_{X_m}(\bar{\mathbb{R}})$. It remains to show that C is a copula. This is, however, time-consuming, so that we shall leave it aside for now. □

The copula of an m-dimensional random vector is invariant under strongly monotone increasing transformations of its components. The following theorem states more generally that the copula of a transformed random vector only depends on which components are transformed monotonically increasingly and which are transformed monotonically decreasingly.

Theorem 3.2 *Let X be an m-dimensional random vector with continuous marginal distributions F_{X_i} and copula C. Suppose that $T_i : \mathbb{R} \to \mathbb{R}$, $i \in \{1, \ldots, m\}$, are continuous functions, which are all strongly monotone, and let*

$$I_i = \begin{cases} \{1\} & \text{if } T_i \text{ is strongly monotone increasing,} \\ \{-1, 0\} & \text{if } T_i \text{ is strongly monotone decreasing.} \end{cases}$$

Then the copula C_T of the transformed random vector

$$X^T = (T_1 \circ X_1, \ldots, T_m \circ X_m)$$

is given by

$$C_T(u_1, \ldots, u_m) = \sum_{\beta \in I_1 \times \cdots \times I_m} (-1)^{N(\beta)} C\big(v(\beta_1, u_1), \ldots, v(\beta_m, u_m)\big),$$

where $N(\beta) = \sum_{i=1}^m 1_{\{\beta_i = -1\}}$ and

$$v(\beta_i, u_i) = \begin{cases} u_i & \text{if } \beta_i = 1, \\ 1 & \text{if } \beta_i = 0, \\ 1 - u_i & \text{if } \beta_i = -1. \end{cases}$$

If, in particular, all T_i are strongly monotone increasing, then $C_T = C$.

Proof We will start by proving the theorem for the special case that the first n transformations are monotone decreasing and the remaining transformations are monotone increasing. In this case, we have

$$I_i = \begin{cases} \{-1, 0\} & \text{if } i \le n, \\ \{1\} & \text{if } i \ge n + 1 \end{cases}$$

and

$$I_1 \times \cdots \times I_m = \{-1, 0\}^n \times \{1\}^{m-n} =: I_{n,m}.$$

Using this simplification and the continuity of the marginals, we obtain

$$\begin{aligned} &F_{X^T}(y_1, \ldots, y_m) \\ &\quad = \mathbf{P}(T_1 \circ X_1 \le y_1, \ldots, T_m \circ X_m \le y_m) \\ &\quad = \mathbf{P}\big(X_1 > T_1^{-1}(y_1), \ldots, X_n > T_n^{-1}(y_n), \\ &\qquad X_{n+1} \le T_{n+1}^{-1}(y_{n+1}), \ldots, X_m \le T_m^{-1}(y_m)\big) \\ &\quad = \mathbf{P}\big(\Omega \setminus \{X_1 \le T^{-1}(y_1)\} \cap \cdots \cap \Omega \setminus \{X_n \le T^{-1}(y_n)\}, \\ &\qquad \cap \{X_{n+1} \le T^{-1}(y_{n+1})\} \cap \cdots \cap \{X_m \le T^{-1}(y_m)\}\big). \end{aligned}$$

For $\beta \in \{-1,0,1\}^m$ and any subsets $A_1,\dots,A_m \subseteq \Omega$ let

$$A(\beta)_i = \begin{cases} \Omega & \text{if } \beta_i = 0, \\ A_i & \text{otherwise.} \end{cases}$$

Then there holds

$$\begin{aligned} &\mathbf{P}(\Omega \setminus A_1 \cap \dots \cap \Omega \setminus A_n \cap A_{n+1} \cap \dots \cap A_m) \\ &= \sum_{\beta \in I_{n,m}} (-1)^{N(\beta)} \mathbf{P}\big(A(\beta)_1 \cap \dots \cap A(\beta)_n \cap A_{n+1} \cap \dots \cap A_m\big) \\ &= \sum_{\beta \in I_{n,m}} (-1)^{N(\beta)} \mathbf{P}\big(A(\beta)_1 \cap \dots \cap A(\beta)_m\big). \end{aligned} \tag{3.1}$$

The second equality in (3.1) is obvious. We prove the first equality by induction over n. The assertion clearly holds for $n=0$. Assume now that the formula has been proven for n. Then we obtain

$$\begin{aligned} &\mathbf{P}(\Omega \setminus A_1 \cap \dots \cap \Omega \setminus A_n \cap \Omega \setminus A_{n+1} \cap A_{n+2} \cap \dots \cap A_m) \\ &= \sum_{\beta \in I_{n,m}} (-1)^{N(\beta)} \mathbf{P}\big(A(\beta)_1 \cap \dots \cap A(\beta)_n \cap \Omega \setminus A_{n+1} \cap A_{n+2} \cap \dots \cap A_m\big) \\ &= \sum_{\beta \in I_{n,m}} (-1)^{N(\beta)} \mathbf{P}\big(A(\beta)_1 \cap \dots \cap A(\beta)_n \cap \Omega \cap A_{n+2} \cap \dots \cap A_m\big) \\ &\quad - \sum_{\beta \in I_{n,m}} (-1)^{N(\beta)} \mathbf{P}\big(A(\beta)_1 \cap \dots \cap A(\beta)_n \cap A_{n+1} \cap A_{n+2} \cap \dots \cap A_m\big) \\ &= \sum_{\beta \in I_{n+1,m}} (-1)^{N(\beta)} \mathbf{P}\big(A(\beta)_1 \cap \dots \cap A(\beta)_{n+1} \cap A_{n+2} \cap \dots \cap A_m\big). \end{aligned}$$

This proves (3.1) for all $n \in \{0,\dots,m\}$.

Setting

$$A_i = \{X_i \le T_i^{-1}(y_i)\} \quad \text{and} \quad a(z,\beta_i) = \begin{cases} \infty & \text{if } \beta_i = 0, \\ z & \text{otherwise,} \end{cases}$$

we get

$$A(\beta)_i = \begin{cases} \Omega & \text{if } \beta_i = 0, \\ \{X_i \le T_i^{-1}(y_i)\} & \text{otherwise,} \end{cases} \quad = \quad \{X_i \le a\big(T_i^{-1}(y_i),\beta_i\big)\}$$

and therefore

$$
\begin{aligned}
F_{X^T}(y_1,\ldots,y_m) &= \mathbf{P}(\Omega\setminus A_1\cap\cdots\cap\Omega\setminus A_n\cap A_{n+1}\cap\ldots A_m)\\
&= \sum_{\beta\in I_{n,m}}(-1)^{N(\beta)}\mathbf{P}\big(A(\beta)_1\cap\cdots\cap A(\beta)_m\big)\\
&= \sum_{\beta\in I_{n,m}}(-1)^{N(\beta)}\mathbf{P}\big(X_1\le a\big(T_1^{-1}(y_1),\beta_1\big),\ldots,\\
&\qquad X_m\le a\big(T_m^{-1}(y_m),\beta_m\big)\big)\\
&= \sum_{\beta\in I_{n,m}}(-1)^{N(\beta)}C\big(F_{X_1}\big(a\big(T_1^{-1}(y_1),\beta_1\big)\big),\ldots,\\
&\qquad F_{X_m}\big(a\big(T_m^{-1}(y_m),\beta_m\big)\big)\big).
\end{aligned}
$$

Observing

$$F_{X_i}\big(T_i^{-1}(y_i)\big)=\mathbf{P}\big(X_i\le T_i^{-1}(y_i)\big)=\mathbf{P}(T_i\circ X_i\ge y_i)=1-F_{T_i\circ X_i}(y_i)$$

for $i\le n$ and $F_{X_i}(T_i^{-1}(y_i))=F_{T_i\circ X_i}(y_i)$ for $i>n$, we obtain

$$
\begin{aligned}
&F_{X^T}(y_1,\ldots,y_m)\\
&\quad=\sum_{\beta\in I_{n,m}}(-1)^{N(\beta)}C\big(v\big(F_{T_1\circ X_1}(y_1),\beta_1\big),\ldots,v\big(F_{T_m\circ X_m}(y_m),\beta_m\big)\big).
\end{aligned}
$$

Now the uniqueness part of Sklar's Theorem 3.1 implies

$$C_T(u_1,\ldots,u_m)=\sum_{\beta\in I_{n,m}}(-1)^{N(\beta)}C\big(v(u_1,\beta_1),\ldots,v(u_m,\beta_m)\big).$$

Hence we have proven the theorem in the special case.

For the general case, observe that there is a permutation σ of $(1,\ldots,m)$ such that the n transformations $T_{\sigma_1},\ldots,T_{\sigma_n}$ are decreasing and the remaining transformations are increasing. We can now apply the formula for C_T, in the special case we know, to the permuted arguments. Permuting back using the inverse of σ yields the formula in the general case since this procedure only involves re-arranging the arguments of C and T. □

Remark 3.2 Let (X_1,X_2) be a two-dimensional random vector with copula C and $T_i:\mathbb{R}\to\mathbb{R}$ strongly monotone transformations. We denote the copula of the transformed random vector $(T_1(X_1),T_2(X_2))$ by C_T. If T_1 and T_2 are increasing, then $C_T=C$. If T_1 is decreasing and T_2 increasing, then

$$C_T(u_1,u_2)=C(1,u_2)-C(1-u_1,u_2).$$

If T_1 and T_2 are decreasing, then

$$C_T(u_1,u_2)=C(1,1)-C(1,1-u_2)-C(1-u_1,1)+C(1-u_1,1-u_2).$$

Example 3.2 Assume that the random variables X_1, X_2 are completely dependent in the sense that there is a monotone increasing function $f: \mathbb{R} \to \mathbb{R}$ such that $X_2 = f \circ X_1$. Then

$$\mathbf{P}(X_1 \leq x_1, X_2 \leq x_2) = \min\{\mathbf{P}(X_1 \leq x_1), \mathbf{P}(X_2 \leq x_2)\}.$$

It follows that their copula is given by $C(u_1, u_2) = \min(u_1, u_2)$. Let $T_1: \mathbb{R} \to \mathbb{R}$ be a strongly monotone decreasing function, and $T_2: \mathbb{R} \to \mathbb{R}$ be a strongly monotone increasing function. Then we get

$$\begin{aligned} C_T(u_1, u_2) &= C(1, u_2) - C(1 - u_1, u_2) \\ &= u_2 - \min(1 - u_1, u_2) = u_2 + \max(u_1 - 1, -u_2) \\ &= \max(u_1 + u_2 - 1, 0). \end{aligned}$$

Assume now that both $\tilde{T}_1$ and $\tilde{T}_2$ are strongly monotone decreasing functions. Then we obtain

$$\begin{aligned} C_{\tilde{T}}(u_1, u_2) &= C(1, 1) - C(1, 1 - u_2) - C(1 - u_1, 1) + C(1 - u_1, 1 - u_2) \\ &= 1 - (1 - u_2) - (1 - u_1) + \min(1 - u_1, 1 - u_2) \\ &= u_1 + u_2 + \min(-u_2, -u_1) = \min(u_1, u_2). \end{aligned}$$

This result is expected because $\tilde{T}_2 \circ X_2 = \tilde{T}_2 \circ f \circ \tilde{T}_1^{-1}(\tilde{T}_1(X_1))$ and the composition $\tilde{T}_2 \circ f \circ \tilde{T}_1^{-1}$ is monotone increasing, so $\tilde{T}_1 \circ X_1$ and $\tilde{T}_2 \circ X_2$ are completely dependent.

Please note that the copula is not always preserved when T_1 and T_2 are strongly monotone decreasing. A counterexample is given by a random vector (X_1, X_2) whose cumulative distribution function is the Gumbel copula with parameter $\theta = 2$, and the transformations $T_i(x) = -x$, $i = 1, 2$.

The linear correlation

$$\operatorname{corr}(X_1, X_2) = \frac{\operatorname{cov}(X_1, X_2)}{\sigma(X_1)\sigma(X_2)}$$

is not an invariant of the copula of the 2-dimensional distribution vector X. There are however other dependent quantities that only depend on the copula.

Definition 3.4 Let X be a 2-dimensional random vector with continuous marginal distributions F_{X_1}, F_{X_2}. *Spearman's rho* is defined by

$$\rho_{\text{Spearman}}(X_1, X_2) = \operatorname{corr}(F_{X_1} \circ X_1, F_{X_2} \circ X_2).$$

Proposition 3.3 *Let X be a 2-dimensional random vector with continuous marginal distributions F_{X_1}, F_{X_2} and copula C. Then we have*

$$\rho_{\text{Spearman}}(X_1, X_2) = 12 \int_0^1 \int_0^1 C(u_1, u_2)\,\mathrm{d}u_1\,\mathrm{d}u_2 - 3.$$

In particular, $\rho_{\text{Spearman}}(X_1, X_2)$ depends on X only through the copula C.

Proof Since $F_{X_i} \circ X_i$ is uniformly distributed

$$\begin{aligned}\operatorname{var}(F_{X_i} \circ X_i) &= \operatorname{var}(U_i) = \mathrm{E}(U_i^2) - \mathrm{E}(U_i)^2 = \int_0^1 u^2\,\mathrm{d}u - \left(\int_0^1 u\,\mathrm{d}u\right)^2 \\ &= \frac{1}{3} - \frac{1}{4} = \frac{1}{12}.\end{aligned}$$

Thus we obtain $\rho_{\text{Spearman}}(X_1, X_2) = 12 \operatorname{cov}(F_{X_1} \circ X_1, F_{X_2} \circ X_2)$. Now, in general, for any random vector Y the equation

$$\operatorname{cov}(Y_1, Y_2) = \int\int \big(F_Y(y_1, y_2) - F_{Y_1}(y_1) F_{Y_2}(y_2)\big)\,\mathrm{d}y_1\,\mathrm{d}y_2$$

holds (see the subsequent Lemma 3.2), so that by definition of the copula and the fact that $F_{X_i} \circ X_i$ is uniformly distributed, it follows that

$$\rho_{\text{Spearman}}(X_1, X_2) = 12 \int_0^1 \int_0^1 \big(C(u_1, u_2) - u_1 u_2\big)\,\mathrm{d}u_1\,\mathrm{d}u_2.$$

The integration $12 \int_0^1 \int_0^1 u_1 u_2\,\mathrm{d}u_1\,\mathrm{d}u_2 = 3$ then delivers the assertion. □

In the proof of Proposition 3.3 we used the following Lemma due to Höffding.

Lemma 3.2 *Let X be a 2-dimensional random vector with continuous marginal distributions F_{X_1}, F_{X_2}. If the covariance of X is finite, then*

$$\operatorname{cov}(X_1, X_2) = \int\int \big(F_X(x_1, x_2) - F_{X_1}(x_1) F_{X_2}(x_2)\big)\,\mathrm{d}x_1\,\mathrm{d}x_2.$$

Proof Let $\tilde{X}$ be a random vector independent of X but with the same distribution. Then we have

$$\begin{aligned}\mathrm{E}\big((X_1 - \tilde{X}_1)(X_2 - \tilde{X}_2)\big) &= \mathrm{E}(X_1 X_2 - X_1 \tilde{X}_2 - \tilde{X}_1 X_2 + \tilde{X}_1 \tilde{X}_2) \\ &= 2\,\mathrm{E}(X_1 X_2) - 2\,\mathrm{E}(X_1)\,\mathrm{E}(X_2) = 2 \operatorname{cov}(X_1, X_2).\end{aligned}$$

For each $a, b \in \mathbb{R}$, obviously $(a - b) = \int_{-\infty}^{\infty} (I_{\{b \le x\}} - I_{\{a \le x\}})\,\mathrm{d}x$, so that we can write:

$$
\begin{aligned}
2\,\mathrm{cov}(X_1, X_2) &= \mathrm{E}\big((X_1 - \tilde{X}_1)(X_2 - \tilde{X}_2)\big) \\
&= \mathrm{E}\bigg(\int_{-\infty}^{\infty}\int_{-\infty}^{\infty} (I_{\{\tilde{X}_1 \le x_1\}} - I_{\{X_1 \le x_1\}}) \\
&\qquad \times (I_{\{\tilde{X}_2 \le x_2\}} - I_{\{X_2 \le x_2\}})\,\mathrm{d}x_1\,\mathrm{d}x_2\bigg) \\
&= \int_{-\infty}^{\infty}\int_{-\infty}^{\infty} \mathrm{E}\big((I_{\{\tilde{X}_1 \le x_1\}} - I_{\{X_1 \le x_1\}}) \\
&\qquad \times (I_{\{\tilde{X}_2 \le x_2\}} - I_{\{X_2 \le x_2\}})\big)\,\mathrm{d}x_1\,\mathrm{d}x_2 \\
&= 2\int_{-\infty}^{\infty}\int_{-\infty}^{\infty} \big(\mathrm{E}(I_{\{X_1 \le x_1\}} I_{\{X_2 \le x_2\}}) \\
&\qquad - \mathrm{E}(I_{\{X_1 \le x_1\}})\,\mathrm{E}(I_{\{X_2 \le x_2\}})\big)\,\mathrm{d}x_1\,\mathrm{d}x_2 \\
&= 2\int_{-\infty}^{\infty}\int_{-\infty}^{\infty} \big(\mathbf{P}(X_1 \le x_1, X_2 \le x_2) \\
&\qquad - \mathbf{P}(X_1 \le x_1)\mathbf{P}(X_2 \le x_2)\big)\,\mathrm{d}x_1\,\mathrm{d}x_2 \\
&= 2\int_{-\infty}^{\infty}\int_{-\infty}^{\infty} \big(F_X(x_1, x_2) - F_{X_1}(x_1)F_{X_2}(x_2)\big)\,\mathrm{d}x_1\,\mathrm{d}x_2. \qquad \square
\end{aligned}
$$

Another measure of dependency that depends only on the copula is Kendall's tau.

Definition 3.5 Let $X = (X_1, X_2)$ be a 2-dimensional random vector with copula C, and $\tilde{X} = (\tilde{X}_1, \tilde{X}_2)$ be another random vector, such that X and $\tilde{X}$ are i.i.d. Then *Kendall's* τ for X is defined by

$$
\tau_{\mathrm{Kendall}}(X_1, X_2) = 2\mathbf{P}\big((X_1 - \tilde{X}_1)(X_2 - \tilde{X}_2) > 0\big) - 1.
$$

Kendall's tau can be interpreted as the extent of synchronicity of the two components.

Lemma 3.3 *Let X be a 2-dimensional random vector with continuous marginal distributions F_{X_1}, F_{X_2}. Then we have*

$$
\tau_{\mathrm{Kendall}}(X_1, X_2) = 4\int_{[0,1]^2} C(u_1, u_2)\,\mathrm{d}C(u_1, u_2) - 1.
$$

In particular, Kendall's τ depends on X only through the copula C.

Proof Let F be the distribution function of X. Then we have

$$
\begin{aligned}
\tau_{\mathrm{Kendall}}(X_1, X_2) &= 2\mathbf{P}\big((X_1 - \tilde{X}_1)(X_2 - \tilde{X}_2) > 0\big) - 1 \\
&= 4\mathbf{P}(X_1 < \tilde{X}_1, X_2 < \tilde{X}_2) - 1
\end{aligned}
$$

$$= 4 \int_{\mathbb{R}^2} \mathbf{P}(X_1 < y_1, X_2 < y_2)\, \mathrm{d}F(y_1, y_2) - 1$$

$$= 4 \int_{\mathbb{R}^2} C\big(F_1(y_1), F_2(y_2)\big)\, \mathrm{d}C\big(F_1(y_1), F_2(y_2)\big) - 1.$$

The assertion now follows with $u_1 = F_1(y_1)$ and $u_2 = F_2(u_2)$. □

If in practice one calculates a distribution, one often uses parametric methods. One considers one or more suitable classes of distributions with a few parameters which can be fitted to the data. What classes are to be chosen depends naturally on the empirical distribution, but ultimately there are only a handful of possible candidates. For copulas one is in a similar situation. There is a whole zoo of copula classes with characteristic properties. In a similar way to the case of distributions, one again needs to estimate only a few parameters after having decided on a class. That the definition of a copula (and, in particular, Condition (ii) in Definition 3.2) is a bit complicated, is therefore not bothersome in insurance practice.

Even if the data for calibrating a copula are not sufficient, comparing several copulas can provide useful sensitivities (see Example 3.3 in the next section).

3.2.1 Examples

3.2.1.1 Gauss Copula

Definition 3.6 Let $\Phi_{0,1}$ be the standard normal distribution function, and $\Phi_{0,\mathrm{corr}}$ the distribution function of the m-dimensional multinormal distribution $X \sim N(0, \mathrm{corr})$, where corr is a correlation matrix. Then the copula defined by

$$C_{\mathrm{corr}}^{\mathrm{Gauss}}(u_1, \dots, u_m) = \Phi_{0,\mathrm{corr}}\big(\Phi_{0,1}^{-1}(u_1), \dots, \Phi_{0,1}^{-1}(u_m)\big)$$

is called the *Gauss copula*.

The Gauss copula $C_{\mathrm{corr}}^{\mathrm{Gauss}}$ is obviously the copula of the multinormally distributed random vector X. From Theorem 3.2 and the transformation

$$x_i = \frac{y_i - \mu_i}{\sigma_i}$$

it follows that it is also the copula of the random vector $Y \sim N(\mu, \Sigma)$, if Σ has the correlation matrix corr.

In two dimensions, with

$$\mathrm{corr} = \begin{pmatrix} 1 & \mathrm{corr}_{12} \\ \mathrm{corr}_{12} & 1 \end{pmatrix}$$

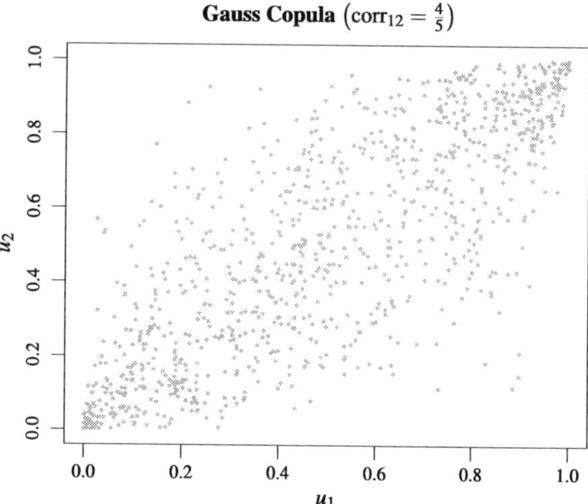

Fig. 3.1 Gauss copula with $\mathrm{corr}_{12} = \frac{4}{5}$; 1000 random points were generated

we obtain explicitly

$$\begin{aligned} C^{\mathrm{Gauss}}_{\mathrm{corr}}(u_1, u_2) &= \Phi_{0,\mathrm{corr}}\big(\Phi^{-1}_{0,1}(u_1), \Phi^{-1}_{0,1}(u_2)\big) \\ &= \frac{1}{2\pi\sqrt{1-(\mathrm{corr}_{12})^2}} \\ &\quad \times \int_{-\infty}^{\Phi^{-1}_{0,1}(u_1)} \int_{-\infty}^{\Phi^{-1}_{0,1}(u_2)} \exp\bigg(-\frac{x^2 - 2\mathrm{corr}_{12}xy + y^2}{2(1-(\mathrm{corr}_{12})^2)}\bigg)\, dx\, dy. \end{aligned}$$

Proposition 3.4 *Let X be a 2-dimensional distribution vector with Gauss copula $C^{\mathrm{Gauss}}_{\mathrm{corr}}$. Then*

$$\rho_{\mathrm{Spearman}}(X_1, X_2) = \frac{6}{\pi} \arcsin\bigg(\frac{\mathrm{corr}_{12}}{2}\bigg),$$

$$\tau_{\mathrm{Kendall}}(X_1, X_2) = \frac{2}{\pi} \arcsin(\mathrm{corr}_{12}).$$

Proof We first show that, for normally distributed 1-dimensional random variables Y_1, Y_2 with $\mathrm{corr}(Y_1, Y_2) = \tilde{\rho}$, the equation

$$\mathbf{P}\big(Y_1 - \mathrm{E}(Y_1) \geq 0, Y_2 - \mathrm{E}(Y_2) \geq 0\big) = \frac{1}{4} + \frac{1}{2\pi} \arcsin \tilde{\rho} \tag{3.2}$$

holds.

Since, for all strictly positive real numbers a_1 and a_2, we have the equation

$$\mathbf{P}\big(Y_1 - \mathrm{E}(Y_1) \geq 0, Y_2 - \mathrm{E}(Y_2) \geq 0\big) = \mathbf{P}\big(a_1\big(Y_1 - \mathrm{E}(Y_1)\big) \geq 0,\ a_2\big(Y_2 - \mathrm{E}(Y_2)\big) \geq 0\big)$$

we can assume, without loss of generality, that Y_1 and Y_2 are each standard normally distributed. $Y := (Y_1, Y_2)$ then has the same distribution as

$$\big(\tilde{Z}_1, \tilde{\rho}\tilde{Z}_1 + \sqrt{1-\tilde{\rho}^2}\tilde{Z}_2\big),$$

where $\tilde{Z}_1, \tilde{Z}_2$ are standard normally distributed independent random variables. Setting $\phi := \arcsin\tilde{\rho}$ we obtain

$$(Y_1, Y_2) \sim (\tilde{Z}_1, \sin\phi\tilde{Z}_1 + \cos\phi\tilde{Z}_2).$$

Let O be a rotation in $\mathbb{R}^2$ and $\tilde{Z} = (\tilde{Z}_1, \tilde{Z}_2)$. Since, for $t \in \mathbb{R}^2$, we have

$$\mathrm{E}(\mathrm{i}t \cdot O\tilde{Z}) = \mathrm{E}\big(\mathrm{i}\,O^\top t \cdot \tilde{Z}\big) = \exp\Big(-\frac{1}{2}\|O^\top t\|^2\Big) = \exp\Big(-\frac{1}{2}\|t\|^2\Big) = \mathrm{E}(\mathrm{i}t \cdot \tilde{Z}).$$

We see that $O\tilde{Z}$ and $\tilde{Z}$ have the same distribution. On symmetry grounds we can therefore write

$$(\tilde{Z}_1, \tilde{Z}_2) = R(\cos\Theta, \sin\Theta)$$

where R is a positive random variable and Θ a uniformly distributed random variable on $[-\pi, \pi[$. It follows that

$$\begin{aligned}
\mathbf{P}(Y_1 \geq 0, Y_2 \geq 0) &= \mathbf{P}(\cos\Theta \geq 0, \sin\phi\cos\Theta + \cos\phi\sin\Theta \geq 0)\\
&= \mathbf{P}\big(\cos\Theta \geq 0, \sin(\phi+\Theta) \geq 0\big)\\
&= \mathbf{P}\Big(\Theta \in \Big[-\frac{\pi}{2}, \frac{\pi}{2}\Big], \phi + \Theta \in [0, \pi]\Big)\\
&= \mathbf{P}\Big(\Theta \in \Big[-\frac{\pi}{2}, \frac{\pi}{2}\Big] \cap [-\phi, \pi - \phi]\Big)\\
&= \frac{1}{2\pi}\Big(\frac{\pi}{2} + \phi\Big),
\end{aligned}$$

and so (3.2) has been proven.

Now we show the assertion for Spearman's rho, ρ_{Spearman}. Denote the density of the standard normal distribution by $\phi_{0,1}$. From Proposition 3.3 it immediately follows that $\rho_{\text{Spearman}}(X_1, X_2)$ depends only on the Gauss copula and not on the marginal distributions X_1, X_2, so that we may assume, without loss of generality, that X_1, X_2 have standard normal distributions. Furthermore, it follows from Proposition 3.3 that

$$\begin{aligned}
\rho_{\text{Spearman}}(X_1, X_2) &= 12\int_0^1\int_0^1 \Phi_{0,\rho}\big(\Phi_{0,1}^{-1}(u_1), \Phi_{0,1}^{-1}(u_2)\big)\,\mathrm{d}u_1\,\mathrm{d}u_2 - 3\\
&\overset{\Phi_{0,1}(x_i)=u_i}{\overbrace{=}}\; 12\int_{-\infty}^{\infty}\int_{-\infty}^{\infty} \Phi_{0,\rho}(x_1, x_2)\phi_{0,1}(x_1)\phi_{0,1}(x_2)\,\mathrm{d}x_1\,\mathrm{d}x_2 - 3
\end{aligned}$$

$$= 12 \int_{-\infty}^{\infty} \int_{-\infty}^{\infty} \mathbf{P}(X_1 \le x_1, X_2 \le x_2)$$
$$\phi_{0,1}(x_1)\phi_{0,1}(x_2)\,\mathrm{d}x_1\,\mathrm{d}x_2 - 3$$
$$= 12\,\mathrm{E}\big(\mathbf{P}\big(X_1 \le Z_1, X_2 \le Z_2 \mid Z_1, Z_2\big)\big) - 3,$$

where Z is a random vector independent of X, with independent standard normally distributed components Z_1, Z_2. Putting $Y = Z - X$, we obtain

$$\rho_{\text{Spearman}}(X_1, X_2) = 12\,\mathbf{P}(Y_1 \ge 0, Y_2 \ge 0) - 3.$$

The random variable Y is, because it is a linear combination of normally distributed random variables, also normally distributed, and has the mean 0 and covariance matrix

$$\tilde{\Sigma} = \begin{pmatrix} 2 & \mathrm{corr}_{12} \\ \mathrm{corr}_{12} & 2 \end{pmatrix}.$$

It follows that the correlation of Y_1 and Y_2 is exactly $\tilde{\rho} = \mathrm{corr}_{12}/2$. Therefore we obtain, using (3.2),

$$\rho_{\text{Spearman}}(X_1, X_2) = 12\left(\frac{1}{4} + \frac{1}{2\pi}\arcsin\tilde{\rho}\right) - 3 = 3 + \frac{6}{\pi}\arcsin\left(\frac{\mathrm{corr}_{12}}{2}\right) - 3.$$

In order to demonstrate the claim for Kendall's τ, we use the formula, obtained *en passant* in Lemma 3.3,

$$\tau_{\text{Kendall}}(X_1, X_2) = 4\,\mathbf{P}(X_1 < \tilde{X}_1, X_2 < \tilde{X}_2) - 1.$$

Since the normal distribution is continuous, this formula can be rewritten, with $Y_i = \tilde{X}_i - X_i$, as

$$\tau_{\text{Kendall}}(X_1, X_2) = 4\,\mathbf{P}(Y_1 \ge 0, Y_2 \ge 0) - 1.$$

From

$$\mathrm{corr}(Y_1, Y_2) = \frac{\mathrm{cov}(\tilde{X}_1 - X_1, \tilde{X}_2 - X_2)}{\sigma(\tilde{X}_1 - X_1)\sigma(\tilde{X}_2 - X_2)}$$
$$= \frac{\mathrm{cov}(\tilde{X}_1, \tilde{X}_2) + \mathrm{cov}(X_1, X_2)}{\sqrt{(\mathrm{var}(\tilde{X}_1) + \mathrm{var}(X_1))}\sqrt{(\mathrm{var}(\tilde{X}_2) + \mathrm{var}(X_2))}}$$
$$= \frac{2\,\mathrm{cov}(X_1, X_2)}{2\sigma(X_1)\sigma(X_2)} = \mathrm{corr}(X_1, X_2)$$

and (3.2) we obtain, with $\tilde{\rho} = \mathrm{corr}_{12}$,

$$\tau_{\text{Kendall}}(X_1, X_2) = 4\left(\frac{1}{4} + \frac{1}{2\pi}\arcsin\mathrm{corr}_{12}\right) - 1,$$

so that we have shown our claim. □

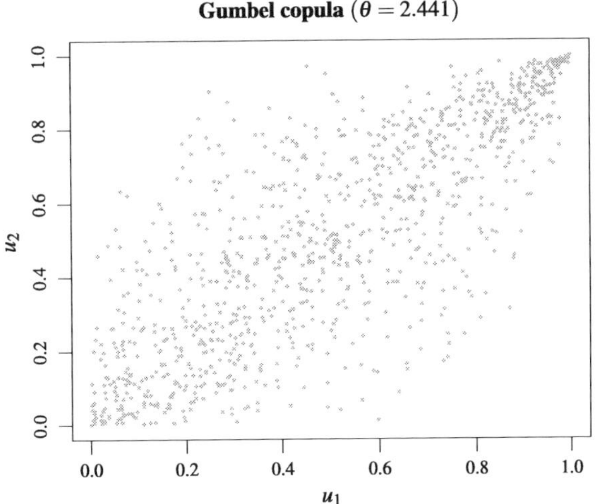

Fig. 3.2 Gumbel copula $C_\theta^{\text{Gumbel}}(u_1, u_2)$ with $\theta = 2.441$. The parameter θ was chosen so that Kendall's τ here has the same value as Kendall's τ for the copula shown in Fig. 3.1. 1000 random points were generated

The Gauss copula can be used for arbitrary marginal distributions to produce a multivariate distribution. Its strength lies in the fact that

- it is easy to use and
- in the special case of a multinormal distribution, it gives the ordinary correlation matrix.

Thus it is often used when there is no more precise information on the dependency structure.

3.2.1.2 Gumbel Copula

The *Gumbel copula* has a particularly simple presentation,

$$C_\theta^{\text{Gumbel}}(u_1, \ldots, u_m) = \mathrm{e}^{-(\sum_{i=1}^m (-\ln u_i)^\theta)^{1/\theta}}.$$

The importance of the Gumbel copula is that it models higher dependency in the tails of distributions which are of special interest to risk management. With it we have a first approach to modeling our hail and car damage example where significant dependence only turns up for large losses. This property of the Gumbel copula can be seen in Fig. 3.2 as the accumulation of points in the upper right corner. Note there is no particular accumulation in the lower left corner.

Proposition 3.5 *For the Gumbel copula C_θ^{Gumbel} we have*

$$\tau_{\text{Kendall}}(X_1, X_2) = 1 - \frac{1}{\theta}.$$

Proof Kendall's τ can be calculated from

$$\begin{aligned}\tau_{\text{Kendall}} &= 4\int_{[0,1]^2} C_\theta^{\text{Gumbel}}(u,v)\,\mathrm{d}C_\theta^{\text{Gumbel}}(u,v) - 1 \\ &= 4\int_{[0,1]^2} C_\theta^{\text{Gumbel}}(u,v)\frac{\partial^2 C_\theta^{\text{Gumbel}}}{\partial u \partial v}\,\mathrm{d}u\,\mathrm{d}v - 1.\end{aligned}$$

Upon putting $f(u) = (-\ln(u))^\theta$, we obtain

$$\begin{aligned}\frac{\partial^2}{\partial u \partial v} C_\theta^{\text{Gumbel}}(u,v) &= \frac{\partial^2}{\partial u \partial v}\exp\left(-\left[f(u)+f(v)\right]^{\frac{1}{\theta}}\right) \\ &= \frac{\partial}{\partial u}\left(-\frac{1}{\theta}C_\theta^{\text{Gumbel}}(u,v)\left[f(u)+f(v)\right]^{\frac{1}{\theta}-1}\frac{\mathrm{d}f}{\mathrm{d}v}\right) \\ &= \frac{1}{\theta^2}C_\theta^{\text{Gumbel}}(u,v)\left[f(u)+f(v)\right]^{\frac{2}{\theta}-2}\frac{\mathrm{d}f}{\mathrm{d}u}\frac{\mathrm{d}f}{\mathrm{d}v} \\ &\quad -\frac{1}{\theta}\left(\frac{1}{\theta}-1\right)C_\theta^{\text{Gumbel}}(u,v)\left[f(u)+f(v)\right]^{\frac{1}{\theta}-2}\frac{\mathrm{d}f}{\mathrm{d}u}\frac{\mathrm{d}f}{\mathrm{d}v} \\ &= \frac{1}{\theta^2}C_\theta^{\text{Gumbel}}(u,v)\left[f(u)+f(v)\right]^{\frac{1}{\theta}-2}\frac{\mathrm{d}f}{\mathrm{d}u}\frac{\mathrm{d}f}{\mathrm{d}v} \\ &\quad \times\left(\left[f(u)+f(v)\right]^{\frac{1}{\theta}}+\theta-1\right).\end{aligned}$$

With $x = f(u)$ and $y = f(v)$, the integral simplifies to

$$\begin{aligned}\tau_{\text{Kendall}} = \frac{4}{\theta^2}\int_0^\infty\int_0^\infty &\exp\left(-2[x+y]^{\frac{1}{\theta}}\right)[x+y]^{\frac{1}{\theta}-2} \\ &\times\left([x+y]^{1/\theta}+\theta-1\right)\mathrm{d}x\,\mathrm{d}y - 1.\end{aligned}$$

Now we consider the transformation

$$\begin{pmatrix} x \\ y \end{pmatrix} \mapsto \begin{pmatrix} a \\ b \end{pmatrix} = \begin{pmatrix} x+y \\ x-y \end{pmatrix}.$$

The inverse function is

$$\varphi(a,b) = \frac{1}{2}\begin{pmatrix} a+b \\ a-b \end{pmatrix}$$

and we have

$$\begin{aligned}\varphi^{-1}\left(]0,\infty[^2\right) &= \left\{\begin{pmatrix} a \\ b \end{pmatrix} : \frac{1}{2}(a+b) > 0, \frac{1}{2}(a-b) > 0\right\} \\ &= \left\{\begin{pmatrix} a \\ b \end{pmatrix} : a > 0, b \in]-a,a[\right\}\end{aligned}$$

as well as

$$\left|\det(\mathrm{D}\varphi)\right| = \left|\det\begin{pmatrix} \frac{1}{2} & \frac{1}{2} \\ \frac{1}{2} & -\frac{1}{2} \end{pmatrix}\right| = \frac{1}{2}.$$

From this follows

$$\begin{aligned}
\tau_{\text{Kendall}} &= \frac{4}{\theta^2}\int_0^\infty \int_{-a}^{a} \exp\{-2a^{1/\theta}\} a^{\frac{1}{\theta}-2}\left(a^{1/\theta}+\theta-1\right) \times \frac{1}{2}\,\mathrm{d}b\,\mathrm{d}a - 1 \\
&= \frac{4}{\theta^2}\int_0^\infty \exp\{-2a^{1/\theta}\} a^{\frac{1}{\theta}-1}\left(a^{\frac{1}{\theta}}+(\theta-1)\right)\mathrm{d}a - 1 \\
&= \frac{4}{\theta}\int_0^\infty \exp(-2z)(z+\theta-1)\,\mathrm{d}z - 1 \\
&= \frac{4}{\theta}\left(-\frac{z}{2}\exp(-2z)\Big|_0^\infty - \left(-\frac{1}{2}\int_0^\infty \exp(-2z)\,\mathrm{d}z\right)\right. \\
&\quad \left. + (\theta-1)\int_0^\infty \exp(-2z)\,\mathrm{d}z\right) - 1 \\
&= \frac{4}{\theta}\left(\frac{1}{4}+(\theta-1)\times\frac{1}{2}\right) - 1 = -\frac{1}{\theta}+1,
\end{aligned}$$

where we made the substitution $z = a^{1/\theta}$, and then integrated by parts. □

Example 3.3 We consider two loss amounts X_1 and X_2, both exponentially distributed with parameter 1, and look at the possible influence of the choice of the copula on the value at risk of $X := X_1 + X_2$ at the level 99 %. We compare the Gauss copula $C_{0.8}^{\text{Gauss}}$ with the Gumbel copula C_θ^{Gumbel} where

$$\theta = \left(1 - \frac{2}{\pi}\arcsin(0.8)\right)^{-1} = 2.441.$$

To make the comparison easier the parameters are chosen such that Kendall's tau is the same for both copulas.

The value at risk is derived from the equation

$$\begin{aligned}
0.99 &= \int_0^{\mathrm{VaR}_{0.99}(X)} \int_0^z f_{(X_1,X_2)}(z-x,x)\,\mathrm{d}x\,\mathrm{d}z \\
&= \int_0^{\mathrm{VaR}_{0.99}(X)} \int_0^z c\left(F_{X_1}(z-x), F_{X_2}(x)\right) f_{X_1}(z-x) f_{X_2}(x)\,\mathrm{d}x\,\mathrm{d}z \\
&= \int_0^{\mathrm{VaR}_{0.99}(X)} \exp(-z)\int_0^z c\left(1-\exp(x-z), 1-\exp(-x)\right)\mathrm{d}x\,\mathrm{d}z,
\end{aligned}$$

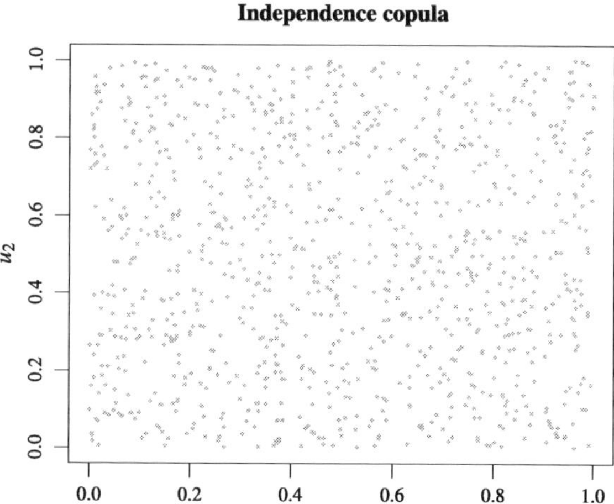

Fig. 3.3 The independence copula $C^{\text{indep}}(u_1, u_2)$. 1000 random points were generated

where for a Gauss copula the density $c = \partial^2 C_{0.8}^{\text{Gauss}} / \partial u \partial v$ is given by

$$c(u, v) = \frac{1}{\sqrt{1-0.8^2}} \exp\left(\frac{(\Phi_{0,1}^{-1}(u))^2 + (\Phi_{0,1}^{-1}(v))^2}{2} + \frac{1.6\,\Phi_{0,1}^{-1}(u)\Phi_{0,1}^{-1}(v) - (\Phi_{0,1}^{-1}(u))^2 - (\Phi_{0,1}^{-1}(v))^2}{2(1-0.8^2)} \right),$$

and for the Gumbel copula the density $c = \partial^2 C_\theta^{\text{Gumbel}} / \partial u \partial v$ is given by

$$c(u, v) = \exp\{-[(-\ln(u))^\theta + (-\ln(v))^\theta]^{1/\theta}\} \frac{1}{uv} [(-\ln(u))^\theta + (-\ln(v))^\theta]^{\frac{1-2\theta}{\theta}} \times (-\ln(u))^{\theta-1} (-\ln(v))^{\theta-1} ([(-\ln(u))^\theta + (-\ln(v))^\theta]^{1/\theta} + \theta - 1).$$

We obtain for the Gauss copula $\text{VaR}_{0.99}(X) = 8.68$, and for the Gumbel copula $\text{VaR}_{0.99}(X) = 9.02$. Since the difference is small of the order of 3%, this sensitivity analysis shows that using the Gauss copula for risk management is unlikely to be problematic for this set of parameters.

3.2.1.3 Independence copula

It follows from Sklar's Theorem 3.1 that the *independence copula* is given by

$$C^{\text{indep}}(F_{X_1}(x_1), \ldots, F_{X_m}(x_m)) = F_X^{\text{indep}}(x_1, \ldots, x_m) = F_{X_1}(x_1) \cdots F_{X_m}(x_m)$$

so that $C^{\text{indep}}(u_1, \ldots, u_m) = u_1 \cdots u_m$.

Proposition 3.6 *For the independence copula we have*

$$\rho_{\text{Spearman}}(X_1, X_2) = 0,$$
$$\tau_{\text{Kendall}}(X_1, X_2) = 0.$$

Proof We calculate

$$\begin{aligned}
\rho_{\text{Spearman}}(X_1, X_2) &= 12 \int_0^1 \int_0^1 C^{\text{indep}}(u, v)\, du\, dv - 3 \\
&= 12 \int_0^1 \int_0^1 uv\, du\, dv - 3 = 0, \\
\tau_{\text{Kendall}}(X_1, X_2) &= 4 \int_0^1 \int_0^1 C^{\text{indep}}(u, v)\, dC^{\text{indep}}(u, v) - 1 \\
&= 4 \int_0^1 \int_0^1 uv\, du\, dv - 1 = 0.
\end{aligned}$$

□

The independence copula is illustrated in Fig. 3.3.

3.2.2 Tail Dependence

In our hail and car damage example the dependence of the two distributions, for hail and for collisions, is only in their tails. A simple possibility of describing tail dependence quantitatively is to compare quantiles. Since we are interested in having a simple indicator, we replace the comparison of quantiles with a limit, and come to the following definition:

Definition 3.7 Let $X = (X_1, X_2)$ be a bivariate random variable. Then its *upper tail dependence (upper marginal dependence)* is given by

$$\lambda_u(X) = \lim_{q \to 1} \mathbf{P}\big(X_2 > \text{VaR}_q(X_2) \mid X_1 > \text{VaR}_q(X_1)\big).$$

The index u in λ_u stands for the word "upper". Upper tail dependence does not necessarily exist for all distributions.

Proposition 3.7 *Let F_X be a bivariate distribution with continuous marginal distributions F_{X_1} and F_{X_2} and copula C. If the upper tail dependence λ_u exists, then*

$$\lambda_u(X_1, X_2) = 2 + \lim_{q \to 1} \frac{C(q, q) - 1}{1 - q}.$$

Proof Obviously we have

$$\begin{aligned}
&\mathbf{P}\big(X_2 > \text{VaR}_q(X_2) \mid X_1 > \text{VaR}_q(X_1)\big) \\
&\quad = \frac{\mathbf{P}(X_2 > \text{VaR}_q(X_2), X_1 > \text{VaR}_q(X_1))}{\mathbf{P}(X > \text{VaR}_q(X_1))}
\end{aligned}$$

$$= \frac{\mathbf{P}(X_1 > \mathrm{VaR}_q(X_1), X_2 > \mathrm{VaR}_q(X_2))}{1-q}$$

$$= \frac{1 - \mathbf{P}(X_1 \le \mathrm{VaR}_q(X_1), X_2 \le \mathrm{VaR}_q(X_2))}{1-q}$$

$$- \frac{\mathbf{P}(X_1 \le \mathrm{VaR}_q(X_1), X_2 > \mathrm{VaR}_q(X_2))}{1-q}$$

$$- \frac{\mathbf{P}(X_1 > \mathrm{VaR}_q(X_1), X_2 \le \mathrm{VaR}_q(X_2))}{1-q}.$$

From

$$\mathbf{P}\big(X_1 \le \mathrm{VaR}_q(X_1), X_2 > \mathrm{VaR}_q(X_2)\big)$$
$$= \mathbf{P}\big(X_1 \le \mathrm{VaR}_q(X_1)\big) - \mathbf{P}\big(X_1 \le \mathrm{VaR}_q(X_1), X_2 \le \mathrm{VaR}_q(X_2)\big)$$

and

$$\mathbf{P}\big(X_1 \le \mathrm{VaR}_q(X_1), X_2 \le \mathrm{VaR}_q(X_2)\big) = C(q,q)$$

follows

$$\mathbf{P}\big(X_2 > \mathrm{VaR}_q(X_2) \mid X_1 > \mathrm{VaR}_q(X_1)\big)$$
$$= \frac{1 - C(q,q) - q + C(q,q) - q + C(q,q)}{1-q}$$
$$= \frac{1 - 2q + C(q,q)}{1-q} = 2 + \frac{C(q,q) - 1}{1-q}.$$

□

Corollary 3.2 *The upper tail dependence for continuous marginal distributions depends only on the copula and not on the marginal distributions.*

We can write $\lambda_u(C)$ instead of $\lambda_u(X)$ in the following.

Proposition 3.8 *For the bivariate Gauss copula with* $\mathrm{corr}_{12} < 1$ *we have*

$$\lambda_u\big(C^{\mathrm{Gauss}}_{\mathrm{corr}}\big) = 0.$$

Proof On the basis of the Corollary 3.2 we can, without loss of generality, assume that we have a joint distribution whose marginals are 1-dimensional standard normal distributions. The joint distribution is then a normal distribution with mean $(0,0)$ and correlation matrix corr. A direct calculation yields

$$\lambda_u(X) = \lim_{q\to 1} \mathbf{P}\big(X_2 > \mathrm{VaR}_q(X_2) \mid X_1 > \mathrm{VaR}_q(X_1)\big)$$
$$= \lim_{q\to 1} \frac{\mathbf{P}(X_2 > \mathrm{VaR}_q(X_2),\ X_1 > \mathrm{VaR}_q(X_1))}{\mathbf{P}(X_1 > \mathrm{VaR}_q(X_1))}$$

$$= \lim_{z\to\infty} \frac{\mathbf{P}(X_2 > z,\, X_1 > z)}{\mathbf{P}(X_1 > z)}$$

$$= \frac{\sqrt{2\pi}}{2\pi\sqrt{1-\mathrm{corr}_{12}{}^2}} \lim_{z\to\infty} \frac{\int_z^\infty \int_z^\infty \exp(-\frac{x^2-2\mathrm{corr}_{12}xy+y^2}{2(1-\mathrm{corr}_{12}{}^2)})\,\mathrm{d}x\,\mathrm{d}y}{\int_z^\infty \exp(-\frac{x^2}{2})\,\mathrm{d}x}.$$

In the following we consider the case $\mathrm{corr}_{12} \geq 0$, and apply the estimate

$$\begin{aligned} x^2 - 2\mathrm{corr}_{12}xy + y^2 &= (1-\mathrm{corr}_{12})\big(x^2+y^2\big) + \mathrm{corr}_{12}(x-y)^2 \\ &\geq (1-\mathrm{corr}_{12})\big(x^2+y^2\big). \end{aligned}$$

In case $\mathrm{corr}_{12} < 0$, the calculation runs similarly using the estimate

$$\begin{aligned} x^2 - 2\mathrm{corr}_{12}xy + y^2 &= (1+\mathrm{corr}_{12})\big(x^2+y^2\big) - \mathrm{corr}_{12}(x+y)^2 \\ &\geq (1+\mathrm{corr}_{12})\big(x^2+y^2\big). \end{aligned}$$

We have

$$\begin{aligned} \lambda_u(X) &\leq \frac{\sqrt{2\pi}}{2\pi\sqrt{1-\mathrm{corr}_{12}{}^2}} \lim_{z\to\infty} \frac{\int_z^\infty \int_z^\infty \exp(-\frac{1-\mathrm{corr}_{12}}{2(1-\mathrm{corr}_{12}{}^2)}(x^2+y^2))\,\mathrm{d}x\,\mathrm{d}y}{\int_z^\infty \exp(-\frac{x^2}{2})\,\mathrm{d}x} \\ &= \frac{1}{\sqrt{2\pi(1-\mathrm{corr}_{12}{}^2)}} \lim_{z\to\infty} \frac{(\int_z^\infty \exp(-\frac{x^2}{\sqrt{2(1+\mathrm{corr}_{12})}})\,\mathrm{d}x)^2}{\int_z^\infty \exp(-\frac{x^2}{2})\,\mathrm{d}x} \\ &= \frac{1}{\sqrt{2\pi(1-\mathrm{corr}_{12}{}^2)}} \\ &\quad \times \lim_{z\to\infty} \frac{2(\int_z^\infty \exp(-\frac{x^2}{\sqrt{2(1+\mathrm{corr}_{12})}})\,\mathrm{d}x)\exp(-\frac{z^2}{\sqrt{2(1+\mathrm{corr}_{12})}})}{\exp(-\frac{z^2}{2})}, \end{aligned}$$

where we used l'Hôpital's rule in the last line. From

$$\sqrt{2(1+\mathrm{corr}_{12})} \leq \sqrt{2(1+1)} = 2$$

follows

$$\frac{z^2}{\sqrt{2(1+\mathrm{corr}_{12})}} \geq \frac{z^2}{2}$$

and thus

$$0 \leq \exp\left(-\frac{z^2}{\sqrt{2(1+\mathrm{corr}_{12})}} + \frac{z^2}{2}\right) \leq 1.$$

Since

$$\int_z^\infty \exp\left(-\frac{x^2}{\sqrt{2(1+\mathrm{corr}_{12})}}\right)\mathrm{d}x$$

converges to 0, and λ_u is non-negative, it has been shown that $\lambda_u(C_{\mathrm{corr}}^{\mathrm{Gauss}}) = 0$. □

Corollary 3.3 *In particular,* $\lambda_u(C^{\text{indep}}) = 0$.

Proposition 3.9 *The upper tail dependence of the bivariate Gumbel copula has the value* $\lambda_u(C_\theta^{\text{Gumbel}}) = 2 - 2^{1/\theta}$.

Proof From $C_\theta^{\text{Gumbel}}(q,q) = e^{-((-\ln q)^\theta + (-\ln q)^\theta)^{1/\theta}} = e^{-2^{1/\theta}(-\ln q)} = q^{2^{1/\theta}}$ follows by l'Hôpital's rule

$$\lambda_u(C_\theta^{\text{Gumbel}}) = 2 + \lim_{q\to 1} \frac{q^{2^{1/\theta}} - 1}{1-q}$$

$$= 2 + \lim_{q\to 1} \frac{2^{1/\theta} q^{2^{1/\theta}-1}}{-1} = 2 - 2^{1/\theta}.$$

□

With this the indicator "upper tail dependence" confirms the intuition, derived from Figs. 3.1 and 3.2 that the dependence in the tail for the Gumbel copula is in fact larger than that for the Gauss copula. Proposition 3.8 even shows that for the Gauss copula the dependence in the tail of the limit distribution vanishes. The dependence structure in our hail and car damage example is therefore qualitatively badly described by the Gauss copula. The Gumbel copula shows itself more suitable in that case.

3.2.3 Modeling with Copulas

In practice, copulas are used in Monte Carlo simulations. Let U be an m-dimensional random vector, with joint distribution function given by the copula C. If $F_1, \ldots, F_m$ are given distribution functions with pseudoinverses $F_i^{\leftarrow}(\alpha) := \inf\{x : F_i(x) \geq \alpha\}$, then

$$X = \begin{pmatrix} F_1^{\leftarrow}(U_1) \\ \vdots \\ F_m^{\leftarrow}(U_m) \end{pmatrix}$$

is a random vector with marginal distributions $F_1, \ldots, F_m$ and copula C. Remark that the pseudoinverses can be determined numerically relatively simply as quantiles. The difficulty of finding U depends on the chosen copula.

Example 3.4 Let $Z \sim N(0, \text{corr})$ be an m dimensional normally distributed random vector with correlation matrix corr, and $\Phi_{0,1}$ the distribution function of a one-dimensional standard normal distribution. Then

$$U = \begin{pmatrix} \Phi_{0,1}(Z_1) \\ \vdots \\ \Phi_{0,1}(Z_m) \end{pmatrix}$$

is a random vector whose distribution function is exactly the Gauss copula $C_{\text{corr}}^{\text{Gauss}}$.

The choice of the copula class can have considerable influence on the calculated risk capital. For instance, the Gumbel copula emphasizes the dependence upon tail risks more than the Gauss copula (see Figs. 3.1 and 3.2).

Because of this, an important use of copulas is often considered to be the possibility of describing the increased dependence on tail risks seen in practice. However, in reality it is hardly ever possible to find the optimal copula for the observed tail dependence. Just as with distributions, only a handfull of well described copulas are at our disposal. The problem of copula estimation arises because to judge the dependence of several risks, as a rule, requires very much more data than is needed to estimate the individual marginal distributions. The theory of copulas does not solve the dependency problem but only provides a structure for it. In spite of this, it is naturally better to describe a dependency noted in the data by the choice of a copula with tail dependency, rather than to ignore it.

3.3 Correlations

One can gauge dependencies of random variables without using copulas. At first glance the simplest indicator seems to be *linear correlation*. For random variables X_1, X_2 this is given by

$$\operatorname{corr}(X_1, X_2) = \frac{\mathrm{E}((X_1 - \mathrm{E}(X_1))(X_2 - \mathrm{E}(X_2)))}{\sigma(X_1)\sigma(X_2)} = \frac{\mathrm{E}(X_1 X_2) - \mathrm{E}(X_1)\,\mathrm{E}(X_2)}{\sigma(X_1)\sigma(X_2)}.$$

This gives an indicator, but does not, however, tell us how one can construct the joint distribution out of X_1 and X_2.

If we now consider m loss functions $X_1, \ldots, X_m$, we can define the correlation matrix corr by

$$\operatorname{corr}_{ij} = \operatorname{corr}(X_i, X_j), \quad i, j = 1, \ldots, m.$$

Proposition 3.10 *Let the random vector $(X_1, \ldots, X_m)$ be multinormally distributed and $X = X_1 + \cdots + X_m$. Then we have*

$$\mathrm{VaR}_\alpha(X) = \mathrm{E}(X) + \sqrt{\sum_{i,j=1}^{m} \operatorname{corr}_{ij}\big(\mathrm{VaR}_\alpha(X_i) - \mathrm{E}(X_i)\big)\big(\mathrm{VaR}_\alpha(X_j) - \mathrm{E}(X_j)\big)}$$

and

$$\mathrm{ES}_\alpha(X) = \mathrm{E}(X) + \sqrt{\sum_{i,j=1}^{m} \operatorname{corr}_{ij}\big(\mathrm{ES}_\alpha(X_i) - \mathrm{E}(X_i)\big)\big(\mathrm{ES}_\alpha(X_j) - \mathrm{E}(X_j)\big)}.$$

Proof Since the random vector $(X_1, \ldots, X_m)$ is multinormally distributed, the linear combination X of its components is normally distributed. Furthermore we have

$$\mathrm{E}(X) = \sum_{i=1}^{m} \mathrm{E}(X_i), \quad \mathrm{var}(X) = \sum_{i,j=1}^{m} \mathrm{corr}_{ij} \sqrt{\mathrm{var}(X_i)\,\mathrm{var}(X_j)}.$$

Proposition 2.1 and Proposition 2.5 imply that there is, for each of $\rho_\alpha = \mathrm{VaR}_\alpha$ and $\rho_\alpha = \mathrm{ES}_\alpha$, a function $f(\alpha)$, such that $\rho_\alpha(Y) = \mathrm{E}(Y) + \sqrt{\mathrm{var}(Y)} f(\alpha)$ for any normally distributed random variable Y. Thus we obtain

$$\begin{aligned}
\rho_\alpha(X) &= \mathrm{E}(X) + \sqrt{\mathrm{var}(X)} f(\alpha) \\
&= \mathrm{E}(X) + \sqrt{\sum_{i,j=1}^{m} \mathrm{corr}_{ij} \sqrt{\mathrm{var}(X_i)\,\mathrm{var}(X_j)}\, f(\alpha)} \\
&= \mathrm{E}(X) + \sqrt{\sum_{i,j=1}^{m} \mathrm{corr}_{ij} \big(\sqrt{\mathrm{var}(X_i)} f(\alpha)\big)\big(\sqrt{\mathrm{var}(X_j)} f(\alpha)\big)} \\
&= \mathrm{E}(X) + \sqrt{\sum_{i,j=1}^{m} \mathrm{corr}_{ij} \big(\rho_\alpha(X_i) - \mathrm{E}(X_i)\big)\big(\rho_\alpha(X_j) - \mathrm{E}(X_j)\big)}.
\end{aligned}$$

□

Under the assumptions of Proposition 3.10 one can therefore determine the risk capital for the (unknown) joint distribution from the risk capitals for the individual loss distributions. While the individual distributions often fail to be normally distributed, in practice this "square root" method is often used to approximate the risk capital $\rho(X)$ through

$$\rho(x) \approx \mathrm{E}(X) + \sqrt{\sum_{i,j=1}^{m} \mathrm{corr}_{ij} \big(\rho(X_i) - \mathrm{E}(X_i)\big)\big(\rho(X_j) - \mathrm{E}(X_j)\big)}.$$

This method has, however, serious weaknesses:

- Risk capital for the joint distribution is only a single value. For risk management the shape of the distribution is at least just as important. About that this method can give no clue.
- In many applications one cannot start with the assumption of a normal distribution or approximately normal distribution.
  - Since for a normal distribution exactly the largest risks tend to be underestimated, such an assumption is especially questionable for risk management.
  - In general, the formula is only an approximation. It should be used only when the order of magnitude of the deviation from the true value is known. Depending on the risk measure and the distribution, the deviation can be arbitrarily

large, and the real risk thus could be arbitrarily badly underestimated (or overestimated).

For internal models the calculation of correlations is not really easier to manage than the calculation of a simple copula (such as, for example, the Gauss copula). But the copula provides the joint distribution, and so a lot more information.

3.4 Functional Dependencies

It is not always the case that each uncertain quantity is driven by its own random variable. In our hail and collision example, let us assume in addition that the insured area is very small, and that, when it hails, every insured is affected to the same extent. Moreover, we will suppose that losses occur proportional to the (normalized) intensity I of the hailstorm and its duration D. The number of hailstorms is modeled by another random variable N. We can model the losses under an insurance policy i with

$$S_i = \sum_{k=1}^{N} \gamma I_k D_k W_i$$

where γ is a proportionality factor, $I_k \sim I$, $D_k \sim D$ holds, and W_i is the value of the insured object. The losses for two insurance policies $i = 1, 2$ are obviously functionally dependent on each other, since $S_1 = S_2 W_1 / W_2$. Admittedly this dependence is trivial in the sense that the two policies are mutually perfectly correlated. We have a non-trivial example if we introduce a new policy $i = 3$ for which the maximal payout is C_3:

$$S_3 = \min\left(\sum_{k=1}^{N} \gamma I_k D_k W_3, C_3\right)$$

It follows that $S_3 = \min(S_2 W_3 / W_2, C_3)$. Clearly the policy $i = 3$ is functionally dependent on the policy $i = 2$. Since this dependency is not linear, the correlation of the two policies is not perfect.

The hail example serves only to illustrate a concept. As a rule, one will admit the possibility that not every insured is affected to the same extent by a hailstorm, and the dependency between policies will be classically described by using correlations or copulas. Functional dependencies are however much used in describing the capital market and in life insurance. Another natural area of application is the description of bonus-malus (debit and credit) systems.

References

1. C. Acerbi, D. Tasche, Expected shortfall: a natural coherent alternative to value at risk. Econ. Notes **31**(2), 379–388 (2002) (p. 73)
2. A. McNeil, R. Frey, P. Embrechts, *Quantitative Risk Management. Concepts, Techniques, Tools* (Princeton University Press, Princeton, 2005) (p. 74)

Chapter 4
Risk Capital

4.1 Risk Capital and Cost of Capital

4.1.1 Risk Capital as a Criterion for Comparing Diverse Risks

In Chap. 2 we encountered risk measures whose values could be interpreted as risk capital. Once one has chosen such a risk measure ρ, one can compare different risks with each other. The measure reflects the risk aversion of the company.

One can view this risk capital simply as a value to be calculated so as to compare risks, but as a rule it is in fact made available by one or more investors, and thus operationally made real. The risk capital is thus a capital buffer that is used up in case of need, if a risk is actually realized. However, under normal circumstances the risk capital will generally remain untouched.

In order to carry out its function, risk capital must be available in case of need. Usually this means that risk capital must be sufficiently fungible. The value of the administration building of an insurance company is, for example, not fungible — or only in extreme scenarios in which the autonomous existence of the enterprise is in question. Complete fungibility, however, does not have to be demanded, since the realization of a risk is not always coupled with an outflow of resources. An example of that would be a stock market crash, that for a life insurer had the result that (without own capital) technical provisions would no longer be covered. In that case, it is enough to reduce the own capital in order to restore coverage without any flow of real capital. The portion of the assets that is now no longer used to cover the own capital, but to offset the technical provisions, also does not have to be fungible, but just has to satisfy the regulatory guidelines for assets that cover technical provisions.

In general, we have the following stratification of assets:

- assets that are covering liabilities and so cannot be used as risk capital,
- risk capital that serves as a defense against risks,
- excess capital that has no business function.

Risk capital is often, but not always and not exclusively, made available by shareholders. In a mutual insurance company the insured offer up the risk capital them-

M. Kriele, J. Wolf, *Value-Oriented Risk Management of Insurance Companies*,
EAA Series, DOI 10.1007/978-1-4471-6305-3_4, © Springer-Verlag London 2014

selves. In life insurance, depending on the jurisdiction and the type of contract, the insured may make available part of the risk capital even in stock corporations. An example would be the terminal bonus fund, which can be used to offset risks and which otherwise is given to policy holders at the end of the durations of the contracts. Another common form of risk capital is subordinate bank loans.

4.1.2 Cost of Capital

If risk capital is made available, opportunity costs arise, which can be interpreted as an interest $s_t + k_t$ on the risk capital C_t, where s_t is the risk-free interest and the *spread* k_t reflects the risk that the risk capital could be (partly) lost. The higher the risk, the higher the spread. The opportunity costs are called *cost of capital.*

Here we are dealing with the same concept as for a zero bond: If the investor, in year t, buys a 1-year-zero bond of a company at the face value N, he expects that at the end of the year he will get back from the company a somewhat higher value $(1 + s_t + k_t)N$. The interest $s_t + k_t$ is larger than the risk-free interest s_t by the spread k_t, since the company could have gone bankrupt in the meantime and the investor would then get back nothing (or only a fraction of the bankruptcy estate). The spread k_t compensates the investor for this risk.

In practice, one of the greatest difficulties is to set the spread k_t. There are various possibilities for this. We first assume for simplicity's sake that we wish to determine a risky interest rate $s_t + k_t$ for the company as a whole.

1. If there is only a single owner of the company, one of the options is that the owner simply sets a risky interest rate $s_t + k_t$ for the management by virtue of his power as the financial resource. The owner will usually have made certain that he cannot invest his money elsewhere, with similar risk, in such a way that the profit exceeds $s_t + k_t$.
2. If the company is actively traded on the stock market, one can determine the so-called β and using the Capital Asset Pricing Model (CAPM, see Appendix A) calculate the corresponding yield $s_t + k_t$. If the company is not actively traded, one can use similar companies as benchmarks. This method, however, makes strong assumptions about the link between real risks and share prices. But share prices are also influenced by the public perception of future profits. Together with the psychological component, the fact that profit and risk are not directly coupled leads to bias.
3. One can determine k_t on the basis of direct modeling. Let X_t be the loss distribution of the company, and let $0 = p_0 < \cdots < p_n = 1$ be a finite increasing sequence of probabilities. We denote by $q_i = \mathrm{VaR}_{p_i}(X_t)$ the p_i-quantile of the loss X_t. The investor assigns to each interval $]q_{i-1}, q_i]$ an expected loss

$$E_{X_t,i} = \mathrm{E}\big(\min(C_t, X_t) | X_t \in]q_{i-1}, q_i]\big)$$

(see Fig. 4.1), where C_t denotes the risk capital. The minimum of the loss distribution and the risk capital is taken since the company is bankrupt after using

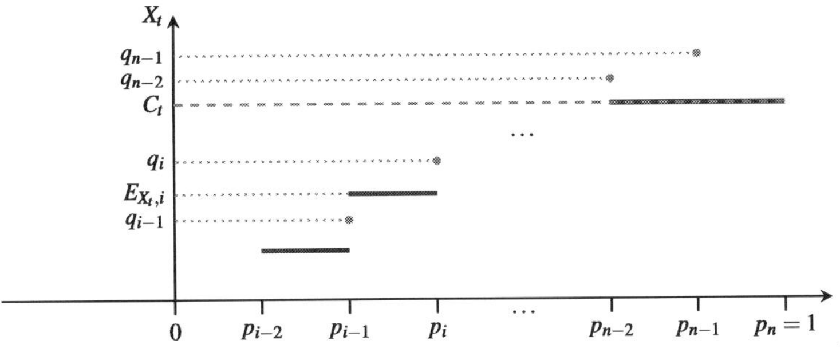

Fig. 4.1 Construction of a spread k_t by the risk profile method. Here $q_{n-2} > C_t$ and so $E_{X_t,n-2} = E_{X_t,n-1} = C_t$

all its risk capital. If the company is part of a group, and has access to group resources that are not part of the risk capital, then the model must be altered correspondingly. The event that with expected probability $1-(p_i+p_{i-1})/2$ a loss of $E_{X_t,i}$ is incurred, can be assigned a spread k_t^i. For example, one can choose k_t^i to be the spread of a zero bond (with full loss given default) whose probability of failure is exactly $1-(p_i+p_{i-1})/2$. To take into account the risk aversion of the investor one could, in a second step, modify k_t^i to be dependent on $E_{X_t,i}$. (The linear relationship between capital invested and profit, which is suggested by the multiplicative formula $(s_t+k_t^i)E_{X_t,i}$, is not always appropriate.) To calculate the spread k_t for the total risk we have only to aggregate the k_t^i over our probability intervals:

$$(s_t+k_t)C_t = (s_t+k_t^0)E_{X_t,0} + \sum_{i=1}^{n}(s_t+k_t^i)(E_{X_t,i} - E_{X_t,i-1}). \tag{4.1}$$

Note that because of (4.1), varying cost of capital arises for companies that put up the same total risk capital C_t, depending on the form of the loss distribution. At present[1] the dependence on the form of the distribution is normally neglected in practice. Neglecting the form of the loss distribution also implies neglecting that hardly ever does a total loss occur. Rather the risk capital is used to absorb partial losses, which do, however, occur and with a greater probability than would be suggested by the desired level of safety.

4.1.2.1 Realistic Cost of Capital to a Company

In practice, cost of capital is not fully borne by the stockholders, but it is possible to draw on certain reserves or hybrid capital to cover (partly) risk capital. (See Sect. 8.2.3.3 in the context of Solvency I).

[1] This text was written in 2013.

The terminal bonus fund in life insurance provides an example of this. The terminal bonus fund can be seen as a type of technical provisions for the terminal bonus, which is to be paid to the policy holder upon termination of the contract. However, the terminal bonus is not a guaranteed benefit. While it cannot be taken away at will by the insurer, in an emergency the insurer can use the terminal bonus fund to cover losses. The terminal bonus would then be reduced. Thus the terminal bonus fund has the character of equity. For this capital the insurance companies pay no costs, but there are restrictions on the conditions under which this capital can be employed. In addition, opportunity costs arise that are difficult to quantify, since a large terminal bonus is less attractive to policy holders than a correspondingly higher guaranteed benefit.

An example of hybrid capital that can be used to cover reserves is given by subordinated loans. Subordinated loans do also engender cost of capital since they bear, as a rule, interest. This interest, however, is fixed by contract. Usually the cost of capital for subordinated loans is less than for equity capital, since in the case of losses first the equity capital will be used up before subordinated loans are brought in to cover losses, so that the equity capital is at greater risk for losses.

Example 4.1 We assume that the available capital of a life-insurance company comprises

- genuine equity provided by shareholders, K_t,
- subordinated debt, D_t,
- the terminal bonus fund, TBF_t,

and that there are no other resources that may be used for coverage. If the capital is optimized in the sense that the available capital exactly offsets the risk capital needed to ensure the desired confidence level, then

$$C_t = K_t + D_t + \mathrm{TBF}_t.$$

To calculate the realistic cost of capital we must also take into account the constraints on the use of the terminal bonus fund and of the subordinated debt. To this end, we assume that in the event of a crisis the company first draws upon the terminal bonus fund, then the equity, and finally the subordinated debt. This order of capital usage implies that the required interest on shareholder's equity should be higher than the interest on subordinated debt.

Let $r_t^K = s_t + k_t^K$ be the interest rate on the equity capital K_t and $r_t^D = s_t + k_t^D$ the interest on the subordinated debt D_t. In this example, there is no interest payable on the terminal bonus fund. Then the average realistic cost of capital $k_t^{\emptyset,\mathrm{real}}$ for the risk capital C_t can be read from

$$\left(s_t + k_t^{\emptyset,\mathrm{real}}\right)C_t = \left(s_t + k_t^K\right)K_t + \left(s_t + k_t^D\right)D_t.$$

Since $C_t \geq K_t + D_t$ and $k_t^K \geq k_t^D$ it holds that

$$\begin{aligned}\left(s_t + k_t^{\emptyset,\mathrm{real}}\right)C_t &\le \left(s_t + k_t^K\right)K_t + \left(s_t + k_t^K\right)D_t \\ &\le \left(s_t + k_t^K\right)(K_t + D_t + \mathrm{TBF}_t) \\ &= \left(s_t + k_t^K\right)C_t\end{aligned}$$

and thus

$$k_t^{\emptyset,\mathrm{real}} \le k_t^K .$$

4.2 Risk-Bearing Capital

While risk capital gives the capital needs of the company, as determined by a risk measure, the *risk-bearing capital* (also known as *available capital*) is the capital that is really available for a company to use in compensating for deviations from the expected course of business.

From an economic perspective the available capital is given by the difference between the market value of the assets and the market value of the liabilities. It thus represents the part of the company's wealth that under realistic assumptions and employing the available information about the market, is not needed to meet liabilities, and so can be used for possible loss compensation.

In order that a company can be said to be *solvent* the available capital must exceed the risk capital. Note that "not being solvent" is not the same as being "insolvent", as *insolvency* means that the assets fail to cover the liabilities. If the liabilities can be reduced in the event of adverse developments, for instance through the reduction of future profit sharing, then the calculation of the risk capital should include such buffering effects, and show the capital character of such liabilities by delivering a correspondingly reduced risk capital. The definition of solvency through comparing available and risk capital thus shows itself to be consistent.

4.3 Types of Risk Capital

There are various types of risk capital in play which are each to be associated with a different perspective. Economic risk capital reflects the purely economic point of view. Rating capital is the capital necessary to achieve a given rating, where in contrast to the economic perspective not financial but only legal insolvency is under consideration. Solvency capital represents the point of view of the supervisory authorities and (to a large extent) policy holders.

4.3.1 Economic Risk Capital

Economic risk capital is the central notion of value-oriented company management.

Definition 4.1 The *economic risk capital* C_t^{EC} is that capital that is necessary to cover possible losses within a given time period and for a given risk tolerance.
We often call economic risk capital simply *economic capital.*

This definition is consciously not so precisely formulated as to allow deciding on the economic risk capital for particular aspects of a company, e.g., "only" for asset management. Thus the economic risk capital is a generic term whose characteristic is the economic viewpoint in considering possible losses.

In practice the economic risk capital is set with the help of a risk measure (Definition 2.1). The risk measure represents the given risk tolerance, so that it has to be ensured that its mathematical form is actually consistent with the views of the management. The given risk tolerance is determined in economic terms by the company's appetite for risk. It is obvious that because of limited resources there is always a residual risk, so that an arbitrarily small willingness for risk cannot be implemented.

One year is the time period often chosen in practice.

The way that economic risk capital functions can best be described with an example (see Fig. 4.2). The economic risk capital is usually calculated at the beginning of a business year and it relates to the whole year. It is influenced both by the situation of the individual company and by the general risk context, which is the same for all similar companies. The *general risk context* reflects external influences such as the volatility of the capital market or the weather, say in relation to hurricanes. The economic capital is a computed quantity (e.g., the expected shortfall with a confidence level of 99.5 %) and must be covered by actual available capital. If the available capital is higher than the economic risk capital (as it was on 1 Jan 2010) then the difference is called *excess capital.* This capital is not needed to keep the business going and so just reduces the risk-adjusted profits. It will be a goal of management to control the available capital and the risks to be accepted so that the excess capital is positive, but by a small amount. The company in the example had reached this goal on 1 Jan 2011 by changing its risk profile. The available capital is naturally on hand not just on 1 January but throughout the year. This is shown in the figure by the background behind the capital bars. In mid-year 2011 the company suffers a considerable loss resulting from unpredictable circumstances. The insolvency of the business can be prevented because there is enough capital available to absorb the loss. This reduces the available capital. In our example neither the particular situation of the company nor the general risk context has changed so that the event has no impact on the economic risk capital.[2] In Fig. 4.2, because of the losses in the 2011 year the risk capital for 2012 can no longer be covered by the available capital. This means that the company is no longer operating in accordance with its risk tolerance. It can now

[2]In principle large catastrophes can lead to a new assessment of the general risk context which then has an effect on the following year's risk capital. An example of this is the re-evaluation of the hurricane risk after Hurricane Katrina devastated New Orleans in 2005.

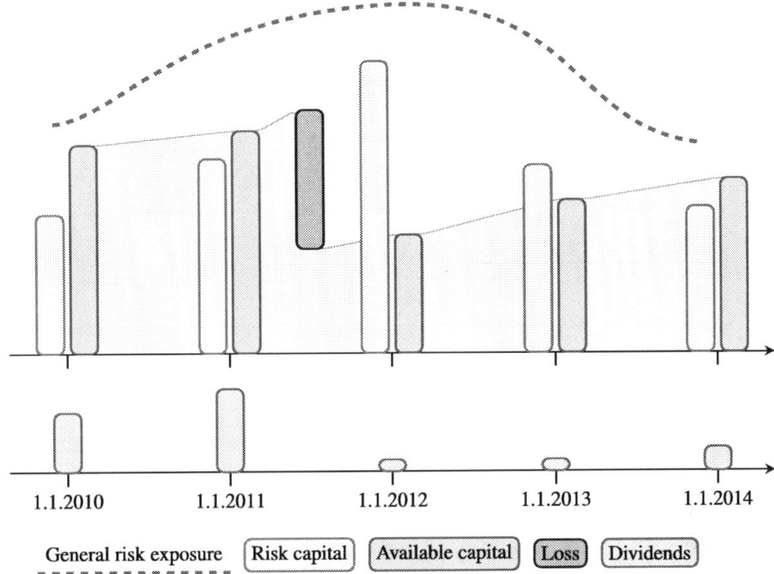

Fig. 4.2 The functioning of economic risk capital

- raise capital from the market, which means additional cost of capital,
- reduce the risks in its portfolio, for example by
  - sale of assets,
  - hedging its investments,
  - increased reinsurance,
  - securitization etc.,
- or by reducing dividends, or retention of earnings, to top up its equity once more in as short a time as possible on its own.

In our example the company chooses the third possibility. In 2014 it again has a level of coverage by available capital that is correct for its risk tolerance. This was possible because it also benefitted from the fact that in the same time period the general risk context improved. In practice this approach is often only successful for small shortfalls, since the risk capital is much larger than the dividends.

4.3.1.1 Operationally Necessary Risk Capital and the Market Value of Technical Reserves

If the economic risk capital for a goal has been calculated so as to protect from economic insolvency with a given high probability then one speaks of operationally necessary risk capital.

Definition 4.2 The *operationally necessary risk capital* at the start of the period t is the capital, needed so that the difference between the market values of the assets and the liabilities remain positive during the given period with a prescribed probability.

The operationally necessary risk capital is thus a special form of the economic necessary risk capital, where the risk measure 'value at risk' is applied to the economic loss in value of a company, that is to the change in the difference between the economic value of its assets and the economic value of its liabilities.

Obviously variations on the operationally necessary risk capital can be imagined that are based on other risk measures, such as the coherent expected shortfall.

The risk capital defined above could be calculated through Monte Carlo simulations of the company. However, because of the complexity of risks and their interactions in the insurance business an implementation of this approach could be difficult. In practice simpler definitions may approximate the economic content of the above definition sufficiently well.

4.3.1.2 Run-off and Going-Concern Problems

Since the amount of operationally necessary risk capital depends on the market value of the liabilities at the end of the period being considered, it is necessary to determine future cash flows in order to calculate the market value of liabilities.

For this there are differing interpretations as to what belongs under "future liabilities" which lead to differing results.

Definition 4.3 The value of the liabilities on a *run-off basis* is the amount that is necessary to liquidate the company and to make all those future payments which are due to the existing liabilities (and consistent with the implicit confidence level).

The idea behind this approach is that the present insurance portfolio seen on its own is secured. The approach reflects the scenario that the company is closed for new business and a trustee is appointed to wind down the existing insurance portfolio.

It is assumed no new business can be entered into. Therefore the relative portion of the fixed costs increases during the course of the projected plan. The costs of doing business go down, however, there are costs in dismantling the business (e.g., financial settlements, costs for dissolving long-standing contracts, etc.) which have to be taken into account. One has to proceed in the same way in a planned reduction of the work-force. In doing so, it has to be taken into account that the number of employees necessary is not proportional to the size of the portfolio. Analogous considerations apply to all other cost factors, such as the buildings used by the business itself. Furthermore, in life insurance the cancellation rates for a portfolio during run-off are different from the cancellation rates for a company conducting new business. In life insurance there is also the obligation to ensure that any bonus funds are correctly disposed of during a run-off phase.

Definition 4.4 The value of the liabilities on a *going-concern basis* is the amount that is necessary to make all those future payments which are due to the existing liabilities (and consistent with the implicit confidence level) on the assumption that the company continues to take on new business.

This approach is based on the notion that the company continues to exist in the future, or that it sells its portfolio to another insurance company.

Since the assumption is that there will be unchanged new business, the relative contribution of the fixed costs remains essentially constant during the course of the projection. In growing or shrinking companies the assumption that new business is taken on at a constant rate can lead to distortions. Therefore it might be recommended to allow for time-dependent new business, even when such assumptions are associated with great uncertainty.

One may take the view that for a company that continues to exist the existing insurance portfolio should be handled separately from new business. In that case, the future new business will merely affect the fixed costs to be modeled. For life insurance there is (in some jurisdictions) the extra consideration that the profit participation for old or new business must be comparable. However, often the acquisition of new business is financed with profits from old business. In this case a strict separation of old and new business is not correct, since one would otherwise overestimate the profits that could be attained from the old business.

Remark 4.1 The going-concern approach is also adopted in calculating embedded values in life insurance and health insurance. In establishing claim reserves the going-concern principle is also implicitly assumed.

Definition 4.5 The value of liabilities on a *reference company basis* (sometimes reference undertaking basis) is so determined that (for an implicit confidence level) it is sufficient to completely wind up the insurance portfolio, on the assumption that the portfolio is transferred to a large, well diversified, insurance company.

In contrast to the going-concern approach it is not the parameters of the undertaking but the reference company that is imputed. In particular, it starts by assuming that we have optimal diversification.

This approach reflects the scenario that the company is no longer adequately solvent, and that the supervisory authority initiates the sale of the portfolio to a large healthy insurance company, in order to preserve the rights of the insured. It can also be seen as the basis for the most objective market valuation possible of the liabilities.

In order to satisfy the first scenario (lack of solvency) one would have also to take into account the costs of transferring the insurance portfolio to the reference company. The costs are however variable (e.g., depending on the management system being used), so that an objective estimate is hardly possible for a virtual reference company. This problem can be avoided if one assumes that the current management system continues to be used. But, in this case, one accepts at least the run-off costs for the management system, which are not simple to determine.

In the second application of market value determination, it should be born in mind that there is an implicit assumption here of a liquid market for the insurance portfolio. In reality, such a market does not exist so that the value determined is only to be viewed as a guide.

4.3.2 Rating Capital

Rating capital C_t^{Rating} is the capitalization that is equivalent to having the rating desired by the company. The basis for this is the empirical default probabilities that are published by the rating agencies, corresponding to the rating categories for each rating company. These default probabilities do not, however, refer to economic insolvency, but rather to legal insolvency. Since in practice the calculations refer to economic insolvency, the rating capital will be somewhat higher than needed. This is because a company that is economicly insolvent only then declares insolvency when there is no longer any genuine hope that bankruptcy can be avoided. This leads to offsetting of the confidence level.

The rating models used by the rating agencies are frequently less detailed than internal risk models, but calculate risk capital for a commonly very high confidence level, e.g., the confidence level corresponding to Standard & Poor's AAA rating. The associated error in the model is so large that a direct interpretation of the rating capital as an individual economic risk capital is not possible. But this is anyway not the function of rating capital. Rating capital must be understood in the context of a well diversified portfolio of corporate bonds which involve several industries and thousands of companies. The rating capital is calibrated for large and broad portfolios, and, by the law of large numbers, a quality rating based on rating capital is appropriate under such circumstances.

Naturally a rating has consequences for an individual company, since it influences its credit costs and the reputation of the undertaking. Especially industrial insurers and reinsurers see themselves as compelled on competitive grounds to have enough available capital for an excellent rating, but even for life insurers ratings are playing an ever larger role, notably in the occupational pension business. Rating agencies are about to include individual risk capital calculations in their ratings. In particular, Standard & Poor's have introduced a separate partial rating for Enterprise Risk Management (ERM), for which internal models are also examined in detail.

4.3.3 Solvency Capital

The solvency capital C_t^{Reg} is the capital mandated by regulations that an insurance company must have on hand in order to be allowed to do business. From the point

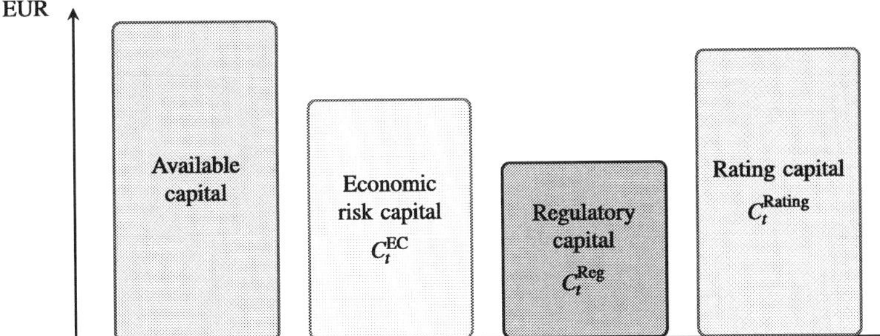

Fig. 4.3 Comparison of different forms of risk capital. The relative capital levels are for illustrative purposes

of view of supervising authorities this serves primarily to protect the insured, who would be harmed if the insurance company became insolvent. Another intended goal is the stabilization of the financial market.

In a similar way to the case of rating capital, the confidence level, relative to which the solvency capital is calculated, is defined relative to the portfolio. However, here the portfolio only includes the national insurance companies and is thus an order of magnitude smaller than the portfolio that is considered by the rating agencies. With Solvency II (Sect. 8.2.4) it will be possible (as it already is in Switzerland with the Swiss Solvency Test) for insurance companies to determine the solvency capital on the basis of individual risk capital calculations.

In Fig. 4.3 we compare the various types of capital, which have been discussed in Sect. 4.3, to the available capital.

4.4 Valuing Insurance Liabilities

4.4.1 Concept and Definition

Both determining risk-bearing capital on economic grounds, and the determination of the economicly necessary capital, depend on the market values of assets and liabilities. While for most investments the price is visible in liquid markets, there is at present no liquid market for trade in insurance portfolios.

Valuations based on prices in a market are distinguished by high transparency and small risk of their being manipulated, and thus are generally accepted.

The fundamental idea behind the *fair value* of an insurance liability is to define a market value for it. The fair value is the price that a third-party expert would ask for taking on the liabilities at time of the valuation. The central challenge is to determine an approximation to this price by some credible procedure.

The market value of financial instruments estimates uncertain future cash flows and includes a risk premium to compensate for this uncertainty. The fair value, as a substitute for this market price, can thus not just be limited purely to expected values. More precisely, the costs of liabilities can be divided into three groups:

1. Expected value of discounted future liabilities.
2. Discounted expected costs for risk capital that is associated with future liabilities. Thus all risks that are covered by the risk capital are included in the valuation of the liabilities.
3. Risk margins for those risks that are not fully covered by the risk capital.

An important motivation for the notion of risk capital is its universality. Ideally one would like all risks to be adequately captured by the risk capital, and that, in particular, the risk premium for uncertain cash flows be represented as a function of the risk capital. In this idealization, the two quantities, 1 and 2 above, are sufficient to determine the fair value. However, this idealization ignores that the risk depends on the whole distribution function and cannot be represented by the risk capital, which is a single real number. For example, there is, in addition to costs and the risk that the settlement of the portfolio is less successfully handled than expected, also the "upside risk" that the expected value of future payments is overestimated. A rational investor would take this possibility, that the run-off of the portfolio goes better than expected, into account during a valuation along with the negative risks. As another concrete example, in Example 4.7 two portfolios are considered which have the same expected value for the discounted future liabilities and the same risk capital, but one is clearly much more risky than the other.

The use of risk margins (item 3 above) takes account of these factors by including further properties of the portfolio in the determination of a fair value. In fact, in modern, market consistent approaches it is preferred to calculate risk margins, wherever possible, without recourse to the risk capital (see Sect. 6.6.4 and the discussion of hedgeable/non-hedgeable risks in Sect. 4.4.2).

4.4.2 Approaches to Valuation of Insurance Liabilities

Insurance liabilities depend on uncertain future cash flows, whose uncertainty is characterized by insurance risk and market risk, but also by how the insurance portfolio is managed, which reflects, e.g., profit sharing, cost structures and claims settlements.

In valuing insurance liabilities the following approaches can be distinguished.

1. *Best estimate perspective.* A critical piece of information for management is the most precise and realistic estimate possible of the liabilities. The expected value of the future cash flow, estimated on the basis of realistic assumptions about the relevant risk factors, provides the *best estimate*. It includes no safety margins and can thus not compensate for taking on the risk in the liabilities.
2. *Economic perspective.* For company management the best estimate is not enough, since the risks associated with future cash flows must be managed and the provisions that are to ensure that future liabilities can be fulfilled must include a risk margin. This risk margin depends on the company's risk tolerance and takes account of the risks from a longer perspective, while short-term extreme scenarios

are captured by the risk capital. Along with the risk tolerance there are a number of relevant factors specific to the company that determine the value of the liabilities from an economic point of view:

(a) The company can influence the cash flows and their risk profile, through their management rules (e.g., profit sharing) and cost structures.
(b) For management decisions, cash flows originating from assets and liabilities need to be assessed while taking their mutual interactions into account.
(c) Mutual insurers may have agreements with their policy holders in place to the extent that the latter have to provide supplementary funds if the mutual insurer is in difficulties. Such agreements would influence the risk profile of the mutual insurer.
(d) If parts of the portfolio are valued individually, for instance when there is a transfer of a portfolio, then the extent of the effects of diversification depends crucially on the structure of the portfolio as a whole, so that the economic value of part of the portfolio depends on the entire portfolio.

3. *Balance sheet perspective.* The appraisal of technical provisions on the balance sheet complies with the legal requirements for accounting. The development of the IFRS (international finance reporting standards) has as a goal to set out accounting regulations from the fair value point of view.
4. *Fair value perspective.* The fair value tries to approximate a market value for liabilities. The basic valuation method must therefore be consistent with the market, i.e., it cannot contradict available market information and should, as far as possible, employ market prices. Since market prices involve risk premiums, the fair value is made up of the best estimate for future cash flows and the *market value margin* (*MVM*), which reflects the market's assessment of the risk of the cash flow. The fair value is closely related to the economic point of view, but cannot take into account particular aspects of the company because they would not carry over to a buyer when liabilities are sold.

The intention to use information about market prices as well as possible motivates the classification into *hedgeable* and *non-hedgeable risks*. A risk falls into the first category when there is a tradable financial instrument, with a uniquely determined market price, using which one can hedge the risk. Since the great majority of insurance risks must be viewed as non-hedgeable, the approach offers itself of dividing cash flows associated with insurance liabilities into their expected values and a remainder. The expected cash flow can be valued by replicating it with financial instruments. The difficulty in obtaining a valuation consistent with the market is then moved entirely to the non-hedgeable remainder.

Example 4.2 For a portfolio of term life insurance contracts the insurance benefits can be projected using realistic mortality rates. The individual payments can be replicated with risk-free zero bonds with the corresponding maturities. The market value of this replicating portfolio then represents the best estimate of the liabilities.

In order to determine the market value margin for mortality risk we note that the entire mortality risk could have been taken on by a reinsurer. The cash flow of the corresponding reinsurance premium can then again be valued by using a replicating portfolio of risk-free zero bonds.

In this approach, the uncertainty in the original cash flow for the insurance payments is transferred, after splitting off the best estimate, to determining the reinsurance premium. As long as there is no liquid market for insurance risks, however, one cannot derive an implicit mortality rate from the market prices in order to calculate the reinsurance premiums.

Example 4.3 In non-life insurance, one can project the cash flow of future payments $(X_1, \dots, X_n)$ from the information in claims triangles. Valuation of the estimators $(x_1, \dots, x_n)$ by means of a replicating portfolio of risk-free zero bonds provides the best estimate of the liabilities.

Conventional methods of setting up claims reserves also give estimators for the variances $\operatorname{var}(X_j)$ of future payments. If one uses $\rho(X_j) = \beta \operatorname{var}(X_j)$ with $\beta > 0$ as a risk measure and assumes that a provider of risk capital expects a relative risk premium of i relative to the risk capital, then one can determine the market value margin of a replicating portfolio of risk-free zero bonds for the cash flow $(i\beta \operatorname{var}(X_1), \dots, i\beta \operatorname{var}(X_n))$ (see Sect. 4.4.3.3).

For this method the question of suitable choices of ρ and i has to be considered in the light of the fact that there is actually no liquid market.

Example 4.4 From time to time catastrophe bonds are issued by insurers or reinsurers, and traded in the capital market. These catastrophe bonds can be used, to a limited extent, to estimate the value of corresponding insurance liabilities.

Suppose R is a reinsurer who has a concentration risk in respect of earthquakes in California. R could, of course, retrocede some of these risks. Another possibility would be to bring these risks to the capital market. Since only standardized securities can be traded, and the investors in the capital market have no particular insight into the reinsurance business and a reinsurer's settlement of claims, it would only with difficulty be possible to assemble reinsurance contracts into a portfolio, and to bring this portfolio to market in small bundles. It is therefore more promising for R to place the earthquake risk before the market separated from concrete reinsurance contracts. Assume that R wishes to protect 100 million €. Then the idea could be implemented as follows: R issues a one-year zero bond with nominal value $N = 100$ million € with the proviso that the loan will not be paid back if there occurs, during that year, in California (or within 100 km of the coast of California) an earthquake of magnitude ≥ 7.5. For this purpose the magnitude of the earthquake is taken to be the value published by the *United States Geological Survey's (USGS) Earthquake Hazards Program*. Such a financial instrument can be interesting to an investor in the capital market for the following reasons:

- There is no moral risk on the part of the issuer.
- When and what must be paid is explicitly specified.
- It is possible to divide up the product into pieces of the same type without substantial administrative effort. Thus it becomes tradable.

- The buyer does not need any deep understanding of insurance mathematics, since the trigger does not depend on the actual damage caused by the earthquake. Outside experts can be brought in to determine the risk of an earthquake.
- The risk is only weakly correlated to the capital market risk in general.

For the reinsurer R the disadvantage is that it has a considerable base risk, since the real damages may turn out to be bigger (or smaller). In addition, R is in no way protected if there is an earthquake of magnitude 7.4 in the center of San Francisco. On the other hand, it is possible that investors are willing, on the grounds of the small correlations with capital market risks, to pay so high a price for this product that this procedure is cheaper than retrocession, in spite of the base risk.

If P is the price for the whole slice that was actually paid, then

$$\frac{N-P}{N} \times \text{amount insured}$$

would be, to a first approximation, the value of the liability for earthquakes in California that can be traced back to earthquakes of magnitude ≥ 7.5. This amount must be adjusted by corrections for the risk-free interest, for the risk of default by the reinsurer, and for the base risk.

A further problem in the valuation of insurance liabilities arises because a hedge against a risk cannot cover the actual dependence of that risk upon other risks in the contract. For example, it seems plausible that one evaluates separately the financial and biometric risks in an index-linked life insurance contract, and then hedges the financial risk using the index. If the rate of cancellation by policy holders does depend on the performance of the index, then how the index evolves can influence the composition of the collective and thus the biometric risk profile. Separate consideration of the risks also neglects the diversification effects that companies do include in their calculations. So, for example, non-life insurers can offer contracts with smaller combined ratios than would be the case if they had to invest only in risk-free assets.

The development of new tradable financial instruments such as insurance derivatives leads to increasing information available on market prices and poses the question of completion of the insurance markets[3] in the sense that risks that have long been non-hedgeable should become hedgeable. Even with a liquid market of tradeable insurance derivatives there remains the problem of base risk, since each insured risk is unique and distinguished by individual properties such as its extent, the size of the guarantees, and the structure of its dependence on other risks.

Since the fair value is supposed to approximate a price at which the transfer of the liabilities could be carried out in the market, it represents a quantity that is particular to a portfolio and so should not depend on the individual business characteristics of the buyer, such as the diversification resulting from acquisition of the portfolio. This

[3]It needs to be checked that the known results of financial mathematics can be carried over to the stochastic processes used for insurance risks. There are results that suggest this is possibly not the case [8].

requirement of the fair value raises the question of how far the factors influencing the cash flow of insurance liabilities are specific to a portfolio, and how the possible company-dependent factors are to be handled.

Though biometric actuarial assumptions for calculation are properties of the insurance portfolio, one generally takes it that during transfer of portfolios the cancellation probabilities and cost rates level off after a certain while to those of the transferee company. Furthermore, the cash flows are influenced by the investment strategy and management practices. If the expectations of policy holders are shaped by the profit sharing strategy of the transferring company, then the profit sharing policy of the transferee company matters, and it becomes relevant how fast the transferee company adjusts the transferred assets according to its own investment strategy.

In order to find the fair value as an approximation to a market price, it is appropriate to replace influence factors that depend on the company with standardized assumptions. For example, the average cancellation probability and cost rates for the industry can be used, as well as any regulatory minimum requirements for profit sharing and the fastest possible reorganization of the assets into an optimal replicating portfolio (as in SST, see [4]).

4.4.3 Implementation Concepts

Determining a fair value calls for, in the first place, a projection of future cash flows on the basis of realistic assumptions. The present value of these cash flows, discounted with the risk-free interest curve, gives the best estimate of the insurance liabilities. The fair value is then derived by adding in the market value margin (MVM), which depends on the risk tolerance of the market, or, in other words, the market price for risk.

Determining the MVM can be based on the market price for other tradable risks, e.g., credit risks. In this connection, the question arises to what extent risks can be compared.

There are three main methods which have been discussed for approximating the fair value:

- Valuation based on a risk measure, e.g., based on a quantile (Sect. 4.4.3.1) or a spectral measure (Sect. 4.4.3.2),
- Valuation based on cost of capital (Sect. 4.4.3.3),
- Valuation based on market consistent methods. We will sketch this approach in Sect. 6.6.4.

4.4.3.1 Quantile Method

In the quantile method, the fair value is defined as the α-quantile of the distribution of the present value of future cash flows. Thus the fair value gives the level of technical provisions that, with probability α, are sufficient to satisfy all future liabilities.

Let V_t be a random variable that describes the present value (discounted with the risk free interest curve) of future liabilities. Then the fair value, according to the quantile method, is given by

$$\mathrm{FV}_\alpha = \mathrm{VaR}_\alpha(V_t).$$

In this sense, the fair value generalizes the concept of value-at-risk to considerations involving multiple periods. This method has the following drawbacks:

1. The quantile method masks high risks on the other side of the α-quantile.
2. The method may lead to negative risk margins. Thus, for example, the 75 % quantile of a sufficiently skew distribution can be smaller than the mean.
3. There is no direct connection between the market price of risky securities and the quantile function. In particular, it turns out to be difficult to fix a plausible value for α.
4. Stochastic simulations to determine quantiles require a great deal of effort.

4.4.3.2 Spectral Measure Method

An alternative to the quantile method is the use of a spectral measure (Definition 2.7),

$$\mathrm{FV}_\phi(X) = \int_0^1 \mathrm{VaR}_p(X)\phi(p)\,\mathrm{d}p,$$

where ϕ is a monotone increasing weight function (see Definition 2.6 and Theorem 2.4).

In comparison to the quantile method, points 3 and 4 of the criticism above continue to stand. Nonetheless, it would be easier than with the quantile method, because of the greater flexibility of spectral measures, to achieve a spectral measure consistent with market prices.

As ϕ is monotonically increasing and does not vanish identically, high risks are not obscured in contrast to the quantile method (see critical point 1 for the quantile method). The following proposition shows that criticism 2 of the quantile method also does not hold for the spectral measure method.

Proposition 4.1 *The spectral measure* FV_ϕ *satisfies* $\mathrm{FV}_\phi(X) \geq \mathrm{E}(X)$, *if the weight function* $\phi \in L^1([0, 1])$ *is (almost everywhere) monotonically increasing.*

Proof As Φ is monotone increasing and $\int_0^1 (\Phi(p) - 1)\,\mathrm{d}p = 0$, there is a $\zeta \in (0, 1)$ with

$$\Phi(p) - 1 = \begin{cases} \leq 0, & p \leq \zeta, \\ \geq 0, & p \geq \zeta. \end{cases}$$

Thus it holds that

$$-\int_0^{\zeta} \big(\Phi(p) - 1\big)\,\mathrm{d}p = \int_{\zeta}^{1} \big(\Phi(p) - 1\big)\,\mathrm{d}p.$$

Since $p \mapsto \mathrm{VaR}_p(X)$ is monotone increasing, it follows that

$$-\int_0^{\zeta} \mathrm{VaR}_p(X)\big(\Phi(p) - 1\big)\,\mathrm{d}p \leq \int_{\zeta}^{1} \mathrm{VaR}_p(X)\big(\Phi(p) - 1\big)\,\mathrm{d}p,$$

or

$$\int_0^{1} \mathrm{VaR}_p(X)\big(\Phi(p) - 1\big)\,\mathrm{d}p \geq 0.$$

From the equation $\mathrm{E}(X) = \int_0^1 \mathrm{VaR}_p(X)\,\mathrm{d}p$ we infer

$$\mathrm{FV}_{\phi}(X) \geq \mathrm{E}(X).$$ □

The simplest example of a spectral measure is the expected shortfall ES_α (Definition 2.4). The expected shortfall with a level of confidence of 70 % is utilized in the USA in the valuation of provisions for "Variable Annuities with Guarantees".

In the following example a generic spectral measure is used to take into account "upside risk".

Example 4.5 Suppose $p_{\mathrm{E}(X)} = F_X(\mathrm{E}(X))$; then, with probability $p_{\mathrm{E}(X)}$, the present value of the future liabilities is smaller than the best estimate. The corresponding events thus represent the performance of a portfolio that went better than expected. Instead of, as for the expected shortfall, not weighting these events at all, we want to give this "upside risk" a weight of $a \in (0, 1)$. Analogously the present value of the liabilities is larger than the best estimate with probability $1 - p_{\mathrm{E}(X)}$. Such events represent performance of a portfolio that went worse than expected, and so represent a "downside risk". We would like to weight this risk with a factor $b > 1$. The resulting weighting function ϕ_a is thus piecewise constant, and the normalization $1 = \int_0^1 \phi_a(p)\,\mathrm{d}p = a p_{\mathrm{E}(X)} + b(1 - p_{\mathrm{E}(X)})$ implies

$$\phi_a = a \mathbb{1}_{[0, p_{\mathrm{E}(X)}[} + \frac{1 - a p_{\mathrm{E}(X)}}{1 - p_{\mathrm{E}(X)}} \mathbb{1}_{[p_{\mathrm{E}(X)}, 1]}.$$

Clearly it holds that $\lim_{a \to 1} \mathrm{FV}_{\phi_a}(X) = \mathrm{E}(\mathrm{X})$. The higher we weight the "upside risk" the closer the fair value lies to the expected value.

Example 4.6 In Example 4.5 we defined the "upside risk" and "downside risk" in relation to the expected value for the sake of intuition. However, this is not necessary. Let $n \geq 1$ and let $\mathbf{a} = (a_1, \ldots, a_n) \in \mathbb{R}^n$ be a vector with $a_1 \leq \cdots \leq a_n$ and $\mathbf{p} = (p_0, \ldots, p_n) \in \mathbb{R}^{n+1}$ a vector with $0 = p_0 < \cdots < p_n = 1$. Then

$$\phi_{\mathbf{a},\mathbf{p}} = \frac{\sum_{i=1}^n a_i \mathbb{1}_{[p_{i-1}, p_i[}}{\sum_{i=1}^n a_i (p_i - p_{i-1})}$$

is a weight function. Clearly we have $\phi_a = \phi_{((1-p_{E(X)})a, 1-ap_{E(X)}),(0,p_{E(X)},1)}$, where we have used that for each $\lambda \in \mathbb{R}^+$ there holds the identity $\phi_{\mathbf{a},\mathbf{p}} = \phi_{\lambda\mathbf{a},\mathbf{p}}$. Now we demonstrate a practical application of this class of weight functions. Insurance company XYZ distinguishes "good", "neutral", and "bad" performance, wherein the boundaries of these classes are defined by the quantiles $\mathrm{VaR}_{0.25}(X)$ and $\mathrm{VaR}_{0.75}(X)$. Therefore $\mathbf{p} = (0, 0.25, 0.75, 1)$. Neutral performance should get the weight 1. If

$$\mathbf{a} = (\alpha, \beta, \gamma),$$

then this means $\beta = 0.25\alpha + 0.5\beta + 0.25\gamma$, so that $\beta = 0.5\alpha + 0.5\gamma$. We have

$$\mathrm{FV}_{\mathbf{a},\mathbf{p}}(X) = \frac{\alpha\,\mathrm{E}(X) + (\beta - \alpha)\,\mathrm{ES}_{0.25}(X) + (\gamma - \beta)\mathrm{ES}_{0.75}(X)}{\beta}.$$

The relative weighting of "bad" and "good" performance follows from the risk strategy, and, in particular, the risk aversion of the company. The larger the ratio of these weightings is the more conservative the company is. The top management of XYZ decides that weights with $\alpha : \gamma = 1 : 3$ are suitable. This gives $\beta = (0.5 + 1.5)\alpha = 2\alpha$, $\gamma = 3\alpha$, and thus

$$\mathrm{FV}_{\mathbf{a},\mathbf{p}}(X) = \frac{\mathrm{E}(X) + \mathrm{ES}_{0.25}(X) + \mathrm{ES}_{0.75}(X)}{2}.$$

4.4.3.3 Cost of Capital Method

The cost of capital method is based on the idea that the market value margin is a sufficient risk premium which enables an investor to raise the risk capital required for the reserves at all future times. For this it has first to be made clear how the required risk capital is to be defined.

Conceptually, the risk capital is determined by applying a risk measure to the distribution of the present value of the future cash flows at time t. This method would take account of the dependencies over time, including trend risks such as a slow continuous worsening of the risk profile of a collective (e.g., as a result of longevity risks), but it would be very complicated. In practice therefore risk capital with a one-year horizon is used.

Economic risk capital C_t^{EC}, which reflects the risk appetite of a company, is however, not appropriate, since each company has a different appetite for risk and the fair value has to be normalized relative to the market. Regulatory capital C_t^{Reg} is normalized with respect to the market, but it may address not only risks which are linked to the liabilities. An example would be regulatory capital for market risks that have developed because of the investment strategy of the company. Since a company to which the liabilities are transferred could invest in risk-free securities matched to the duration of the liabilities, market risk is not relevant to the fair value. In addition, market risk is already considered in valuing the assets. Because of this a special risk capital C_t^{FV} is defined for determining fair values. In practice this is often the regulatory capital where the contribution from those risks, which are not relevant to the fair value of liabilities, is deducted. If the regulatory capital C_t^{Reg} is not appropriate,[4] then one can use a rating capital C_t^{Rating} for a suitable rating, e.g., BBB.

Denoting by V_t the present value of the future liabilities discounted with a risk-free interest curve s_t, by C_t^{FV} the risk capital for the calculation of a fair value, and by k_t the relative cost of capital, the fair value results as

$$\text{FV}_t = \text{E}(V_t) + \text{MVM}_t,$$

where the market value margin MVM_t is defined as the present value of the future cost of capital

$$\text{MVM}_t = \sum_{\tau=t+1}^{\infty} \frac{k_\tau C_\tau^{\text{FV}}}{\prod_{\tilde{\tau}=t+1}^{\tau}(1+s_{\tilde{\tau}})}.$$

Using a risk-free interest curve brings out that, in any case, risk capital has to be provided until the complete disposal of the insurance portfolio, even if the original insurance company was not solvent.

The relative cost of capital rate k_t is determined as the spread above the risk-free interest rate which compensates for the risk of default of C_t^{FV} as a result of adverse developments in the cash flow. Making a determination of k_t that meets the amount of risk requires the analysis of the distribution of defaults in C_t^{FV}. (Compare this with the decomposition of C_t into tranches which hold different risks in Sect. 4.1.2, bullet point 3). A pragmatic way to avoid the difficulties of such an analysis is to choose for k_t a universal percentage that corresponds to the average risk content in a business sector.

A further simplification is to establish the risk capital C_t^{FV} only at the start and to use at later times a portion weighted by the volume

$$C_\tau^{\text{FV}} = \frac{\text{E}(V_\tau)}{\text{E}(V_t)} C_t^{\text{FV}}. \tag{4.2}$$

This procedure can be refined by using for each of the various drivers of risk in the calculation of C_t^{FV} an adjusted volume weight.

For example, for Solvency II a whole hierarchy of possible simplifications has been proposed [7, Paragraph TP.5.32]:

1. complete projection without simplification of the future risk capital for all risks
2. approximation of the risks for individual risk classes and calculation of the prospective risk capital for the simplified risk description
3. scaling of C_t^{FV} over the various risk drivers with suitable respective volume weights
4. use of the best estimate of future liabilities as the only driver of risk, as in (4.2)

---

[4]For example, solvency capital in line with Solvency I is not, since this capital notion takes too sweeping a view of risks.

5. determining all future SCR (solvency capital requirements) by a duration method
6. approximation of the market value margin as a fixed percentage of the best estimate of the future liabilities.

The "cost of capital" approach is implemented in the Swiss Solvency Test and in Solvency II. In both cases k_t is set at a constant 6 %, and the calculation of the risk capital C_t^{FV} follows the calculation of the solvency capital, where it is, however, assumed that the assets are to be restructured as quickly as possible into a portfolio that optimally replicates the liabilities.

In the following example we show that the cost of capital method cannot capture all aspects of risk, since risk is accounted for only through the value of a single risk measure.

Example 4.7 Given $z > 0$ and $z_t = (1 - 10\,\%\,(t-1))z$ we consider the following two portfolios:

1. *Portfolio A*. In the year t ($t \in \{1, \dots, 10\}$) the portfolio generated the following payments:

$$Z_t^A = \begin{cases} Z_t^{(1)} = z_t & \text{with a probability of } 99\,\%, \\ Z_t^{(2)} = (1+5)z_t & \text{with a probability of } 1\,\%. \end{cases}$$

2. *Portfolio B*. In the year t ($t \in \{1, \dots, 10\}$) the portfolio generated the following payments:

$$Z_t^B = \begin{cases} Z_t^{(1)} = z_t & \text{with a probability of } 79\,\%, \\ Z_t^{(2)} = (1+5)z_t & \text{with a probability of } 1\,\%, \\ Z_t^{(3)} = (1+0.5)z_t & \text{with a probability of } 10\,\%, \\ Z_t^{(3)} = (1-0.5)z_t & \text{with a probability of } 10\,\%. \end{cases}$$

We assume that the risk capital is the value at risk at confidence level 99.5 %. Clearly we have

$$\mathrm{E}(Z_t^A) = \mathrm{E}(Z_t^B) = 0.99z_t + 0.01 \times 6z_t = 1.05z_t$$

and

$$C_t^{\mathrm{FV}}(Z_t^A) = \mathrm{VaR}_{99.5\,\%}(Z_t^A) = 6z_t = \mathrm{VaR}_{99.5\,\%}(Z_t^B) = C_t^{\mathrm{FV}}(Z_t^B).$$

With the same choice of k_t and s_t one obtains the same fair value using the cost of capital method. Nonetheless, portfolio B can be held to be more risky as it has a higher variance, since

$$\begin{aligned} \mathrm{var}(Z_t^A) &= \mathrm{E}((Z_t^A)^2) - \mathrm{E}(Z_t^A)^2 \\ &= 0.99{z_t}^2 + 0.01 \times 36{z_t}^2 - 1.05^2{z_t}^2 \\ &= 0.2475{z_t}^2 \end{aligned}$$

and

$$\begin{aligned}\mathrm{var}(Z_t^B) &= \mathrm{E}((Z_t^B)^2) - \mathrm{E}(Z_t^B)^2\\ &= 0.79{z_t}^2 + 0.1 \times \frac{9}{4}{z_t}^2 + 0.1 \times \frac{1}{4}{z_t}^2 + 0.01 \times 36{z_t}^2 - 1.05^2{z_t}^2\\ &= 0.2975{z_t}^2.\end{aligned}$$

An additional point of criticism of the classic cost of capital method is that it only takes account of the "downside risk".

4.4.4 Valuing Technical Provisions According to IFRS

IFRS stands for the International Financial Reporting Standards. With its "Exposure Draft Insurance Contracts" in July 2010 the IASB[5] deviated from its previous guiding principle of fair value for a fictional market transaction. They specified that insurance contracts were to be valued on the basis of the company-specific situation that is expected for the time when the contracts are fulfilled. If there are replicating instruments for parts of the contract, then the market value of these replicating instruments is to be used for these parts. Otherwise the basic principle is that technical liabilities are to be valued as the sum of a "current estimate" plus a risk margin. By a current estimate is meant the present value, discounted with a risk-free interest curve, of the expected payment streams of the insurance contract. The payment streams are to be determined from the point of view of the company and on the basis of company-specific circumstances using realistic probabilities. Determining the risk margin can be done with the quantile method, by establishing the expected shortfall, or with the cost of capital method, and should be carried out in the context of an insurance portfolio in which similar contracts with largely similar risks are grouped and where these groups are managed separately. In this way the structure of diversification in the particular company is taken into consideration. If, as a result of the valuation using the current estimate and a risk margin, there is a profit at inception of the contract then an additional provision in the amount of the profit has to be set up, the so-called residual margin. This residual margin is depleted during the term of the contract *pro rata temporis* or in accordance with the expected evolution of losses and payments, and booked as income.

4.5 Approaches to Modeling Risk Capital

Risk capital can be modeled in many ways. We will briefly sketch some of the most popular ones in what follows. A real understanding for the general relationships can

[5] International Accounting Standards Board.

only be achieved when one has studied a good, but not too complex, model in detail. For that purpose the Swiss Solvency Test [4] (SST) is particularly well suited, and we introduce it in Sect. 4.6.1.

4.5.1 Factor-Based Models

Factor-based models are very pragmatic. The risk capital is to be calculated using the simplest possible formulas. In the simplest case, one scales a normed base loss with a volume parameter. If, for example, one is considering the risk that share prices fall, one can calculate the value at risk

$$\mathrm{VaR}_{99.5\,\%}(A_{\mathrm{norm}}) =: f$$

for a parcel of shares A_{norm} with starting value $W(A_{\mathrm{norm}}) = 1€$. For a parcel A with a different value $W(A)$ one simply puts $\mathrm{VaR}_{99.5\,\%}(A) = f\,W(A)$ without explicitly doing the calculation of the value at risk. This method can now be generalized to other risks, so that one finally arrives at a system of general predetermined factors and corresponding, company-dependent volume parameters, which are multiplied to obtain the risk capitals. These risk capitals are often simply added up, or aggregated under some predetermined assumption for their correlation (Sect. 3.3).

Once the risk factors are known, factor-based models have the advantage that the risk capital is extremely simple to calculate. However, the linear relationship implied by a factor and the volume parameter is not always a good approximation. If there is a quantitative change in the risks, the whole factor system must be calibrated anew in order to calculate the risk capital. Factor-based models can easily create wrong business incentives. For instance, the risk capital in the Solvency I framework was calculated for life insurance companies with the formula

$$4\,\%\ \text{technical provisions} + 0.3\,\%\ \text{capital at risk}.$$

Here one of the justifications for the first factor was that the technical provisions should be a good proxy for the investment volume, and that investment risk is accounted for in the factor of 4 %.[6] An unexpected side-effect was that a company that had more cautious (and thus higher) technical provisions, had to have more solvency capital, even though it had achieved a higher degree of safety. The risk capital required by Solvency I also does not depend on the type of investment, so that both risk-free and risky investments lead to the same solvency capital.

For computational simplicity the standard model for Solvency II is in part factor-based (Sect. 4.6.2). However, the model was constructed in such a way that false incentives are minimized.

---

[6] 3 % of the technical provisions was reckoned for the investment risks and the remaining 1 % for operational risks.

4.5.2 Analytic Models

The idea of an analytic model is to employ actuarial theory to derive as simple formulas as possible that are easy to work with and can easily be interpreted. An important advantage of this procedure is that the simplifications necessary are motivated by theory, so that in general it is possible to estimate the range of applicability of the model.

Example 4.8 A very simple example of an (in practice often over-simplified) analytical model is the use of a normal distribution $N(\mu, \sigma)$ for the total loss. We choose as the risk measure the value at risk for confidence level α. By Proposition 2.1 it then holds that $\mathrm{VaR}_\alpha(N(\mu,\sigma)) = \mu + \sigma {\Phi_{0,1}}^{-1}(\alpha)$. If we take for example $\alpha = 99.5\ \%$, then ${\Phi_{0,1}}^{-1}(\alpha) = 2.58$, so that we have

$$\mathrm{VaR}_{99.5\ \%}\big(N(\mu,\sigma)\big) = \mu + 2.58\sigma$$

and only need to estimate the expectation and standard deviation to carry out the calculation of the risk capital. The assumption of a normal distribution is completely transparent so its applicability is easy to check. Since assuming a normal distribution for an insurance portfolio with cumulative losses and large losses is, as a rule, not justified, this special model must be rejected in that case.

In Sect. 4.6.1 we will meet a much more complicated analytic model that can actually be used in practice.

4.5.3 Scenario-Based Models and Stress Tests

4.5.3.1 Conception

The fundamental goal of stress tests is to investigate the effects of extreme, but not implausible shocks on the financial well-being of an insurance company. The conception rests on the assumption that the value V of an insurance company or a portfolio can be expressed as a deterministic function f of a number of stochastic risk factors:

$$V = f(R_1, \ldots, R_n).$$

A stress test describes unusual situations that occur as a result of sudden changes in one or more risk factors. In this connection there are the following perspectives:

1. What effects does an extreme change in risk factors have?
2. What changes in which risk factors give rise to extreme changes in values? (Looking for the company's specific worst-case scenario.)
3. To what extent does the mapping f (the model) remain adequate under unusual prevailing circumstances?

The predictive capacity of stress tests for risk management depends crucially on the extent to which the scenarios are plausible and can be taken seriously. It would be ideal if one could reproducibly attribute a probability to each stress scenario. In that case the results of the stress test could be, weighted with probabilities, included in existing models of risk capital. Otherwise stress tests are a qualitative supplement to quantitative capital models, that can deliver additional knowledge about the capital buffers needed in extreme situations that are possibly not covered in the capital models (e.g., events on the far side of quantiles in VaR models).

Stress scenarios can be distinguished according to several criteria.

Historical scenarios are oriented toward extreme events in the past. The advantage is that the relevance of already observed changes in the risk factors cannot be neglected. However, the danger is that future extreme events may well dwarf past extremes. In addition, the historical scenarios for the insurance company are not necessarily the scenarios with the greatest effects.

Hypothetical scenarios address the influence of possible future adverse developments of relevant risk factors, e.g., structural changes resulting from a changed financial environment or from a changed risk profile consequent upon a new business strategy.

When one investigates the effects of changes in an individual risk factor, one speaks of a *single scenario* in contrast to a *multiple scenario*, that describes the influence of the simultaneous changes of several risk factors. In the multiple scenario approach the dependence structure of the risk factors has to be taken into account in order to avoid implausible scenarios.

Standard scenarios focus on the possible extreme changes of risk factors in the market as a whole and are mostly prescribed externally, e.g., by the regulators.

To identify *company-specific scenarios* with extreme consequences is a task of risk management. Interesting questions in this regard concern, for example, the combinations of changes in risk factors that lead to the heaviest losses with a given probability (individual worst-case scenario) or the size of a shock that the company could just cope with (inverse stress test).

4.5.3.2 Regulatory Stress Tests

The regulatory stress tests are standard stress tests. The regulators prescribe scenarios for which the insurance company must check whether its assets cover the liabilities and the solvency requirements.

As an example, we will describe the German Federal Financial Supervisory Authority's stress test.[7] The basis of this stress test is the balance from the foregoing year that is first brought forward to the balance sheet date of the current year.

For the valuation of the assets the fundamental principle is that market values are to be reported as in the appendix to the balance sheet. There are two exceptions to

[7]BaFin-Stresstest.

this rule. Firstly, fixed-interest securities which are classified as "held to maturity" since legally (according to § 341 b HGB) there is no risk that its value changes in the balance sheet. Secondly, an increased risk of change in market value is taken into account by multiplying the present market value with a factor (according to § 51 paragraph 2 InvG).

In bringing forward the balance sheet, it is assumed that the ongoing investment income is reinvested using the same mix of investment categories. On the liability side it is assumed that new business and contractual maturities completely cancel each other out. The technical provisions are assigned an interest rate that equals the sum of the guaranteed rate and the direct surplus rate.[8] The allocated part of the reserve for premium refund ("gebundene RfB")[9] is assigned an interest of half the guaranteed rate. For loss provisions one must take account of inflation.

The updated balance is then subjected to the following stress scenarios:

- price losses for fixed-interest bonds of 10 %
- decline of the stock market by 35 %
- price losses of 20 % for stocks and of 5 % for bonds
- price declines in the stock market of 20 % and real estate market losses of 10 %

In a multiple scenario a stress correlation of +1 is assumed. Credit risk is taken into account through additionally prescribed adjustments depending on rating classes. Hedging and buffering on the liability side (equity capital, reserves for free premium refunds, terminal bonus fund) are accounted for.

A stress test can be considered as passed in a particular scenario if the assets exceed the sum of the liabilities and the solvency requirements.

While the design of these stress scenarios relies on historical experience, the regulators do not assign probabilities to them.

A stress test that is not survived is an indicator of an insufficient capacity to bear risk on the part of the insurance company. The deficit cannot, however, be interpreted as a concrete requirement for capital, since the stress test employs rather sweeping assumptions and perhaps overestimates the potential losses. A stress test which is not passed triggers a dialog with regulators and provides a motivation to develop the risk management system further, after more careful analysis.

Possible enhancements of the regulatory stress tests could consist in the inclusion of actuarial risks, assignment of probabilities to the stress scenarios, or in the task for risk management of determining the company's particular worst-case scenarios (endogenesis of scenarios) and investigating the robustness of the models' assumptions.

[8]The direct surplus rate ("Direktgutschrift") is a peculiarity of the German insurance market. It describes a part of the bonus which is immediately allocated to policy holders.

[9]The RfB is a reserve concept specific to the German market.

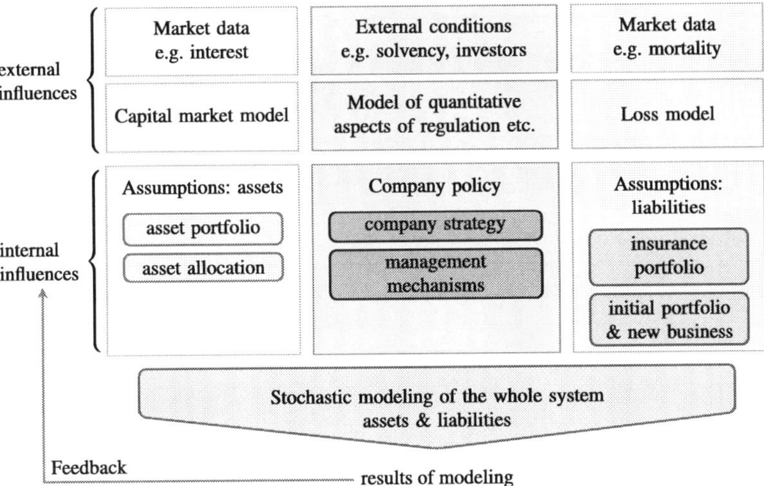

Fig. 4.4 A schematic for Monte Carlo modeling

4.5.4 Monte Carlo Models

Monte Carlo models are closely related to scenario-based models. In Monte Carlo models risk scenarios are automatically generated from preset distributions or stochastic processes, and their effects on the company are directly simulated. In this way one gets a discrete approximation to the overall distribution of risk for the company, so that arbitrary risk measures can be calculated. We illustrate this with the value of risk for the confidence level α. Assume that the Monte Carlo simulation consists of N scenarios and let $\lfloor r \rfloor = \sup_{n\in\mathbb{Z}}\{n \leq r\}$ for $r \in \mathbb{R}$. To work out the value at risk one first sorts the N losses obtained from the Monte Carlo simulation in descending order, i.e, $\text{Loss}_i \geq \text{Loss}_{i+1}$. Then one picks the $(\lfloor N(1-\alpha)\rfloor + 1)$-th loss from this list. If N is large, then the average of the first $\lfloor N(1-\alpha)\rfloor$ losses from the sorted list is a good approximation of the expected shortfall (see Theorem 2.5).

Because of the flexibility of the Monte Carlo method, it can be used to calculate the risk capital for arbitrarily complex undertakings. However, the effort required can be very great, and since the aggregate distribution is given in numerical form interpretation is more difficult than with analytic models.

The structure of a Monte Carlo model for insurance is shown in the schematic Fig. 4.4.

4.5.5 The Problem of Modeling Reinsurance

Some of the methods that are demonstrated here have difficulties with taking into account non-proportional reinsurance. In reinsurance there are two countervaling components to risk:

1. risk reduction as a primary effect
2. limitation of the risk reduction because of the risk of default on the part of the reinsurer

For a non-proportional reinsurance contract the factor-based method is unsuitable, since its foundation is a proportionality between losses and volume. Analytic models rapidly become so complex as a result of including reinsurance that no easy to use formulas can be derived. Scenario-based models and Monte Carlo models, on the other hand, can in principle model the risk reduction due to non-proportional reinsurance contracts well because of their flexibility.

The effect of the risk of default by the reinsurer is often nearly impossible to estimate because of retrocession. Ratings are not a good proxy for this since they take into account the solvency of the reinsurer and not just the fulfillment of insurance contracts. In addition, default of reinsurers is well correlated with catastrophic losses, an effect that is missed by using ratings.

4.5.6 Feedback of Investment Risk on Capital

If an insurance company borrows risk capital it will invest this capital. As a rule this investment will not be a risk-free security, but be in line with the investment strategy for other assets. To take into account both the investment income from the risk capital as well as the associating investment risk one has to solve an implicit equation. Let V_{t-1} be the (economic) liability at the end of year $t-1$, which has, at a minimum, to be covered by capital at the start of year t. The starting capital A is a parameter in the loss distribution X for the year, so we shall write $X(A)$. If there were no feedback effect, one could simply set $A = V_{t-1}$ and determine the risk capital $\tilde{C}_t$ from the equation

$$\tilde{C}_t = \frac{\rho(X(V_{t-1}))}{1+s_t}, \tag{4.3}$$

where we have noted that $\rho(X)$ is the (accumulated) loss at the end of the year and thus must be discounted at the risk-free interest rate. If the capital $\tilde{C}_t$ is invested without risk, then it is constant with respect to ω (but not in time), and we obtain

$$\rho\big(X(V_{t-1}+\tilde{C}_t)\big) - \rho\big(X(V_{t-1})\big) - (1+s_t)\tilde{C}_t.$$

Equation (4.3) for $\tilde{C}_t$ is therefore, with this assumption on the capital, equivalent to the implicit equation determining C_t

$$0 = \rho\big(X(V_{t-1}+C_t)\big). \tag{4.4}$$

But (4.4), in contrast to (4.3), continues to hold if there is feedback as described above.

The implicit (4.4) can usually be solved by an iterative procedure. For this we remark that (in realistic circumstances) for fixed ω the map $A \mapsto X(A)$ is monotone decreasing.[10] So generally $C_t \mapsto \rho(X(V_{t-1}+C_t))$ is also monotone decreasing, and we can employ a simple Newton method to solve (4.4). Nonetheless, this method is computationally very complex, since ρ itself is determined stochastically. If, for example, the Monte Carlo method is used, one has to calculate not one but hundreds of thousand of scenarios to determine C_t. In practice, at present[11] (4.3) is often used.

4.6 Risk Capital Models in Practice

4.6.1 The Swiss Solvency Test (SST)[12]

As far as we are aware, the SST is the first complete economicly based regulatory risk capital model. It was developed at the beginning of the millennium in cooperation between the Eidgenössische Finanzmarktaufsicht (FINMA), leading insurance companies, and Swiss universities. It has been in use in practice since 2008, and since 2011 the results from the SST are regulatorily binding. The SST can either be calculated using a regulatory *Standard Model* or using an internal (or partially internal) model that is developed by the insurer and certified by the Swiss regulator. In this section we concentrate on the Standard Model and give a simplified description of both the P&C Standard Model and the Life Standard Model. In particular, we will ignore all reinsurance effects.

The SST Standard Model has the following two seminal and unique properties:

- The calculation results in probability distributions rather than merely in capital figures.
- Concrete extreme scenarios are used in order to correct the tails of the probability distributions (see Sect. 4.6.1.3).

These properties aid in testing the Standard Model and make it possible to compare easily different levels of confidence.

[10] An example for a situation where the map is not monotone decreasing could be a financial investment in the company, in which small investors are less than proportionally liable for the risk. Such a product does not accord with the rules of a free market. However, for example, in privatising important public utility companies it could be politically desirable to, on the one hand, spread the ownership across the public, and, on the other hand, to protect these non-professional investors, by and large, from market fluctuations.

[11] This text is written in 2013.

[12] This section contains a *simplified* description of the Swiss Solvency Test as an example for a risk capital model that is applied in practice. Please note that this description is not meant to be 100 % faithful due to our simplifications and the fact that the SST is constantly developed further. Readers who are interested in a correct description and its official interpretation are referred to the official website of the Swiss financial supervisor (FINMA).

It is the aim of the SST to first estimate the magnitude of risks, and second to estimate the financial capacity of the company to bear these risks. The magnitude of the risk is measured using the target capital (Definition 4.7). The risk capacity of the company is quantified through the risk bearing capital (Definition 4.6).

In this section, a time index with square brackets ($^{[t]}$) indicates that the corresponding quantity is based on information which is available at time t. We denote the risk-free interest curve at time t (at the beginning of period $t+1$) by $\tau \mapsto s_\tau^{[t]}$ ($\tau > t$). Here $s_\tau^{[t]}$ is the risk-free 1-year interest for period τ, based on information available at time t. To simplify notation we denote the risk-free one-year interest rate for period t by $s_t = s_t^{[t-1]}$.

Remark 4.2 Observe that the risk-free interest curve $s^{[t]}$ is a stochastic quantity if t denotes a future point in time. Otherwise this quantity is deterministic.

The risk-bearing capital can loosely be thought of as the difference between the market value of assets and the market value of liabilities. The risk capital is then derived from the distribution of the risk-bearing capital. In order to make this more precise, we first need to decompose the market value of liabilities into three components,

- the best estimate of the present value of future liabilities,
- the price of hedgeable risks,
- the price of non-hedgeable risks.

The sum of the present value of future liabilities and the price of hedgeable risks is often determined directly using a Monte Carlo approach. The price of non-hedgeable risks is approximated in a separate step using a cost of capital approach (cf. Sect. 4.4.3.3). As one needs to know the risk capital for the cost of capital approach and as the risk capital depends on the risk bearing capital, we would obtain a complex implicit equation for the risk capital. In order to avoid this complication, the SST defines the risk-bearing capital without the price of non-hedgeable risks, and then corrects for this omission in a secondary step once risk capital has been calculated.

Definition 4.6 Let A_t^{Start} be the market consistent value of assets at the beginning of period t and V_t^{Start} the sum of

- the expected present value of future liabilities at the beginning of period t, where the risk-free interest curve is used for discounting,
- and the price of hedgeable risks inherent in the liabilities.

The *risk-bearing capital* ($\text{RTK}^{\text{Start}}$) is given by

$$\text{RTK}_t^{\text{Start}} = A_t^{\text{Start}} - V_t^{\text{Start}}.$$

The corresponding value at the end of period t is given by

$$\text{RTK}_t^{\text{End}} = A_t^{\text{End}} - V_t^{\text{End}}.$$

Remark 4.3 Our abbreviation for "risk-bearing capital", RTK, stands for the German word "Risikotragendes Kapital". We use the German abbreviation as "RBC" usually stands for "risk-based capital" which is another name for economic capital.

Definition 4.7 The *loss function* is given by the negative change (loss) in the risk-bearing capital during one year,

$$-\Delta \mathrm{RTK}_t = -\left(\frac{\mathrm{RTK}_t^{\mathrm{End}}}{1 + s_t^{[t-1]}} - \mathrm{RTK}_t^{\mathrm{Start}} \right).$$

The risk capital at the beginning of period t is defined as the expected shortfall of the loss function for the 99 % level of confidence,

$$C_t^{\mathrm{Reg}} = \mathrm{ES}_{99\,\%}(-\Delta \mathrm{RTK}_t).$$

The *target capital* at time t is the sum of the risk capital and the *market value margin* MVM:

$$\mathrm{TC}_t = C_t^{\mathrm{Reg}} + \frac{\mathrm{MVM}_t}{1 + s_t^{[t-1]}}.$$

Remark 4.4 The market value margin represents the price of non-hedgeable risks inherent in the liabilities. Consequently the market value of liabilities is approximated by

$$V_t^{\mathrm{Start}} + \frac{\mathrm{MVM}_t}{1 + s_t^{[t-1]}}.$$

Neither the risk-bearing capital RTK nor the target capital TC_t has a direct economic interpretation due to the transferral of the market value margin from the value of liabilities to the target capital.

In the following sections we will employ (a simplified version of) the SST Standard Model in order to determine the distribution for $-\Delta \mathrm{RTK}_t$. The risk capital C_t^{Reg} is the expected shortfall of this distribution. The change $-\Delta \mathrm{RTK}_t$ takes into account the following risks:

- market risk,
- insurance risk,
- credit risk,
- the effect of extreme scenarios.

Operational risk is not quantitatively accounted for in the Swiss Solvency Test.

4.6.1.1 Credit Risk

The SST Standard Model uses a simple factor-based approach following Basel II. In order to account for the result K in the distribution of $-\Delta \mathrm{RTK}_t$, the latter is simply shifted by K. Alternatively, one could add K to the expected shortfall.

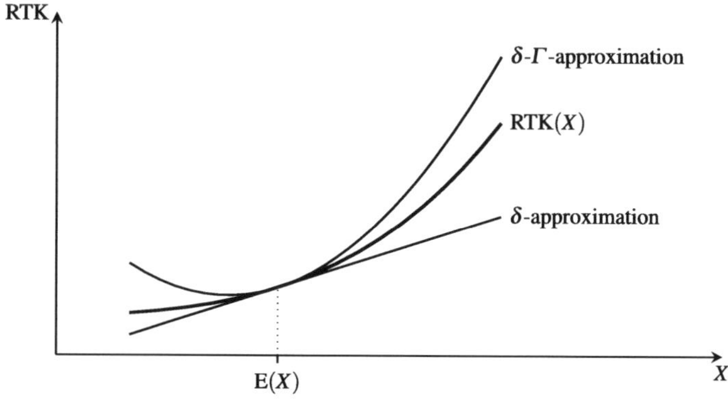

Fig. 4.5 The δ-Γ-approximation

4.6.1.2 The Quadratic Model for Both Market Risk and Life Risk

Market risk for both life insurance and non-life insurance, as well as insurance risk for life insurance, are estimated using the so-called δ-Γ-model. The risk bearing capital, RTK, is viewed as a function of risk factors, X, which are random. The idea of the δ-Γ-model is to approximate $\mathrm{RTK}(X)$ through a quadratic function of X and then to determine the distribution of this quadratic approximation.[13] Assume that the vector X consists of n risk factors X_i which represent market risk and (where appropriate) life insurance risks. FINMA prescribes a sensitivity h_i for each risk factor X_i. Assuming that $X \mapsto \mathrm{RTK}(X)$ is twice continuously differentiable the Taylor approximation gives

$$\mathrm{RTK}_t^{\mathrm{End}}\big(\mathrm{E}(X)+h\big) = \mathrm{RTK}_t^{\mathrm{End}}\big(\mathrm{E}(X)\big) + \sum_{i=1}^{n}\left(\frac{\partial \mathrm{RTK}_t^{\mathrm{End}}}{\partial X_i}\right)_{|\mathrm{E}(X)} h_i + \frac{1}{2}\sum_{i,j=1}^{n}\left(\frac{\partial^2 \mathrm{RTK}_t^{\mathrm{End}}}{\partial X_i \partial X_j}\right)_{|\mathrm{E}(X)} h_i h_j + o_2(h), \tag{4.5}$$

where $x \mapsto o_2(x)$ is a continuous function with $\lim_{x\to 0} o_2(x)/|x|^2 = 0$.

Let $e_i = (0,\ldots,0,1,0,\ldots,0)^\top \in \mathbb{R}^n$ be the i-th standard unit vector and define δ and Γ as the finite approximations with respect to the sensitivities $h \in \mathbb{R}^n$ of the first- and second-order derivatives of RTK,

$$\delta_i = \frac{\mathrm{RTK}_t^{\mathrm{End}}(\mathrm{E}(X)+h_i e_i) - \mathrm{RTK}_t^{\mathrm{End}}(\mathrm{E}(X)-h_i e_i)}{2h_i},$$

[13] In the original literature [3] only a linear approximation has been described. The extension to a quadratic model is fairly recent [6].

$$\Gamma_{ii} = \frac{\mathrm{RTK}_t^{\mathrm{End}}(\mathrm{E}(X) + h_i e_i) - \mathrm{RTK}_t^{\mathrm{End}}(\mathrm{E}(X))}{h_i^2} + \frac{\mathrm{RTK}_t^{\mathrm{End}}(\mathrm{E}(X) - h_i e_i) - \mathrm{RTK}_t^{\mathrm{End}}(\mathrm{E}(X))}{h_i^2},$$

$$\Gamma_{ij} = \frac{\mathrm{RTK}_t^{\mathrm{End}}(\mathrm{E}(X) + h_i e_i + h_j e_j) - \mathrm{RTK}_t^{\mathrm{End}}(\mathrm{E}(X) + h_i e_i - h_j e_j)}{4 h_i h_j} + \frac{\mathrm{RTK}_t^{\mathrm{End}}(\mathrm{E}(X) - h_i e_i - h_j e_j) - \mathrm{RTK}_t^{\mathrm{End}}(\mathrm{E}(X) - h_i e_i + h_j e_j)}{4 h_i h_j}$$

(for $i \neq j$).

Writing $\Delta X = X - \mathrm{E}(X)$ we obtain the approximation

$$\mathrm{RTK}_t^{\mathrm{End}}\big(\mathrm{E}(X) + \Delta X\big) - \mathrm{RTK}_t^{\mathrm{End}}\big(\mathrm{E}(X)\big) \approx \delta \cdot \Delta X + \frac{1}{2} \Delta X^\top \Gamma \Delta X \tag{4.6}$$

(see Fig. 4.5), which in turn implies

$$\begin{aligned} -\Delta \mathrm{RTK}_t(X) &= -\left(\frac{\mathrm{RTK}_t^{\mathrm{End}}(X)}{1 + s_t^{[t-1]}} - \mathrm{RTK}_t^{\mathrm{Start}} \right) \\ &\approx -\frac{\delta \cdot \Delta X + \frac{1}{2} \Delta X^\top \Gamma \Delta X}{1 + s_t^{[t-1]}} - \left(\frac{\mathrm{RTK}_t^{\mathrm{End}}(\mathrm{E}(X))}{1 + s_t^{[t-1]}} - \mathrm{RTK}_t^{\mathrm{Start}} \right). \end{aligned} \tag{4.7}$$

The first summand is the random variable describing the risk while the second summand is simply the discounted, expected loss.[14] The risk factors are normally distributed by assumption, so ΔX is also normally distributed. FINMA prescribes the standard deviation $\sigma(\Delta X)$ and the correlation matrix of ΔX, corr. Then

$$\Delta X \sim N\big(0, \mathrm{corr} \odot \big(\sigma(\Delta X) \sigma(\Delta X)^\top\big)\big),$$

where multiplication $\odot$ is meant component-wise. While $\Delta\mathrm{RTK}$ is not normally distributed due to the quadratic term, it is easy to obtain the distribution for $-\Delta \mathrm{RTK}_t(X)$ numerically, based on the known distribution of ΔX.

Remark 4.5 The δ-Γ-model has the advantage that it is conceptually simple and easy to implement if only a few risk factors are considered. On the other hand, while approximation (4.7) is good for $|\Delta X| \ll 1$ it may be unsuitable for $|\Delta X| \gg 1$. In the context of the SST we regularly have $|\Delta X| \gg 1$ because the SST is concerned with the 99 % quantile. Since the SST is concerned with large deviations from the

[14] It is a loss because of the minus sign in front of the parenthesis. Of course, a negative loss would be a gain.

expected value, the δ-Γ-model entails severe model risk. This model risk could be reduced by using the approximation

$$\begin{aligned}\mathrm{RTK}_t^{\mathrm{End}}\big(\mathrm{VaR}_{99.5\,\%}(X)+h\big) &= \mathrm{RTK}_t^{\mathrm{End}}\big(\mathrm{VaR}_{99.5\,\%}(X)\big) \\ &\quad+\sum_{i=1}^{n}\left(\frac{\partial \mathrm{RTK}_t^{\mathrm{End}}}{\partial X_i}\right)_{|\mathrm{VaR}_{99.5\,\%}(X)} h_i \\ &\quad+\frac{1}{2}\sum_{i,j=1}^{n}\left(\frac{\partial^2 \mathrm{RTK}_t^{\mathrm{End}}}{\partial X_i \partial X_j}\right)_{|\mathrm{VaR}_{99.5\,\%}(X)} h_i h_j + o_2(h),\end{aligned}$$

rather than (4.5). However, in order to use this approach one would have to estimate $\mathrm{RTK}_t^{\mathrm{End}}$ and its derivative at $\mathrm{VaR}_{99.5\,\%}(X)$ rather than $\mathrm{E}(X)$. In other words, we would increase the parameter risk in order to decrease the model risk.

Example 4.9 (Life risk for the 2012 SST) Both fluctuation risk and parameter risk are addressed. FINMA defines 7 risk factors for each category:

1: mortality risk,
2: longevity risk,
3: disability risk,
4: reactivation risk,
5: cost risk,
6: lapse risk,
7: exercise risk of policy holder options.

Denote by $P=(P_1,\dots,P_7)^\top$ the risk factors for parameter risk, where for each i the risk factor P_i corresponds to the risk type i. FINMA further prescribes a δ-Γ-model and the parameters

$$h=(10\,\%,10\,\%,10\,\%,10\,\%,10\,\%,10\,\%,10\,\%)^\top,$$

$$\sigma(\Delta P)=(5\,\%,10\,\%,20\,\%,10\,\%,10\,\%,25\,\%,10\,\%)^\top,$$

$$\mathrm{corr}(\Delta P)=\begin{pmatrix}1&0&0&0&0&0&0\\0&1&0&0&0&0&0\\0&0&1&0&0&0&0\\0&0&0&1&0&0&0\\0&0&0&0&1&0&0\\0&0&0&0&0&1&0.75\\0&0&0&0&0&0.75&1\end{pmatrix}.$$

While most of the risk factors P are self-explanatory, risk factor P_2 for longevity risk warrants a little more explanation. This risk factor applies to annuities in the pay-out phase. The mortality probabilities for these annuities are expected to be of the form

$$q_x(t)=\mathrm{e}^{-\lambda_x t}q_x(0),$$

where x is the age of the insured, t represents time, and the parameter λ_x determines the expected gain in life expectancy. The longevity risk factor pertains to this parameter.

FINMA prescribes for fluctuation risk $F = (F_1, \dots, F_7)^\top$ the same risk types as for parameter risk. The random variable $\mathrm{RTK}_t^{\mathrm{End}}(F)$ is assumed to be normally distributed with correlation matrix

$$\mathrm{corr}\big(\mathrm{RTK}_t^{\mathrm{End}}(F)\big) = \begin{pmatrix} 1 & 0 & 0 & 0 & 0 & 0 & 0 \\ 0 & 1 & 0 & 0 & 0 & 0 & 0 \\ 0 & 0 & 1 & 0 & 0 & 0 & 0 \\ 0 & 0 & 0 & 1 & 0 & 0 & 0 \\ 0 & 0 & 0 & 0 & 1 & 0 & 0 \\ 0 & 0 & 0 & 0 & 0 & 1 & 0.75 \\ 0 & 0 & 0 & 0 & 0 & 0.75 & 1 \end{pmatrix}.$$

The standard deviations $\sigma(\mathrm{RTK}_t^{\mathrm{End}}(F))$ for fluctuation risk are estimated directly by the insurer. Fluctuation risk is usually small, and in most portfolios there are some fluctuation risk factors (e.g., F_3) which significantly dominate the others. As an example consider fluctuation risk for mortality, F_1. Assume there are n policies in the portfolio which provide mortality benefit during the year under consideration. Let S_i be the insured mortality benefit of the policy $i \in \{1, \dots, n\}$ and V_i be the corresponding technical provision. If the policy holder dies, the technical provision is released and the company makes a net loss of $S_i - V_i$. Assuming that each policy holder has only a single policy and that their probabilities of death are independent, we obtain for the standard deviation of the fluctuation risk

$$\sigma\big(\mathrm{RTK}_t^{\mathrm{End}}(F_1)\big) = \sqrt{\sum_{i=1}^{n} q_{x_i}(1 - q_{x_i})(S_i - V_i)^2},$$

where q_{x_i} is the mortality probability of the policy holder with policy i.

Having determined $\sigma(\mathrm{RTK}_t^{\mathrm{End}}(F))$ it is numerically easy to determine

$$\mathrm{RTK}_t^{\mathrm{End}}(F) - \mathrm{RTK}_t^{\mathrm{End}}\big(\mathrm{E}(F)\big) \sim N\big(0, \mathrm{cov}\big(\mathrm{RTK}_t^{\mathrm{End}}(F)\big)\big).$$

Parameter risk and fluctuation risk are assumed to be uncorrelated. This implies

$$\begin{aligned} &\mathrm{RTK}_t^{\mathrm{End}}(F, P) - \mathrm{RTK}_t^{\mathrm{End}}\big(\mathrm{E}(F), \mathrm{E}(P)\big) \\ &\quad = \mathrm{RTK}_t^{\mathrm{End}}(F, P) - \mathrm{RTK}_t^{\mathrm{End}}\big(\mathrm{E}(F), P\big) \\ &\qquad + \mathrm{RTK}_t^{\mathrm{End}}\big(\mathrm{E}(F), P\big) - \mathrm{RTK}_t^{\mathrm{End}}\big(\mathrm{E}(F), \mathrm{E}(P)\big) \\ &\quad \approx \overbrace{\mathrm{RTK}_t^{\mathrm{End}}(F) - \mathrm{RTK}_t^{\mathrm{End}}\big(\mathrm{E}(F)\big)}^{\sim N(0, \mathrm{cov}(\mathrm{RTK}_t^{\mathrm{End}}(F)))} \\ &\qquad + \delta \cdot \Delta P + \frac{1}{2} \Delta P^\top \Gamma \Delta P, \end{aligned}$$

where $\Delta P \sim N(0, \mathrm{cov}(\Delta P))$.

Example 4.10 market risk for the 2012 SST FINMA defines $n = 77$ risk factors for market risk:

$i \in \{1, \dots, 13\}$: risk factors, which describe the risk-free CHF interest curve, i.e.,

$$(X_1, \dots, X_{13}) = \left(s_t^{[t-1]}, \dots, s_{t-1+10}^{[t-1]}, s_{t-1+15}^{[t-1]}, s_{t-1+20}^{[t-1]}, s_{t-1+30}^{[t-1]}\right),$$

$i \in \{14, \dots, 26\}$: risk factors, which describe the risk-free EUR interest curve,
$i \in \{27, \dots, 39\}$: risk factors, which describe the risk-free USD interest curve,
$i \in \{40, \dots, 52\}$: risk factors, which describe the risk-free GBP interest curve,
$i \in \{53, \dots, 56\}$: risk factors for spreads: AAA, AA, A, BBB,
$i \in \{57, \dots, 60\}$: risk factors for FX risk, CHF-EUR, CHF-USD, CHF-GBP, CHF-JPY,
$i \in \{61\}$: risk factor for implied volatility: USD/CHF 3M,
$i \in \{62, \dots, 68\}$: risk factors for stock indices: MSCI CHF, MSCI EMU, MSCI US, MSCI UK, MSCI JP, Pacific excl. Japan, Small cap EMU,
$i \in \{69\}$: risk factor for implied volatility: VIX (Chicago Board Options Exchange Volatility Index),
$i \in \{70, \dots, 73\}$: risk factors for (Swiss) real estate indices, SWX IAZI Performance, Rüd Blass, WUPIX A, as well as commercial direct property,
$i \in \{74\}$: risk factor for hedge funds,
$i \in \{75\}$: risk factor for private equity,
$i \in \{76\}$: risk factor for holdings,
$i \in \{77\}$: risk factor for implied volatility: yield.

FINMA also prescribes the standard deviation $\sigma(\Delta X)$ and the correlation matrix corr of ΔX.[15] The sensitivities h_i have the value 100 bp for $i \in \{1, \dots, 56\}$ and $10\,\%\ \Delta X_i$ for $i \in \{57, \dots, 77\}$.

4.6.1.3 Extreme Scenarios

The SST Standard Model is based on an analytic distribution for the change

$$\Delta \mathrm{RTK}_t = \frac{\mathrm{RTK}_t^{\mathrm{End}}}{1 + s_t^{[t-1]}} - \mathrm{RTK}_t^{\mathrm{Start}}$$

of the discounted risk-bearing capital. One design element of the SST Standard Model is its comparative simplicity. As a consequence, the analytic distribution for market risk and insurance risk underestimates tail risk that dominates the target capital. In addition, there are only few experience data in the tail. Hence the construction and parametrization of an accurate model for the tail would, in any case, be very difficult.

---

[15] We refer to the FINMA website for the actual values.

The solution of the SST is to estimate the effect on the risk-bearing capital of several representative extreme scenarios and then to aggregate the results with the distribution. These extreme scenarios model events with a very small probability of happening (typically one in a hundred or one in a thousand years). Typical examples are a pandemic or a financial market crisis.

Such extreme scenarios can affect negatively both the assets and the liabilities. For instance, a very severe pandemic would affect production severely, and therefore would have a significant effect on the capital markets. A list of extreme scenarios, which were considered initially in 2006, can be found in Sect. 5.2.1 of [4]. However, the list of required scenarios is constantly updated.

Example 4.11 The scenario describes an explosion in a chemical plant. The probability of such an event is estimated to be 0.5 %. A real event that could serve as inspirations would be the accident in Seveso in 1976. It is assumed that the plant leaks a toxic gas. 10 % of the inhabitants of a neighboring town (20,000 inhabitants) are affected. 1 % of the affected population die and 10 % will be permanently disabled which precludes them from work. The remaining 89 % of the affected population need to be treated in a hospital for several days, and afterwards are not fit for work for several weeks. 20 % of the 500 employees are affected. Since they were closer to the accident, they are affected worse. 10 % of the affected employees die and 30 % are permanently disabled. The remaining 60 % have to be treated in a hospital, and cannot work afterwards for several weeks. In addition to the human toll, the plant is totally destroyed and there is considerable damage in the neighborhood of the plant. These damages include the poisoning of a river, damage to neighboring buildings and cars as well as liability claims for the suffering endured. There are also damages due to salary loss and business interruption both at the plant and at neighboring facilities.

Aggregation of the Analytical Distribution, the Credit Risk, and the Extreme Scenarios for the Calculation of the Distribution of $\Delta \mathrm{RTK}_t$ The extreme scenarios are formally designated as events $S_1, \dots, S_m$ (in the SST Standard Model we have $m \approx 15$). Let S_0 be the event that none of the extreme scenarios has happened. This event corresponds to the hypothesis "normal year" under which the cumulative distribution function $\tilde{F}_0$ representing $-\Delta\mathrm{RTK}_t \mid S_0$ has been derived. (We will derive $\tilde{F}_0$ for non-life insurance in Sect. 4.6.1.5 and for a simple life insurance example in Sect. 4.6.1.4.) Formally we have

$$\tilde{F}_0(x) = \mathbf{P}[-\Delta\mathrm{RTK}_t \leq x \mid S_0].$$

The probability space does not contain credit risk as we have obtained the necessary capital K covering credit risk separately in Sect. 4.6.1.1. As indicated in that section, credit risk is incorporated through a shift of the distribution function,

$$F_0(x) = \tilde{F}_0(x - K).$$

Since the probability $p_j = \mathbf{P}[S_j] \ll 1$ of any extreme scenario is between 0.1 % and 1 %, it is so small that the simplification is justified that only one extreme scenario can happen in any given year, i.e., $\mathbf{P}[S_i \cap S_j] = 0$ for $i \neq j$ and $i, j \geq 1$. Hence the events $S_0, \ldots, S_m$ provide a partition of the probability space, and we obtain $p_0 = \mathbf{P}[S_0] = 1 - \sum_{j=1}^{m} \mathbf{P}[S_j]$.

The company determines for each extreme scenario the corresponding additional loss $c_j \geq 0$. The extreme scenario is aggregated to the distribution function F_0 through an appropriate shift by c_j

$$F_j(x) = \mathbf{P}[-\Delta \mathrm{RTK}_t \leq x \mid S_j] = F_0(x - c_j) = \tilde{F}_0(x - c_j - K).$$

This ansatz is a simplification. In reality the occurrence of an extreme scenario would also affect other risk characteristics, for instance the variance of the short term interest rate, which in turn affects the distribution of $\Delta \mathrm{RTK}_t$.

Finally, the unconditional distribution F of $-\Delta \mathrm{RTK}_t$ results from mixing the distributions F_j,

$$F(x) = \mathbf{P}[-\Delta \mathrm{RTK}_t \leq x] = \sum_{j=0}^{m} p_j F_j(x).$$

We can now calculate the risk capital

$$C_t^{\mathrm{Reg}} = \mathrm{ES}_{99\,\%}(-\Delta \mathrm{RTK}_t).$$

and therefore the target capital TC_t.

4.6.1.4 A Strongly Simplified Numerical Life Example

Consider a life insurer which has 10 single insurance contracts for $n = 10$ different policy holders in its portfolio at time $t = 0$: We assume that all the 10 policy holders have exactly the same risk profile and that each policy holder gets $S = 100$ if she or he survives the next $T = 3$ years and nothing otherwise. Assume that the yearly probability of death is given by $q_t = 2\,\%$. Since the contract does not involve any financial option it is possible to value it by simply discounting the liability cash flow with the risk-free rate and adding the market value margin for mortality risk. We assume that government zero bonds are risk-free and that the spot rate curve is given by[16]

Year t:	1	2	3
Spot rate $s_t^{[0]}$:	0.155 %	0.013 %	−0.011 %

[16]This is the first part of the CHF risk-free spot rate curve provided by FINMA for the SST 2011. The unusual pattern (including negative interest for the third year) reflects the turmoil in financial markets at that time.

We obtain

$$\begin{aligned} V_1^{\text{Start}} &= \left(1+s_T^{[0]}\right)^{-T}\left(\prod_{t=1}^{T}(1-q_t)\right)\cdot n\cdot S \\ &= (1-0.011\ \%)^{-3}\cdot(98\ \%)^3\cdot 1000 = 941.5. \end{aligned}$$

The asset portfolio at time $t=0$ at the beginning of the first year consists of the following investments:

Zero Swiss Government bonds	Maturity:	none	1	2	3
	Nominal value B_t:	–	0	300	500
Swiss stock index	Market value V_{stock}:	300	–	–	–

The market value of the (risk-free) Swiss government bonds is simply given by discounting with the risk-free rate. Hence we obtain for the value of the asset portfolio

$$\begin{aligned} A_1^{\text{Start}} &= V_{\text{stock}} + \sum_{t=1}^{T}\left(1+s_t^{[0]}\right)^{-t} B_t \\ &= 300 + (1+0.013\ \%)^{-2}\cdot 300 + (1-0.011\ \%)^{-3}\cdot 500 = 1100.1 \end{aligned}$$

and thus

$$\text{RTK}_1^{\text{Start}} = A_1^{\text{Start}} - V_1^{\text{Start}} = 158.6.$$

In this example we ignore insurance fluctuation risk. Assuming that the expected appreciation of the stock index is 5 % the expected risk bearing capital at the end of the period is given by

$$\begin{aligned} \text{E}\left(\text{RTK}_1^{\text{End}}\right) &= \text{E}\left(A_1^{\text{End}} - V_1^{\text{End}}\right) \\ &= V_{\text{stock}}\cdot(1+5\ \%) + \left(1+s_1^{[0]}\right)\sum_{t=2}^{T}\left(1+s_t^{[0]}\right)^{-t} B_t \\ &\quad - \left(1+s_1^{[0]}\right)\left(1+s_T^{[0]}\right)^{-T}\left(\prod_{t=1}^{T}(1-q_t)\right) n\cdot S \\ &= 300\cdot 1.05 + 1.00155\left(\frac{300}{1.00013^2}+\frac{500}{0.99989^3}\right) \\ &\quad - \frac{1.00155}{0.99989^3}\cdot 0.98^3\cdot 1000 \\ &= 1116.3 - 943.0 = 173.4, \end{aligned}$$

where all values were rounded after the calculation was performed. The corresponding random variable is

$$\mathrm{RTK}_1^{\mathrm{End}}(x) = V_{\mathrm{stock}} \cdot (1 + 5\ \%) \cdot x_4 + (1 + x_1) \sum_{t=2}^{T} (1 + x_t)^{-t} B_t$$
$$- (1 + x_1)(1 + x_3) \left(\prod_{t=1}^{T} (1 - q_t \cdot x_5) \right) n \cdot S$$

where x_4 describes investment risk for the Swiss stock index, x_5 mortality risk and

$$\mathrm{E}(x) = (0.155\ \%, 0.013\ \%, -0.011\ \%, 1, 1)^\top$$

holds. Replacing x by $\mathrm{E}(x) \pm \Delta x$ with

$$\Delta x_i = \left(\delta_i^1 \cdot 1\ \%, \delta_i^2 \cdot 1\ \%, \delta_i^3 \cdot 1\ \%, \delta_i^4 \cdot 10\ \%, \delta_i^5 \cdot 10\ \%\right)^\top$$

and

$$\delta_i^j = \begin{cases} 1 & \text{if } i = j \\ 0 & \text{otherwise} \end{cases}$$

we obtain

$$\left(\mathrm{RTK}_1^{\mathrm{End}}(\mathrm{E}(x) + \Delta x_i)\right)_{i=1,\dots,5} = (171.95, 167.45, 186.37, 204.87, 179.13)^\top,$$
$$\left(\mathrm{RTK}_1^{\mathrm{End}}(\mathrm{E}(x) - \Delta x_i)\right)_{i=1,\dots,5} = (174.78, 179.46, 159.83, 141.87, 167.58)^\top.$$

Hence the approximate derivative

$$\delta_i = \frac{\mathrm{RTK}_1^{\mathrm{End}}(\mathrm{E}(x) + \Delta x_i) - \mathrm{RTK}_1^{\mathrm{End}}(\mathrm{E}(x) - \Delta x_i)}{2\Delta x_i}$$

is given by

$$\delta = (-141.42, -600.82, 1326.65, 315.00, 57.73)^\top.$$

The Γ-matrix, given by

$$\Gamma_{ij} = \frac{\mathrm{RTK}_1^{\mathrm{End}}(\mathrm{E}(x) + \Delta x_i + \Delta x_j) - \mathrm{RTK}_1^{\mathrm{End}}(\mathrm{E}(x) + \Delta x_i - \Delta x_j)}{4\Delta x_i \Delta x_j}$$
$$+ \frac{\mathrm{RTK}_1^{\mathrm{End}}(\mathrm{E}(x) - \Delta x_i - \Delta x_j) - \mathrm{RTK}_1^{\mathrm{End}}(\mathrm{E}(x) + \Delta x_i + \Delta x_j)}{4\Delta x_i \Delta x_j},$$

is calculated similarly,

$$\Gamma = \begin{pmatrix} 0.00 & -599.89 & 1324.60 & 0.00 & 57.64 \\ -599.89 & 1803.05 & 0.00 & 0.00 & 0.00 \\ 1324.60 & 0.00 & -5310.74 & 0.00 & -173.27 \\ 0.00 & 0.00 & 0.00 & 315.00 & 0.00 \\ 57.64 & 0.00 & -173.27 & 0.00 & 55.38 \end{pmatrix}.$$

This enables us to approximately calculate

$$\mathrm{RTK}_1^{\mathrm{End}}(x) - \mathrm{E}\big(\mathrm{RTK}_1^{\mathrm{End}}\big) \approx \delta^\top \Delta x + \frac{1}{2}(\Delta x)^\top \Gamma \Delta x$$

for any random draw $x(\omega) = \mathrm{E}(x) + \Delta x(\omega)$. The random variable x is assumed to be normally distributed where FINMA prescribes

$$\sigma(x) = (0.00603, 0.00606, 0.00633, 0.15052, 0.05000)^\top$$

and

$$\mathrm{corr}(x) = \begin{pmatrix} 1.00000 & 0.72156 & 0.54556 & 0.40224 & 0.00000 \\ 0.72156 & 1.00000 & 0.95319 & 0.43339 & 0.00000 \\ 0.54556 & 0.95319 & 1.00000 & 0.41304 & 0.00000 \\ 0.40224 & 0.43339 & 0.41304 & 1.00000 & 0.00000 \\ 0.00000 & 0.00000 & 0.00000 & 0.00000 & 1.00000 \end{pmatrix}.$$

So the random variable $\mathrm{RTK}_1^{\mathrm{End}}$ is completely known, and therefore also $-\Delta\mathrm{RTK}_0$ and its distribution function $\tilde{F}_0$.

We add the impact of the following extreme scenarios:

Scenario	Δy_1	Δy_2	Δy_3	Δy_4	Δy_5
SZ01: Equity drop −60:	0.000 %	0.000 %	0.000 %	−60 %	0 %
SZ03: Stock market crash (1987):	−0.155 %	−0.013 %	0.000 %	−23 %	0 %
SZ04: Nikkei crash (1989/90):	1.563 %	1.098 %	1.177 %	−26 %	0 %
SZ05: European currency crisis (1992):	−0.155 %	−0.013 %	0.000 %	−6 %	0 %
SZ06: US interest crisis (1994):	1.111 %	1.406 %	1.509 %	−19 %	0 %
SZ07: Russian crisis / LTCM (1998):	−0.155 %	−0.013 %	0.000 %	−28 %	0 %
SZ08: Stock market crash (2000/2001):	−0.155 %	−0.013 %	0.000 %	−36 %	0 %
SZ11: Financial crisis (2008):	−0.155 %	−0.013 %	0.000 %	−39 %	0 %

where Δy_i are the sensitivities describing the impact upon our risk factors of each scenario. These sensitivities are prescribed by FINMA (and are updated each year). Had we applied the historic interest rate shocks for scenarios SZ03, SZ05, SZ07, SZ08, SZ11 to the current spot rate curve in our example, then we would have obtained negative interest rates for the first two years, and the 3-year spot rate would have become even more negative. We have therefore capped the shocks so that the shocked spot rate curve satisfies $s^{[0]}_{1,\text{shocked}} = s^{[0]}_{2,\text{shocked}} = 0$ and $s^{[0]}_{3,\text{shocked}} = s^{[0]}_3 < 0$. Each scenario is supposed to have the probability 0.1 % of happening. Using the quadratic approximation $\delta^\top \Delta y + \frac{1}{2}(\Delta y)^\top \Gamma \Delta y$ we can calculate the effect of each scenario and obtain

Scenario	Probability	ΔRTK
SZ01:	0.001	−132.3
SZ03:	0.001	−64.4
SZ04:	0.001	−65.6
SZ05:	0.001	−17.4
SZ06:	0.001	−43.2
SZ07:	0.001	−76.5
SZ08:	0.001	−92.0
SZ11:	0.001	−98.2

We could account for these scenarios using the method given on page 139. However, since we need to perform Monte Carlo simulations in order to obtain our random variable $\delta^\top \Delta x + \frac{1}{2}(\Delta x)^\top \Gamma \Delta x$, we can use a simpler method. Assume that the Monte Carlo simulation provides the outcomes $X_1, \dots, X_N$. Assuming N is large enough we can simply add

$$-132.3 \text{ to } X_1, \dots, X_{0.001 \cdot N}, \quad -64.4 \text{ to } X_{0.001\cdot N+1}, \dots, X_{2\cdot 0.001\cdot N} \quad \text{and so on.}$$

Observe that this method uses the same approximation as the method given on page 139, namely that no two extreme scenarios can happen at the same time. We denote the resulting random variable by

$$\text{Aggr}\left(\delta^\top \Delta x + \frac{1}{2}(\Delta x)^\top \Gamma \Delta x, \text{Scen}\right).$$

Putting everything together we obtain the following risk capitals:

Insurance:

$$\text{ES}_{1\,\%}\left(\delta_5 \Delta x_5 + \frac{1}{2}\Gamma_{55}(\Delta x_5)^2\right) = -13.4.$$

Market:

$$\text{ES}_{1\,\%}\left(\delta^\top_{[1,\dots,4]} \Delta x_{[1,\dots,4]} + \frac{1}{2}(\Delta x_{[1,\dots,4]})^\top \Gamma_{[1,\dots,4][1,\dots,4]} \Delta x_{[1,\dots,4]}\right) = -105.6.$$

Market and insurance:

$$\mathrm{ES}_{1\,\%}\left(\delta^\top \Delta x + \frac{1}{2}(\Delta x)^\top\right) = -106.3.$$

Market, insurance, and scenario:

$$\mathrm{ES}_{1\,\%}\left(\mathrm{Aggr}\left(\delta^\top \Delta x + \frac{1}{2}(\Delta x)^\top \Gamma \Delta x, \mathrm{Scen}\right)\right) = -116.0.$$

In order to calculate the market value margin MVM we observe that in our example the only non-hedgeable risks are the biometric risk and the scenario risk. We assume that the risk capitals for these risks at later times are well approximated by the risk capital at the beginning, scaled by the value of liabilities. We obtain

$$\begin{aligned}
\mathrm{MVM} &= 6\,\% \cdot \big(-13.4 - (116.0 - 106.3)\big) \\
&\quad \cdot \left(1 + \left(1 + s_1^{[0]}\right)^{-1} \frac{V_2^{\mathrm{Start}}}{V_1^{\mathrm{Start}}} + \left(1 + s_2^{[0]}\right)^{-2} \frac{V_3^{\mathrm{Start}}}{V_1^{\mathrm{Start}}}\right) \\
&= 6\,\% \cdot (-13.3 - 116.0 + 106.3) \\
&\quad \cdot \left(1 + \frac{1}{1.00155} \cdot \frac{943.0}{941.5} + \frac{1}{1.00013^2} \cdot \frac{941.7}{941.5}\right) \\
&= -4.17.
\end{aligned}$$

Hence the target capital is given by

$$\begin{aligned}
\mathrm{TC}_1 &= \frac{\mathrm{ES}_{1\,\%}(\mathrm{Aggr}(\delta^\top \Delta \mathrm{x} + \frac{1}{2}(\Delta \mathrm{x})^\top \Gamma \Delta \mathrm{x}, \mathrm{Scen}))}{1 + s_1^{[0]}} \\
&\quad + \frac{-s_1^{[0]} \cdot \mathrm{RTK}_1^{\mathrm{Start}} + \mathrm{MVM}}{1 + s_1^{[0]}} \\
&= \frac{-116.0 - 0.00155 \cdot 158.6 - 4.17}{1.00155} = -120.2
\end{aligned}$$

and the SST-ratio is

$$-\frac{\mathrm{RTK}_1^{\mathrm{Start}}}{\mathrm{TC}_1} = \frac{158.6}{120.2} = 132\,\%,$$

well about the required threshold of 100 %.

4.6.1.5 The Distribution $\tilde{F}_0$ for Non-life Risk

In the following pages we will give a simplified account of the SST model for non-life insurance [4]. In particular we will ignore reinsurance.

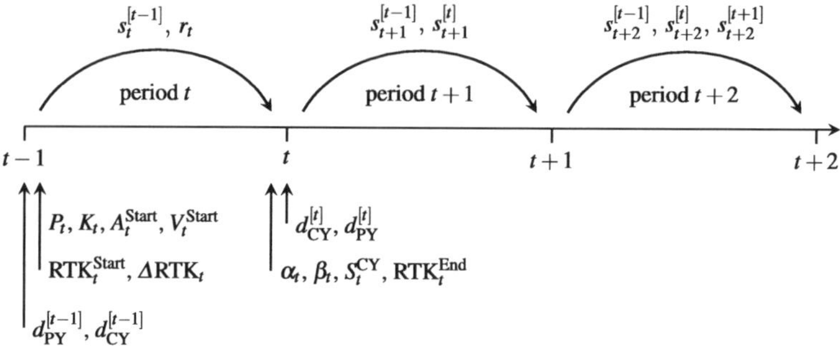

Fig. 4.6 The time model of the SST for non-life

In each model, simplifications have to be made, and the SST model is no exception. As each simplification leads to additional model error, it is necessary to weigh the benefit derived from the simplification against the corresponding loss of accuracy. We will therefore pay particular attention to these additional error terms.

A description of the model can be found in Sect. 4.4 of the Technical Document [4].[17] It should be noted that the Technical Document was written in 2006 and may not fully reflect the current status of the SST model.

We consider the year t corresponding to the time interval $\text{CY} = [t-1, t[$ (CY: Current Year). Losses from events that happen during CY are referred to as CY-losses or "new losses". The years before CY correspond to the time interval $\text{PY} =]-\infty, t-1[$ (PY: Previous Years). Losses from events that have occurred during PY are called PY-losses. The time model of the SST is depicted in Fig. 4.6.

We consider the following risks.

- Investment risk modeled through the random variable r_t that corresponds to the (relative) investment result.
- Valuation risk from the uncertainty of future interest rates ($s_\tau^{[t]}$, $\tau \geq t$). This risk affects both the assets A_τ^{Start} and the liabilities V_τ^{Start}.
- CY-losses.
- Adjustments to the expected run-off result of PY-losses.

We will derive the distribution $\tilde{F}_0$ in four steps:

1. Decomposition of the risk bearing capital into the terms

$$\Delta\text{RTK}_t = (T_{\text{risk invest}} + T_{\text{exp invest}} + T_{\text{risk ins}} + T_{\text{exp ins}} + T_{\text{error}}),$$

where

- $T_{\text{exp invest}}$ denotes the expected investment result,

[17] We do not fully comply with the notation in [4] in order to be more consistent with the rest of this book. For instance, in our description the payment pattern β_τ refers to the end, rather than the beginning, of year τ.

- $T_{\text{risk invest}}$ denotes the risk result from investments,
- $T_{\text{exp ins}}$ denotes the expected insurance results,
- $T_{\text{risk ins}}$ denotes the risk result from insurance,
- T_{error} denotes the error term.

(Proposition 4.2 on page 155). The SST model neglects the error term T_{error}.

2. Calculation of $T_{\text{exp invest}} + T_{\text{risk invest}}$ following Example 4.10 on page 138.
3. Calculation of $T_{\text{exp ins}} + T_{\text{risk ins}}$ in the following steps:

   (a) Result from CY-losses.

      (i) In a first step we calculate normal losses (high frequency, low severity losses) $S_t^{\text{CY,nl}}$. They are assumed to have a lognormal distribution and will be calculated in Corollary 4.2, page 163. In Lemma 4.3 on page 161 we derive the variation coefficient for the normal losses of each line of business. This result will be used in Corollary 4.2.

      (ii) In a second step we will determine the distribution for large losses. The general form of this distribution is given on page 164. We will explain the calculation in detail for the example of losses due to hail (Example 4.12 on page 164).

   (b) Calculation of the run-off result for PY-losses. The resulting distribution is given in Proposition 4.3.

4. Aggregation of the distributions. On page 169 we will combine the previous results to obtain

$$T_{\text{exp invest}} + T_{\text{risk invest}} + T_{\text{exp invest}} + T_{\text{risk invest}}.$$

In order to decompose the risk-bearing capital $T_{\text{risk invest}}$, $T_{\text{exp ins}}$, $T_{\text{risk ins}}$, T_{error} (Proposition 4.2) we need some preparation.

Let $R_{\text{PY}}^{[t-1]}$ be the deterministic best estimate of non-discounted PY-loss reserves at the beginning of period t. These reserves include the IBNR (incurred but not (yet) reported) reserves, reserves for future loss administration expenses which consist of allocated loss adjustment expenses (ALAE), and unallocated loss adjustment expenses (ULAE). For more details see Sect. 4.4.2 in [4]. We denote by $(\beta_\tau)_{\tau \geq t}$ the expected payment pattern for PY-losses which has been determined at the beginning of period t. This payment pattern is normalized so that it satisfies the constraint $\sum_{\tau=t}^{\infty} \beta_\tau = 1$.

The discount factor for the cash flow at the end of period τ back to time t (at the beginning of period $(t+1)$ is given by

$$v_\tau^{[t]} = \prod_{i=t+1}^{\tau} \left(1 + s_i^{[t]}\right)^{-1}$$

(see Fig. 4.7). The present value at the beginning of period t of PY-loss reserves is then given by

$$d_{\text{PY}}^{[t-1]} R_{\text{PY}}^{[t-1]}$$

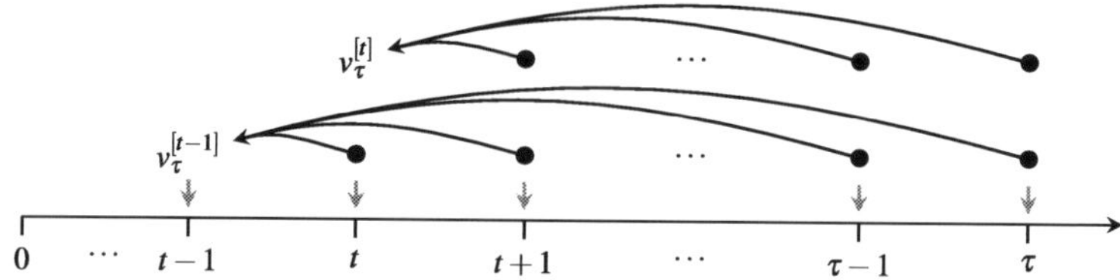

Fig. 4.7 Definition of discount factors

where we have set

$$d_{\text{PY}}^{[t-1]} = \sum_{\tau=t}^{\infty} \beta_\tau v_\tau^{[t-1]}.$$

Let P_t be the expected premium at time $t-1$ which will be earned during period t. This estimate is composed of the unearned premium reserve upr_{t-1} from the preceding year and the remainder, $P_t - \text{upr}_{t-1}$, which is modeled as if this amount were paid at the beginning of period t. Observe that upr_{t-1} is contained in the assets A_t^{Start} at the beginning of period t. (See also Sect. 4.4.4.1 in [4].) New business that covers risks after the end of period t is not modeled, which implies that there is no term upr_t.

The present value of liabilities at the beginning of period t is now given by

$$V_t^{\text{Start}} = \text{upr}_{t-1} + d_{\text{PY}}^{[t-1]} R_{\text{PY}}^{[t-1]}. \tag{4.8}$$

A_t^{Start} can be read directly off the economic balance sheet. This value does not account for incoming premium or expenses at the beginning of period t. Premium and expenses that occur during period t are accounted for in the end-of year balance sheet value for assets, A_t^{End}.

At the beginning of period t both the assets at time t, A_t^{End}, and the present value of liabilities at time t, V_t^{End}, are random variables; and it is our goal to calculate them. Let K_t be the expected costs for administration expenses during period t. For simplicity we assume that these costs are incurred at the beginning of period t. We denote by S_t^{CY} the stochastic non-discounted loss amount from those insurance contracts which are sold during period t. For simplicity we assume that all actual payments for such losses occur at the end of the corresponding period. Let $(\alpha_\tau)_{\tau \geq t}$ be the payment pattern for these losses. Without loss of generality, we can assume that this payment pattern satisfies the constraint $\sum_{\tau=t}^{\infty} \alpha_\tau = 1$. We also assume that $(\alpha_\tau)_{\tau \geq t}$ is known at time $t-1$. Observe that this simplification only affects the timing when these loss amounts are paid, but—apart from discounting effects—not their total amount. The non-discounted cash flow at time τ for CY-losses is then given by $\alpha_\tau S_t^{\text{CY}}$. Writing

$$d_{\text{CY}}^{[t]} = \sum_{\tau=t+1}^{\infty} \alpha_\tau v_\tau^{[t]}$$

the present value at time t of future payments for CY-losses is therefore equal to

$$\left(\alpha_t + d_{\mathrm{CY}}^{[t]}\right) S_t^{\mathrm{CY}}.$$

The first summand $\alpha_t S_t^{\mathrm{CY}}$ is paid immediately at the end of period t whereas the remaining part is accounted for in the liabilities V_t^{End}. At time t we will also re-evaluate the PY-loss reserves based on new experience from period t. This re-evaluation is modeled through a stochastic correction factor c_t^{PY}, where we set for the non-discounted run-off result from PY-losses

$$\left(1 - c_t^{\mathrm{PY}}\right) R_{\mathrm{PY}}^{[t-1]}.$$

The correction factor satisfies $\mathrm{E}(c_t^{\mathrm{PY}}) = 1$, since $R_{\mathrm{PY}}^{[t-1]}$ is defined as the best estimate at time t of future run-off costs.

We obtain

$$A_t^{\mathrm{End}} = (1 + r_t)(A_t + P_t - \mathrm{upr}_{t-1} - K_t) - \alpha_t S_t^{\mathrm{CY}} - \beta_t c_t^{\mathrm{PY}} R_{\mathrm{PY}}^{[t-1]}, \qquad (4.9)$$

$$V_t^{\mathrm{End}} = d_{\mathrm{CY}}^{[t]} S_t^{\mathrm{CY}} + d_{\mathrm{PY}}^{[t]} c_t^{\mathrm{PY}} R_{\mathrm{PY}}^{[t-1]}. \qquad (4.10)$$

Proposition 4.2 will provide an approximation for the loss function $-\Delta\mathrm{RTK}_t$, and it is one of the goals of this section to calculate the corresponding approximation error as explicitly as possible. All expected values and variances will be calculated based on information known at time $t-1$. Hence all quantities with an index $^{[t-1]}$ can be viewed as constant.

The following lemma will be used in the proof of Proposition 4.2.

Lemma 4.1 *Assume* $s_t^{[t-1]} \le s_{\max}$ *and*

$$0 \le s_\tau^{[t-1]}, \quad \left|s_\tau^{[t]} - s_\tau^{[t-1]}\right| \le \delta < 1$$

for all $\tau \ge t$.

Then there are constants c_τ *with* $|c_\tau| \le 1$ *and*

$$\mathrm{E}\left(d_{\mathrm{CY}}^{[t]}\right) = \left(1 + s_t^{[t-1]}\right) d_{\mathrm{CY}}^{[t-1]} - \alpha_t + \sum_{\tau=t+1}^{\infty} \alpha_\tau c_\tau \sum_{k=t+1}^{\tau} a_{k,\tau}(\delta)\,\mathrm{var}\left(s_k^{[t]}\right),$$

$$\mathrm{E}\left(d_{\mathrm{PY}}^{[t]}\right) = \left(1 + s_t^{[t-1]}\right) d_{\mathrm{PY}}^{[t-1]} - \beta_t + \sum_{\tau=t+1}^{\infty} \beta_\tau c_\tau \sum_{k=t+1}^{\tau} a_{k,\tau}(\delta)\,\mathrm{var}\left(s_k^{[t]}\right),$$

where

$$a_{i,\tau}(\delta) = \frac{1 + s_{\max}}{2\delta} \left(\frac{\delta(\delta_i^{t+1} + i - t) + (1-\delta)(1 - (1-\delta)^{\tau-i})}{(1-\delta)^{\tau-i+1}} \right)$$

and δ_i^{t+1} *denotes the Kronecker delta.*

Remark 4.6 The terms $c_\tau \sum_{k=t+1}^{\tau} a_{k,\tau}(\delta)\operatorname{var}(s_k^{[t]})$, $c_\tau \sum_{k=t+1}^{\tau} a_{k,\tau}(\delta)\operatorname{var}(s_k^{[t]})$ describe the error that is incurred, if the expected value of the discount factor

$$\mathrm{E}\left(\prod_{k=t+1}^{\tau}\left(1+s_k^{[t]}\right)^{-1}\right)$$

is replaced by the discount factor calculated from the expected interest rate,

$$\prod_{k=t+1}^{\tau}\left(1+\mathrm{E}(s_k^{[t]})\right)^{-1}=\prod_{k=t+1}^{\tau}\left(1+s_k^{[t-1]}\right)^{-1}.$$

It is therefore not a surprising result that the smaller the interest risk, $\operatorname{var}(s_k^{[t]})$, the smaller the error of the approximation.

For the proof of Lemma 4.1 we will need the following technical lemma:

Lemma 4.2 *Let $a, b, x_k \geq 0$ and assume that g_n satisfies the inequality*

$$g_n \leq a\left(g_{n-1}+\sum_{k=1}^{n-1} x_k + n x_n\right).$$

Then $g_0 = b$ implies

$$g_n \leq a^n b + \sum_{k=1}^{n}\left(k a^{n-k+1} + \frac{a}{a-1}\left(a^{n-k}-1\right)\right)x_k.$$

Proof We first use an induction argument to show

$$g_n \leq h_n := a^n b + \sum_{k=1}^{n} k x_k a^{n-k+1} + \sum_{k=1}^{n-1} x_k \sum_{i=1}^{n-k} a^i.$$

We clearly have $h_0 = b = g_0$ and

$$g_1 \leq a g_0 + a x_1 = ab + a x_1 = h_1,$$

where we have used the convention that a summation over the empty index set has the value 0. Now assume that the inequality $g_{n-1} \leq h_{n-1}$ holds for $n \geq 2$. Then

$$\begin{aligned}
g_n &\leq a\left(h_{n-1}+\sum_{k=1}^{n-1} x_k + n x_n\right)\\
&= a\left(a^{n-1} b + \sum_{k=1}^{n-1} k x_k a^{n-1-k+1} + \sum_{k=1}^{n-2} x_k \sum_{i=1}^{n-1-k} a^i + \sum_{k=1}^{n-1} x_k + n x_n\right)
\end{aligned}$$

$$
\begin{aligned}
&= a^n b + \sum_{k=1}^{n-1} k x_k a^{n-k+1} + \sum_{k=1}^{n-2} x_k \sum_{i=1}^{n-1-k} a^{i+1} + a \sum_{k=1}^{n-1} x_k + a n x_n \\
&= a^n b + \sum_{k=1}^{n} k x_k a^{n-k+1} + \sum_{k=1}^{n-2} x_k \sum_{i=2}^{n-k} a^i + a \sum_{k=1}^{n-1} x_k \\
&= a^n b + \sum_{k=1}^{n} k x_k a^{n-k+1} + \sum_{k=1}^{n-2} x_k \sum_{i=1}^{n-k} a^i + a x_{n-1} \\
&\le a^n b + \sum_{k=1}^{n} k x_k a^{n-k+1} + \sum_{k=1}^{n-1} x_k \sum_{i=1}^{n-k} a^i \\
&= h_n.
\end{aligned}
$$

Now our assertion follows from

$$
\sum_{i=1}^{n-k} a^i = \frac{a^{n-k+1} - a}{a - 1} = \frac{a}{a-1}\left(a^{n-k} - 1\right).
$$

□

We can now prove Lemma 4.1:

Proof Writing $\Delta_\tau = s_\tau^{[t]} - s_\tau^{[t-1]}$ and

$$
v_\tau^{[t]} = \prod_{i=t+1}^{\tau} \left(1 + s_i^{[t]}\right)^{-1}
$$

we use the identity

$$
\frac{\partial}{\partial x_j} \prod_{i=t+1}^{\tau} (1 + x_i)^{-1} = -\frac{1}{1 + x_j} \prod_{i=t+1}^{\tau} (1 + x_i)^{-1}
$$

to obtain the estimate

$$
\begin{aligned}
v_\tau^{[t]} &= \prod_{i=t+1}^{\tau} \left(1 + s_i^{[t]}\right)^{-1} \\
&= \prod_{i=t+1}^{\tau} (1 + s_i^{[t-1]})^{-1} - \sum_{j=t+1}^{\tau} \frac{\Delta_j}{1 + s_j^{[t-1]}} \prod_{i=t+1}^{\tau} (1 + s_i^{[t-1]})^{-1} \\
&\quad + f_\tau(\Delta_{t+1}, \ldots, \Delta_\tau) \\
&= \prod_{i=t+1}^{\tau} (1 + s_i^{[t-1]})^{-1} \left(1 - \sum_{j=t+1}^{\tau} \frac{\Delta_j}{1 + s_j^{[t-1]}}\right) + f_\tau(\Delta_{t+1}, \ldots, \Delta_\tau)
\end{aligned}
$$

$$= v_\tau^{[t-1]}(1+s_t^{[t-1]})\left(1-\sum_{j=t+1}^{\tau}\frac{\Delta_j}{1+s_j^{[t-1]}}\right)+f_\tau(\Delta_{t+1},\ldots,\Delta_\tau)$$

$$= v_{\tau-1}^{[t-1]}\frac{1+s_t^{[t-1]}}{1+s_\tau^{[t-1]}}\left(1-\sum_{j=t+1}^{\tau}\frac{\Delta_j}{1+s_j^{[t-1]}}\right)+f_\tau(\Delta_{t+1},\ldots,\Delta_\tau), \tag{4.11}$$

where f_τ are functions that satisfy $\lim_{\|x\|\to 0} f_\tau(x)/\|x\| = 0$. Using (4.11) for $\tau - 1$ instead of τ,

$$v_\tau^{[t]} = \frac{v_{\tau-1}^{[t]}}{1+s_\tau^{[t]}} = v_{\tau-1}^{[t-1]}\frac{1+s_t^{[t-1]}}{1+s_\tau^{[t]}}\left(1-\sum_{j=t+1}^{\tau-1}\frac{\Delta_j}{1+s_j^{[t-1]}}\right)+\frac{f_{\tau-1}}{1+s_\tau^{[t]}},$$

we obtain a recursion formula for the error term f_τ. Multiplying

$$0 = \left(v_{\tau-1}^{[t-1]}\frac{1+s_t^{[t-1]}}{1+s_\tau^{[t-1]}}\left(1-\sum_{j=t+1}^{\tau}\frac{\Delta_j}{1+s_j^{[t-1]}}\right)+f_\tau\right)$$
$$-\left(v_{\tau-1}^{[t-1]}\frac{1+s_t^{[t-1]}}{1+s_\tau^{[t]}}\left(1-\sum_{j=t+1}^{\tau-1}\frac{\Delta_j}{1+s_j^{[t-1]}}\right)+\frac{f_{\tau-1}}{1+s_\tau^{[t]}}\right)$$

with

$$\frac{(1+s_\tau^{[t-1]})(1+s_\tau^{[t]})}{1+s_t^{[t-1]}}$$

we get

$$0 = v_{\tau-1}^{[t-1]}(1+s_\tau^{[t]})\left(1-\sum_{j=t+1}^{\tau}\frac{\Delta_j}{1+s_j^{[t-1]}}\right)+\frac{(1+s_\tau^{[t-1]})(1+s_\tau^{[t]})}{1+s_t^{[t-1]}}f_\tau$$
$$-v_{\tau-1}^{[t-1]}(1+s_\tau^{[t-1]})\left(1-\sum_{j=t+1}^{\tau-1}\frac{\Delta_j}{1+s_j^{[t-1]}}\right)-\frac{(1+s_\tau^{[t-1]})}{1+s_t^{[t-1]}}f_{\tau-1}$$
$$= v_{\tau-1}^{[t-1]}\left((1+s_\tau^{[t-1]}+\Delta_\tau)\left(1-\sum_{j=t+1}^{\tau-1}\frac{\Delta_j}{1+s_j^{[t-1]}}-\frac{\Delta_\tau}{1+s_\tau^{[t-1]}}\right)\right.$$
$$\left.-(1+s_\tau^{[t-1]})\left(1-\sum_{j=t+1}^{\tau-1}\frac{\Delta_j}{1+s_j^{[t-1]}}\right)\right)$$
$$+\frac{(1+s_\tau^{[t-1]})}{1+s_t^{[t-1]}}\left((1+s_\tau^{[t]})f_\tau - f_{\tau-1}\right)$$

$$= v_{\tau-1}^{[t-1]}\Delta_\tau\left(1 - \sum_{j=t+1}^{\tau-1}\frac{\Delta_j}{1+s_j^{[t-1]}} - \frac{1+s_\tau^{[t-1]}+\Delta_\tau}{1+s_\tau^{[t-1]}}\right)$$

$$+\frac{\left(1+s_\tau^{[t-1]}\right)}{1+s_t^{[t-1]}}\left(\left(1+s_\tau^{[t]}\right)f_\tau - f_{\tau-1}\right)$$

$$= v_{\tau-1}^{[t-1]}\Delta_\tau\sum_{j=t+1}^{\tau}\frac{\Delta_j}{1+s_j^{[t-1]}} + \frac{\left(1+s_\tau^{[t-1]}\right)}{1+s_t^{[t-1]}}\left(\left(1+s_\tau^{[t-1]}+\Delta_\tau\right)f_\tau - f_{\tau-1}\right).$$

This implies

$$f_\tau = \frac{1}{1+s_\tau^{[t-1]}+\Delta_\tau}\left(f_{\tau-1} - \frac{1+s_t^{[t-1]}}{1+s_\tau^{[t-1]}}v_{\tau-1}^{[t-1]}\Delta_\tau\sum_{j=t+1}^{\tau}\frac{\Delta_j}{1+s_j^{[t-1]}}\right).$$

From our assumption $0 \le s_t^{[t-1]} \le s_{\max}$, $|\Delta_j| \le \delta$ for all j we get

$$f_\tau \le \frac{1}{1-\delta}\left(f_{\tau-1} + (1+s_{\max})|\Delta_\tau|\sum_{j=t+1}^{\tau}\frac{|\Delta_j|}{1+s_j^{[t-1]}}\right)$$

$$\le \frac{1}{1-\delta}\left(f_{\tau-1} + \frac{1+s_{\max}}{2}\sum_{j=t+1}^{\tau}\frac{\Delta_\tau^2+\Delta_j^2}{1+s_j^{[t-1]}}\right)$$

$$\le \frac{1}{1-\delta}\left(f_{\tau-1} + \frac{1+s_{\max}}{2}(\tau-t)\Delta_\tau^2 + \frac{1+s_{\max}}{2}\sum_{j=t+1}^{\tau-1}\Delta_j^2\right).$$

Setting

$$n = \tau - t, \qquad a = \frac{1}{1-\delta}, \qquad x_k = \Delta_{t+k}^2, \qquad g_k = \frac{2}{1+s_{\max}}f_{t+k} \quad (k \ge 1)$$

and using (4.11) as well as $|\Delta_\tau| \le \delta$ we obtain the estimates

$$g_1 = \frac{2}{1+s_{\max}}f_{t+1}$$

$$= \frac{2}{1+s_{\max}}\left(v_{t+1}^{[t]} - v_{t+1}^{[t-1]}(1+s_t^{[t-1]})\left(1 - \frac{\Delta_{t+1}}{1+s_{t+1}^{[t-1]}}\right)\right)$$

$$= \frac{2}{1+s_{\max}}\left(\frac{1}{1+s_{t+1}^{[t-1]}+\Delta_{t+1}} - \frac{1}{1+s_{t+1}^{[t-1]}}\frac{1+s_{t+1}^{[t-1]}-\Delta_{t+1}}{1+s_{t+1}^{[t-1]}}\right)$$

$$= \frac{2}{1+s_{\max}}\frac{(1+s_{t+1}^{[t-1]})^2 - (1+s_{t+1}^{[t-1]}+\Delta_{t+1})(1+s_{t+1}^{[t-1]}-\Delta_{t+1})}{(1+s_{t+1}^{[t-1]}+\Delta_{t+1})(1+s_{t+1}^{[t-1]})^2}$$

$$= \frac{2}{1+s_{\max}} \frac{\Delta_{t+1}^2}{(1+s_{t+1}^{[t-1]}+\Delta_{t+1})(1+s_{t+1}^{[t-1]})^2}$$
$$\le \frac{\Delta_{t+1}^2+\Delta_{t+1}^2}{1+\Delta_{t+1}} \le \frac{\Delta_{t+1}^2+x_1}{1-\delta} = a\big(\Delta_{t+1}^2+x_1\big)$$

and

$$g_n = \frac{2}{1+s_{\max}} f_{t+n}$$
$$\le \frac{2}{1+s_{\max}} \frac{1}{1-\delta}\left(f_{t+n-1} + \frac{1+s_{\max}}{2} n\Delta_{t+n}^2 + \frac{1+s_{\max}}{2}\sum_{j=t+1}^{t+n-1}\Delta_j^2\right)$$
$$= \frac{1}{1-\delta}\left(g_{n-1}+nx_n+\sum_{k=1}^{n-1}x_k\right).$$

This inequality is also satisfied for $n=1$, if we set $g_0=\Delta_{t+1}^2$. Lemma 4.2 implies the estimate

$$f_\tau \le \frac{1+s_{\max}}{2}\left(\frac{\Delta_{t+1}^2}{(1-\delta)^{\tau-t}} + \sum_{i=t+1}^{\tau}\left(\frac{i-t}{(1-\delta)^{\tau-i+1}} + \frac{1}{\delta}\left(\frac{1}{(1-\delta)^{\tau-i}}-1\right)\right)\Delta_i^2\right)$$
$$= \frac{1+s_{\max}}{2\delta}\sum_{i=t+1}^{\tau}\frac{\delta(\delta_i^{t+1}+i-t)+(1-\delta)(1-(1-\delta)^{\tau-i})}{(1-\delta)^{\tau-i+1}}\Delta_i^2$$
$$= \sum_{i=t+1}^{\tau} a_{i,\tau}(\delta)\Delta_i^2.$$

Taking the expected value of (4.11) results in

$$\mathrm{E}\big(v_\tau^{[t]}\big) - \big(1+s_i^{[t-1]}\big)v_i^{[t-1]} = -v_\tau^{[t-1]}\big(1+s_t^{[t-1]}\big)\sum_{j=t+1}^{\tau}\frac{\overbrace{\mathrm{E}(\Delta_j)}^{=0}}{1+s_j^{[t\ 1]}} + \mathrm{E}(f_\tau)$$
$$= \mathrm{E}(f_\tau) \le \sum_{i=t+1}^{\tau} a_{i,\tau}(\delta)\,\mathrm{E}\big(\Delta_i^2\big)$$
$$= \sum_{i=t+1}^{\tau} a_{i,\tau}(\delta)\,\mathrm{var}(\Delta_i) = \sum_{i=t+1}^{\tau} a_{i,\tau}(\delta)\mathrm{var}\big(s_i^{[t]}\big).$$

We have shown that there are constants c_τ with $|c_\tau| \leq 1$ and

$$\mathrm{E}\big(v_\tau^{[t]}\big) = \big(1 + s_t^{[t-1]}\big) v_\tau^{[t-1]} + c_\tau \sum_{i=t+1}^{\tau} a_{i,\tau}(\delta)\,\mathrm{var}\big(s_i^{[t]}\big).$$

This implies the equation

$$\begin{aligned}
\mathrm{E}\big(d_{\mathrm{CY}}^{[t]}\big) &= \sum_{\tau=t+1}^{\infty} \alpha_\tau\,\mathrm{E}\big(v_\tau^{[t]}\big) \\
&= \sum_{\tau=t+1}^{\infty} \alpha_\tau \big(1 + s_t^{[t-1]}\big) v_\tau^{[t-1]} + \sum_{\tau=t+1}^{\infty} \alpha_\tau c_\tau \sum_{i=t+1}^{\tau} a_{i,\tau}(\delta)\,\mathrm{var}\big(s_i^{[t]}\big) \\
&= \sum_{\tau=t}^{\infty} \alpha_\tau \big(1 + s_t^{[t-1]}\big) v_\tau^{[t-1]} - \alpha_t \big(1 + s_t^{[t-1]}\big) v_t^{[t-1]} \\
&\quad + \sum_{\tau=t+1}^{\infty} \alpha_\tau c_\tau \sum_{i=t+1}^{\tau} a_{i,\tau}(\delta)\,\mathrm{var}\big(s_i^{[t]}\big) \\
&= \big(1 + s_t^{[t-1]}\big) d_{\mathrm{CY}}^{[t-1]} - \alpha_t + \sum_{\tau=t+1}^{\infty} \alpha_\tau c_\tau \sum_{i=t+1}^{\tau} a_{i,\tau}(\delta)\,\mathrm{var}\big(s_i^{[t]}\big)
\end{aligned}$$

and, analogously,

$$\mathrm{E}\big(d_{\mathrm{PY}}^{[t]}\big) = \big(1 + s_t^{[t-1]}\big) d_{\mathrm{PY}}^{[t-1]} - \beta_t + \sum_{\tau=t+1}^{\infty} \beta_\tau c_\tau \sum_{i=t+1}^{\tau} a_{i,\tau}(\delta)\,\mathrm{var}\big(s_i^{[t]}\big).$$

□

We are now in a position to give an estimate for the loss function $-\Delta\mathrm{RTK}_t$:

Proposition 4.2 *Assume* $s_t^{[t-1]} \leq s_{\max}$ *and*

$$0 \leq s_\tau^{[t-1]}, \quad \big|s_\tau^{[t]} - s_\tau^{[t-1]}\big| \leq \delta < 1$$

for all $\tau \geq t$.

Then the loss function is given by

$$-\Delta\mathrm{RTK}_t = -(T_{\text{risk invest}} + T_{\text{exp invest}} + T_{\text{risk ins}} + T_{\text{exp ins}} + T_{\text{error}}),$$

where

$$\begin{aligned}
T_{\text{risk invest}} &= \frac{r_t - \mathrm{E}(r_t)}{1 + s_t^{[t-1]}} \big(A_t^{\text{Start}} + P_t - \mathrm{upr}_{t-1} - K_t\big) \\
&\quad - \left(\frac{\alpha_t + d_{\mathrm{CY}}^{[t]}}{1 + s_t^{[t-1]}} - d_{\mathrm{CY}}^{[t-1]}\right) \mathrm{E}\big(S_t^{\mathrm{CY}}\big) - \left(\frac{\beta_t + d_{\mathrm{PY}}^{[t]}}{1 + s_t^{[t-1]}} - d_{\mathrm{PY}}^{[t-1]}\right) R_{\mathrm{PY}}^{[t-1]},
\end{aligned}$$

$$\begin{aligned}
T_{\text{exp invest}} &= \frac{\mathrm{E}(r_t) - s_t^{[t-1]}}{1 + s_t^{[t-1]}} \big(A_t^{\text{Start}} + P_t - \text{upr}_{t-1} - K_t\big), \\
T_{\text{risk ins}} &= -d_{\text{CY}}^{[t-1]} \big(S_t^{\text{CY}} - \mathrm{E}\big(S_t^{\text{CY}}\big)\big) - d_{\text{PY}}^{[t-1]} \big(c_t^{\text{PY}} - 1\big) R_{\text{PY}}^{[t-1]}, \\
T_{\text{exp ins}} &= P_t - K_t - d_{\text{CY}}^{[t-1]} \mathrm{E}\big(S_t^{\text{CY}}\big), \\
T_{\text{error}} &= -\left(\frac{d_{\text{CY}}^{[t]} - \mathrm{E}(d_{\text{CY}}^{[t]})}{1 + s_t^{[t-1]}} + \frac{\sum_{\tau=t+1}^{\infty} \alpha_\tau \tilde{f}_\tau \sum_{k=t+1}^{\tau} a_{k,\tau}(\delta) \operatorname{var}(s_k^{[t]})}{1 + s_t^{[t-1]}} \right) \\
&\quad \times \big(S_t^{\text{CY}} - \mathrm{E}\big(S_t^{\text{CY}}\big)\big) \\
&\quad - \left(\frac{d_{\text{PY}}^{[t]} - \mathrm{E}(d_{\text{PY}}^{[t]})}{1 + s_t^{[t-1]}} + \frac{\sum_{\tau=t+1}^{\infty} \beta_\tau \tilde{f}_\tau \sum_{k=t+1}^{\tau} a_{k,\tau}(\delta) \operatorname{var}(s_k^{[t]})}{1 + s_t^{[t-1]}} \right) \\
&\quad \times \big(c_t^{\text{PY}} - 1\big) R_{\text{PY}}^{[t-1]}
\end{aligned}$$

and $a_{k,\tau}(\delta)$ *are defined as in Lemma* 4.1.

Proof Equations (4.8), (4.9), (4.10) imply

$$\begin{aligned}
\text{RTK}_t^{\text{Start}} &= A_t^{\text{Start}} - \text{upr}_{t-1} - d_{\text{PY}}^{[t-1]} R_{\text{PY}}^{[t-1]}, \\
\text{RTK}_t^{\text{End}} &= (1 + r_t)\big(A_t^{\text{Start}} + P_t - \text{upr}_{t-1} - K_t\big) \\
&\quad - \big(\alpha_t + d_{\text{CY}}^{[t]}\big) S_t^{\text{CY}} - \big(\beta_t + d_{\text{PY}}^{[t]}\big) c_t^{\text{PY}} R_{\text{PY}}^{[t-1]}.
\end{aligned}$$

Hence the loss is given by

$$\begin{aligned}
\Delta\text{RTK}_t &= \frac{\text{RTK}_t^{\text{End}}}{1 + s_t^{[t-1]}} - \text{RTK}_t^{\text{Start}} \\
&= \frac{1 + r_t}{1 + s_t^{[t-1]}} \big(A_t^{\text{Start}} + P_t - \text{upr}_{t-1} - K_t\big) \\
&\quad - \frac{(\alpha_t + d_{\text{CY}}^{[t]}) S_t^{\text{CY}} + (\beta_t + d_{\text{PY}}^{[t]}) c_t^{\text{PY}} R_{\text{PY}}^{[t-1]}}{1 + s_t^{[t-1]}} \\
&\quad \overbrace{- \frac{1 + s_t^{[t-1]}}{1 + s_t^{[t-1]}} \big(A_t^{\text{Start}} + P_t - \text{upr}_{t-1} - K_t\big) + P_t - K_t + d_{\text{PY}}^{[t-1]} R_{\text{PY}}^{[t-1]}}^{=-\text{RTK}_t^{\text{Start}}} \\
&= \frac{r_t - s_t^{[t-1]}}{1 + s_t^{[t-1]}} \big(A_t^{\text{Start}} + P_t - \text{upr}_{t-1} - K_t\big) + P_t - K_t + d_{\text{PY}}^{[t-1]} R_{\text{PY}}^{[t-1]} \\
&\quad - \frac{(\alpha_t + d_{\text{CY}}^{[t]}) S_t^{\text{CY}} + (\beta_t + d_{\text{PY}}^{[t]}) c_t^{\text{PY}} R_{\text{PY}}^{[t-1]}}{1 + s_t^{[t-1]}}.
\end{aligned}$$

We can break down the right-hand side of this equation into contributions from

- the expected investment result, $I_{\text{exp invest}}$,
- the investment risk result, $I_{\text{risk invest}}$,
- the expected insurance result, $I_{\text{exp ins}}$,
- and three remaining terms, $I_{\text{R},1}$, $I_{\text{R},2}$, $I_{\text{R},3}$:

$$\begin{aligned}
\Delta\text{RTK}_t = {} & \overbrace{\frac{\text{E}(r_t) - s_t^{[t-1]}}{1+s_t^{[t-1]}}\left(A_t^{\text{Start}} + P_t - \text{upr}_{t-1} - K_t\right)}^{=I_{\text{exp invest}}} \\
& + \overbrace{\frac{r_t - \text{E}(r_t)}{1+s_t^{[t-1]}}\left(A_t^{\text{Start}} + P_t - \text{upr}_{t-1} - K_t\right)}^{=I_{\text{risk invest}}} + \overbrace{P_t - K_t}^{=I_{\text{exp ins}}} \\
& + \overbrace{d_{\text{PY}}^{[t-1]} R_{\text{PY}}^{[t-1]}}^{=I_{\text{R},1}} - \overbrace{\frac{(\alpha_t + d_{\text{CY}}^{[t]}) S_t^{\text{CY}}}{1+s_t^{[t-1]}}}^{=I_{\text{R},2}} - \overbrace{\frac{(\beta_t + d_{\text{PY}}^{[t]}) c_t^{\text{PY}} R_{\text{PY}}^{[t-1]}}{1+s_t^{[t-1]}}}^{=I_{\text{R},3}}. \qquad (4.12)
\end{aligned}$$

We write $\Delta S_t^{\text{CY}} = S_t^{\text{CY}} - \text{E}(\Delta S_t^{\text{CY}})$, and we use Lemma 4.1 in order to simplify $I_{\text{R},2}$:

$$\begin{aligned}
I_{\text{R},2} = {} & \frac{\alpha_t + d_{\text{CY}}^{[t]}}{1+s_t^{[t-1]}} S_t^{\text{CY}} \\
= {} & \frac{\alpha_t + \text{E}(d_{\text{CY}}^{[t]})}{1+s_t^{[t-1]}} \text{E}\left(S_t^{\text{CY}}\right) + \frac{\alpha_t + \text{E}(d_{\text{CY}}^{[t]})}{1+s_t^{[t-1]}} \Delta S_t^{\text{CY}} \\
& + \frac{d_{\text{CY}}^{[t]} - \text{E}(d_{\text{CY}}^{[t]})}{1+s_t^{[t-1]}} \text{E}\left(S_t^{\text{CY}}\right) + \frac{d_{\text{CY}}^{[t]} - \text{E}(d_{\text{CY}}^{[t]})}{1+s_t^{[t-1]}} \Delta S_t^{\text{CY}} \\
= {} & \frac{\alpha_t + d_{\text{CY}}^{[t]}}{1+s_t^{[t-1]}} \text{E}\left(S_t^{\text{CY}}\right) + \frac{d_{\text{CY}}^{[t]} - \text{E}(d_{\text{CY}}^{[t]})}{1+s_t^{[t-1]}} \Delta S_t^{\text{CY}} \\
& + \frac{(1+s_t^{[t-1]}) d_{\text{CY}}^{[t-1]} + \sum_{\tau=t+1}^{\infty} \alpha_\tau c_\tau \sum_{k=t+1}^{\tau} a_{k,\tau}(\delta) \operatorname{var}(s_k^{[t]})}{1+s_t^{[t-1]}} \Delta S_t^{\text{CY}} \\
= {} & \overbrace{d_{\text{CY}}^{[t-1]} \text{E}\left(S_t^{\text{CY}}\right)}^{=I_{\text{CY,exp ins}}} + \overbrace{\left(\frac{\alpha_t + d_{\text{CY}}^{[t]}}{1+s_t^{[t-1]}} - d_{\text{CY}}^{[t-1]}\right) \text{E}\left(S_t^{\text{CY}}\right)}^{=I_{\text{CY,risk invest}}} + \overbrace{d_{\text{CY}}^{[t-1]} \Delta S_t^{\text{CY}}}^{=I_{\text{CY,risk ins}}} \\
& + \overbrace{\frac{d_{\text{CY}}^{[t]} - \text{E}(d_{\text{CY}}^{[t]}) + \sum_{\tau=t+1}^{\infty} \alpha_\tau c_\tau \sum_{k=t+1}^{\tau} a_{k,\tau}(\delta) \operatorname{var}(s_k^{[t]})}{1+s_t^{[t-1]}} \Delta S_t^{\text{CY}}}^{=I_{\text{CY,error}}}
\end{aligned}$$

In the same way we get

$$
\begin{aligned}
I_{\mathrm{R},3} &= \frac{\beta_t + d_{\mathrm{PY}}^{[t]}}{1+s_t^{[t-1]}} c_t^{\mathrm{PY}} R_{\mathrm{PY}}^{[t-1]} \\
&= \overbrace{d_{\mathrm{PY}}^{[t-1]} R_{\mathrm{PY}}^{[t-1]}}^{=I_{\mathrm{R},1}} + \overbrace{\left(\frac{\beta_t + d_{\mathrm{PY}}^{[t]}}{1+s_t^{[t-1]}} - d_{\mathrm{PY}}^{[t-1]} \right) R_{\mathrm{PY}}^{[t-1]}}^{=I_{\mathrm{PY,risk\ invest}}} + \overbrace{d_{\mathrm{PY}}^{[t-1]} \left(c_t^{\mathrm{PY}} - 1\right) R_{\mathrm{PY}}^{[t-1]}}^{=I_{\mathrm{PY,risk\ invest}}} \\
&\quad + \overbrace{\frac{d_{\mathrm{PY}}^{[t]} - \mathrm{E}(d_{\mathrm{PY}}^{[t]}) + \sum_{\tau=t+1}^{\infty} \beta_\tau c_\tau \sum_{k=t+1}^{\tau} a_{k,\tau}(\delta) \operatorname{var}(s_k^{[t]})}{1+s_t^{[t-1]}} \left(c_t^{\mathrm{PY}} - 1\right) R_{\mathrm{PY}}^{[t-1]}}^{=I_{\mathrm{PY,error}}}.
\end{aligned}
$$

With

$$
\begin{aligned}
&T_{\mathrm{exp\ invest}} = I_{\mathrm{exp\ invest}}, \qquad T_{\mathrm{risk\ invest}} = I_{\mathrm{risk\ invest}} - I_{\mathrm{CY,risk\ invest}} - I_{\mathrm{PY,risk\ invest}}, \\
&T_{\mathrm{risk\ ins}} = -I_{\mathrm{CY,risk\ ins}} - I_{\mathrm{PY,risk\ ins}}, \qquad T_{\mathrm{exp\ ins}} = I_{\mathrm{exp\ ins}} - I_{\mathrm{CY,exp\ ins}}, \\
&T_{\mathrm{eror}} = -I_{\mathrm{CY,error}} - I_{\mathrm{PY,error}}.
\end{aligned}
$$

we have proved the assertion. □

Corollary 4.1 *Assume that S_t^{CY} and c_t^{PY} are independent of $s_\tau^{[t]}$ for all $\tau \geq t+1$. Then*

$$
\begin{aligned}
\mathrm{E}(\Delta\mathrm{RTK}_t) &= \frac{\mathrm{E}(r_t) - s_t^{[t-1]}}{1+s_t^{[t-1]}} \left(A_t^{\mathrm{Start}} + P_t - \mathrm{upr}_{t-1} - K_t\right) \\
&\quad + P_t - K_t - d_{\mathrm{CY}}^{[t-1]} \mathrm{E}\left(S_t^{\mathrm{CY}}\right). \\
&\quad + c \sum_{\tau=t+1}^{\infty} \frac{\alpha_\tau \mathrm{E}(S_t^{\mathrm{CY}}) + \beta_\tau R_{\mathrm{PY}}^{[t-1]}}{1+s_t^{[t-1]}} \sum_{k=t+1}^{\tau} a_{k,\tau}(\delta) \operatorname{var}\left(s_k^{[t]}\right),
\end{aligned}
$$

holds with $|c| \leq 1$.

Proof The expected value of T_{error} vanishes since any two independent random variables a, b satisfy the identity

$$
\mathrm{E}(ab) = \mathrm{E}(a)\,\mathrm{E}(b).
$$

Hence we get, using Lemma 4.1,

$$
\begin{aligned}
\mathrm{E}(\Delta\mathrm{RTK}_t) &= \frac{\mathrm{E}(r_t) - s_t^{[t-1]}}{1+s_t^{[t-1]}} \left(A_t^{\mathrm{Start}} + P_t - \mathrm{upr}_{t-1} - K_t\right) \\
&\quad + P_t - K_t - d_{\mathrm{CY}}^{[t-1]} \mathrm{E}\left(S_t^{\mathrm{CY}}\right)
\end{aligned}
$$

$$
\begin{aligned}
&- \mathrm{E}\left(\frac{\alpha_t + d_{\mathrm{CY}}^{[t]}}{1+s_t^{[t-1]}} - d_{\mathrm{CY}}^{[t-1]}\right)\mathrm{E}\big(S_t^{\mathrm{CY}}\big) - \mathrm{E}\left(\frac{\beta_t + d_{\mathrm{PY}}^{[t]}}{1+s_t^{[t-1]}} - d_{\mathrm{PY}}^{[t-1]}\right)R_{\mathrm{PY}}^{[t-1]} \\
&= \frac{\mathrm{E}(r_t) - s_t^{[t-1]}}{1+s_t^{[t-1]}}\big(A_t^{\mathrm{Start}} + P_t - \mathrm{upr}_{t-1} - K_t\big) \\
&\quad + P_t - K_t - d_{\mathrm{CY}}^{[t-1]}\mathrm{E}\big(S_t^{\mathrm{CY}}\big) \\
&\quad - \sum_{\tau=t+1}^{\infty} \frac{\alpha_\tau\,\mathrm{E}(S_t^{\mathrm{CY}}) + \beta_\tau R_{\mathrm{PY}}^{[t-1]}}{1+s_t^{[t-1]}} c_\tau \sum_{k=t+1}^{\tau} a_{k,\tau}(\delta)\,\mathrm{var}\big(s_k^{[t]}\big).
\end{aligned}
$$

The assertion follows now from

$$
\frac{\alpha_\tau\,\mathrm{E}(S_t^{\mathrm{CY}}) + \beta_\tau R_{\mathrm{PY}}^{[t-1]}}{1+s_t^{[t-1]}} \sum_{k=t+1}^{\tau} a_{k,\tau}(\delta)\,\mathrm{var}\big(s_k^{[t]}\big) \geq 0, \quad |c_\tau| \leq 1
$$

for all τ. □

Remark 4.7 The term $(\mathrm{E}(r_t) - s_t^{[t-1]})(1+s_t^{[t-1]})^{-1}(A_t^{\mathrm{Start}} + P_t - \mathrm{upr}_{t-1} - K_t)$ represents the expected investment result and the term $P_t - K_t - d_{\mathrm{CY}}^{[t-1]}\mathrm{E}(S_t^{\mathrm{CY}})$ represents the expected insurance result. The uncertainty at time t regarding the prognosis for the interest rate curve is expressed by

$$
c \sum_{\tau=t+1}^{\infty} \frac{\alpha_\tau\,\mathrm{E}(S_t^{\mathrm{CY}}) + \beta_\tau R_{\mathrm{PY}}^{[t-1]}}{1+s_t^{[t-1]}} \sum_{k=t+1}^{\tau} a_{k,\tau}(\delta)\,\mathrm{var}\big(s_k^{[t]}\big).
$$

Remark 4.8 The error term T_{error} in Proposition 4.2 can be decomposed into a term that is dominated by the variance of the risk-free interest curve,

$$
\begin{aligned}
T_{\mathrm{error},1} = &- \sum_{\tau=t+1}^{\infty} \Big(\alpha_\tau\big(S_t^{\mathrm{CY}} - \mathrm{E}\big(S_t^{\mathrm{CY}}\big)\big) + \beta_\tau\big(c_t^{\mathrm{PY}} - 1\big)R_{\mathrm{PY}}^{[t-1]}\Big) \\
&\times \frac{c_\tau \sum_{k=t+1}^{\tau} a_{k,\tau}(\delta)\,\mathrm{var}(s_k^{[t]})}{1+s_t^{[t-1]}},
\end{aligned}
$$

and a term that is dominated by the quadratic deviation from expected value,

$$
T_{\mathrm{error},2} = -\frac{d_{\mathrm{CY}}^{[t]} - \mathrm{E}(d_{\mathrm{CY}}^{[t]})}{1+s_t^{[t-1]}}\big(S_t^{\mathrm{CY}} - \mathrm{E}\big(S_t^{\mathrm{CY}}\big)\big) - \frac{d_{\mathrm{PY}}^{[t]} - \mathrm{E}(d_{\mathrm{PY}}^{[t]})}{1+s_t^{[t-1]}}\big(c_t^{\mathrm{PY}} - 1\big)R_{\mathrm{PY}}^{[t-1]}.
$$

The error term $T_{\mathrm{error},1}$ is not given explicitly, because c_τ is unknown apart from the property $|c_\tau| \leq 1$. Since the term

$$
\Big(\alpha_\tau\big(S_t^{\mathrm{CY}} - \mathrm{E}\big(S_t^{\mathrm{CY}}\big)\big) + \beta_\tau\big(c_t^{\mathrm{PY}} - 1\big)R_{\mathrm{PY}}^{[t-1]}\Big)
$$

fails to be small in comparison to the other terms in Proposition 4.2, it is not possible to neglect $T_{\mathrm{error},1}$, unless $\mathrm{var}(s_\tau^{[t]}) \ll 1$ for all $\tau \geq t$. This cannot be assumed in general.

The term $T_{\mathrm{error},2}$ is small relative to the other terms in Proposition 4.2 if the deviation from expected value is small. In general, this cannot be assumed either.

The SST model neglects T_{error} for simplicity.

Remark 4.9 For stochastic processes which are usually used to describe the interest rate curve, there are *not any* constants $s_{\max}$ and $\delta < 1$ which satisfy $0 \leq s_\tau^{[t-1]} \leq s_{\max}$ and $|s_\tau^{[t]} - s_\tau^{[t-1]}| \leq \delta$. In order to be able to apply Proposition 4.2 it would therefore be necessary to ensure in a further step that these inequalities are violated only on a "small" set, and that this violation can be neglected for the calculation of $\Delta\mathrm{RTK}_t$.

Calculation of $T_{\mathrm{exp\ ins}} + T_{\mathrm{risk\ ins}}$ (Part 1): The Model for CY Losses The non-discounted yearly loss can be seen as the stochastic sum

$$S_t^{\mathrm{CY}} = \sum_{j=1}^{N} Y_j$$

of individual losses Y_j, where N is the random number of losses during the period t. Hence the distribution of S_t^{CY} could be modeled through the normal approximation (central limit theorem) or using the Panjer algorithm. However, both approaches are problematic. The problem with the normal approximation is that the distributions Y_j are fat-tailed. The Panjer algorithm does not work in practice because of the high number of losses (typically 10^6 or more). Hence another modeling method has to be chosen.

The yearly loss is split into the sum of high frequency low severity "normal losses" and the sum of low frequency high severity "major losses",

$$S_t^{\mathrm{CY}} = S_t^{\mathrm{CY,nl}} + S_t^{\mathrm{CY,ml}}.$$

The SST allows the choice between two different thresholds between the severity of normal losses and major losses, CHF 1M and CHF 5M. The number of major losses is small so that it is possible to employ the Panjer algorithm. While the number of normal losses is still high, they are thin-tailed and it is possible to use the central limit theorem. The SST model assumes that the sum of normal losses and the sum of large losses are independent. Hence both types of losses are aggregated through the convolution of the distributions for $S_t^{\mathrm{CY,nl}}$ and $S_t^{\mathrm{CY,ml}}$.

Distribution of Normal Losses We consider first the distribution for normal losses of a fixed line of business k. The distribution depends on both the individual risk of random fluctuation and external risks which can be assumed to be the same for all insurers. In order to describe the latter we define a discrete random variable $\Theta\colon \Omega \to \mathbb{N}$. For any random variable $a\colon \Omega \to \mathbb{R}$ we write

$$[a \mid \Theta = \vartheta]\colon \Theta^{-1}(\vartheta) \to \mathbb{R}$$

to denote a conditioned on $\{\Theta = \vartheta\}$.

For the proof of Lemma 4.4 we will need the "law of the total variance" which we will prove first.

Lemma 4.3 *Let a, b be random variables. Then the "law of total variance",*

$$\operatorname{var}(a) = \operatorname{var}\bigl(\mathrm{E}(a \mid b)\bigr) + \mathrm{E}\bigl(\operatorname{var}(a \mid b)\bigr),$$

holds, where the conditional variance is defined by $\operatorname{var}(a \mid b) = \mathrm{E}(a^2 \mid b) - \mathrm{E}(a \mid b)^2$.

Proof It is sufficient to calculate

$$\begin{aligned}\operatorname{var}(a) &= \mathrm{E}\bigl(a^2\bigr) - \mathrm{E}(a)^2 = \mathrm{E}\bigl(\mathrm{E}\bigl(a^2 \mid b\bigr)\bigr) - \mathrm{E}\bigl(\mathrm{E}(a \mid b)\bigr)^2 \\ &= \mathrm{E}\bigl(\operatorname{var}(a \mid b) + \mathrm{E}(a \mid b)^2\bigr) - \mathrm{E}\bigl(\mathrm{E}(a \mid b)\bigr)^2 \\ &= \mathrm{E}\bigl(\operatorname{var}(a \mid b)\bigr) + \operatorname{var}\bigl(\mathrm{E}(a \mid b)\bigr).\end{aligned}$$

□

Lemma 4.4 *Let k be a line of business and assume that its external risks are described by a discrete random variable Θ^k. Assume further that the loss relating to this line of business is described by a random variable of the form*

$$S_t^{\mathrm{CY,nl},k} = \sum_{j=1}^{N^{\mathrm{nl},k}} Y_j^{\mathrm{nl},k}$$

such that the following holds:

(i) *For each ϑ the conditional random variables*

$$\bigl[Y_j^{\mathrm{nl},k} \mid \Theta^k = \vartheta\bigr] \sim \bigl[Y_1^{\mathrm{nl},k} \mid \Theta^k = \vartheta\bigr] \quad (j \in \mathbb{N})$$

are independent and identically distributed (iid);

(ii) *There is a measurable map $\lambda_k^{\mathrm{nl}} \colon \Theta^k(\Omega) \to \mathbb{R}$, such that*

$$\bigl[N^{\mathrm{nl},k} \mid \Theta^k = \vartheta\bigr] \sim \operatorname{Poisson}\Bigl(\lambda_k^{\mathrm{nl}}(\vartheta)\Bigr)$$

holds for all $\vartheta \in \Theta^k(\Omega)$;

(iii) *for each $i \in \mathbb{N}$ and each $\vartheta \in \Theta^k(\Omega)$ the random variables*

$$\bigl[N^{\mathrm{nl},k} \mid \Theta^k = \vartheta\bigr] \quad \text{and} \quad \bigl[Y_i^{\mathrm{nl},k} \mid \Theta^k = \vartheta\bigr]$$

are independent;

(iv) *the random variables $\lambda_k^{\mathrm{nl}} \circ \Theta^k$ and $\mathrm{E}(Y_1^{\mathrm{nl},k} \mid \Theta^k)$ are independent;*

(v) *the random variables $\lambda_k^{\mathrm{nl}} \circ \Theta^k$ and $\mathrm{E}((Y_1^{\mathrm{nl},k})^2 \mid \Theta^k)$ are independent.*

Then the variation coefficient vc *of the loss distribution* $S_t^{\mathrm{CY,nl},k}$ *is given by*

$$\mathrm{vc}^2\big(S_t^{\mathrm{CY,nl},k}\big) = \frac{\mathrm{var}(\mathrm{E}(S_t^{\mathrm{CY,nl},k} \mid \Theta^k))}{\mathrm{E}^2(S_t^{\mathrm{CY,nl},k})} + \frac{\mathrm{vc}^2(Y_1^{\mathrm{nl},k}) - 1}{\mathrm{E}(N^{\mathrm{nl},k})}. \tag{4.13}$$

Remark 4.10 Independence of $\lambda_k^{\mathrm{nl}} \circ \Theta^k$ and $\mathrm{E}((Y_1^{\mathrm{nl},k})^2 \mid \Theta^k)$ is ensured if $\Theta^k = (\Theta_N^k, \Theta_Y^k)$ can be split into two independent parts Θ_N^k, Θ_Y^k such that $\lambda_k^{\mathrm{nl}} \circ \Theta^k = \lambda_k^{\mathrm{nl}} \circ \Theta_N^k$ and $\mathrm{E}((Y_1^{\mathrm{nl},k})^2 \mid \Theta^k) = \mathrm{E}((Y_1^{\mathrm{nl},k})^2 \mid \Theta_Y^k)$ hold.

Proof The law of total variance (Lemma 4.3) implies

$$\begin{aligned}\mathrm{vc}^2\big(S_t^{\mathrm{CY,nl},k}\big) &= \frac{\mathrm{var}(S_t^{\mathrm{CY,nl},k})}{\mathrm{E}(S_t^{\mathrm{CY,nl},k})^2} \\ &= \frac{\mathrm{var}(\mathrm{E}(S_t^{\mathrm{CY,nl},k} \mid \Theta^k))}{\mathrm{E}(S_t^{\mathrm{CY,nl},k})^2} + \frac{\mathrm{E}(\mathrm{var}(S_t^{\mathrm{CY,nl},k} \mid \Theta^k))}{\mathrm{E}(S_t^{\mathrm{CY,nl},k})^2}.\end{aligned}$$

For any random variable a we have $\mathrm{E}(a \mid \Theta^k)_{|\{\Theta^k=\vartheta\}} = \mathrm{E}([a \mid \Theta^k = \vartheta])$. Hence

$$\mathrm{E}\big(N^{\mathrm{nl},k} \mid \Theta^k\big) = \mathrm{var}\big(N^{\mathrm{nl},k} \mid \Theta^k\big) = \lambda_k^{\mathrm{nl}}(\vartheta)$$

holds and we obtain

$$\begin{aligned}\mathrm{var}\big(S_t^{\mathrm{CY,nl},k} \mid \Theta^k\big)_{|\{\Theta^k=\vartheta\}} &= \mathrm{var}\big(\big[S_t^{\mathrm{CY,nl},k} \mid \Theta^k = \vartheta\big]\big) \\ &= \mathrm{E}\big(\big[N^{\mathrm{nl},k} \mid \Theta^k = \vartheta\big]\big)\,\mathrm{var}\big(\big[Y_1^{\mathrm{nl},k} \mid \Theta^k = \vartheta\big]\big) \\ &\quad + \mathrm{var}\big(\big[N^{\mathrm{nl},k} \mid \Theta^k = \vartheta\big]\big)\,\mathrm{E}\big(\big[Y_1^{\mathrm{nl},k} \mid \Theta^k = \vartheta\big]\big)^2 \\ &= \lambda_k^{\mathrm{nl}}(\vartheta)\,\mathrm{E}\big(\big[Y_1^{\mathrm{nl},k} \mid \Theta^k = \vartheta\big]^2\big).\end{aligned}$$

Taking the expected value on both sides we get

$$\mathrm{E}\big(\mathrm{var}\big(S_t^{\mathrm{CY,nl},k} \mid \Theta^k\big)\big) = \mathrm{E}\big(N^{\mathrm{nl},k}\big)\,\mathrm{E}\big(\big(Y_1^{\mathrm{nl},k}\big)^2\big),$$

where we have used assumption 4.4. Assumption 4.4 implies

$$\begin{aligned}\mathrm{E}\big(S_t^{\mathrm{CY,nl},k}\big) &= \mathrm{E}\big(\mathrm{E}\big(S_t^{\mathrm{CY,nl},k} \mid \Theta^k\big)\big) = \mathrm{E}\big(\lambda_k^{\mathrm{nl}} \circ \Theta^k\, \mathrm{E}\big(Y_1^{\mathrm{nl},k}\big)\big)\\ &= \mathrm{E}\big(N^{\mathrm{nl},k}\big)\,\mathrm{E}\big(Y_1^{\mathrm{nl},k}\big),\end{aligned}$$

which in turn gives

$$\begin{aligned}\mathrm{vc}^2\big(S_t^{\mathrm{CY,nl},k}\big) &= \frac{\mathrm{var}(\mathrm{E}(S_t^{\mathrm{CY,nl},k} \mid \Theta^k))}{\mathrm{E}(S_t^{\mathrm{CY,nl},k})^2} + \frac{\mathrm{E}(N^{\mathrm{nl},k})\,\mathrm{E}((Y_1^{\mathrm{nl},k})^2)}{\mathrm{E}(N^{\mathrm{nl},k})^2\,\mathrm{E}(Y_1^{\mathrm{nl},k})^2}\\ &= \mathrm{vc}^2\big(\mathrm{E}\big(S_t^{\mathrm{CY,nl},k} \mid \Theta^k\big)\big)\\ &\quad + \frac{\mathrm{E}((Y_1^{\mathrm{nl},k})^2) - \mathrm{E}(Y_1^{\mathrm{nl},k})^2 + \mathrm{E}(Y_1^{\mathrm{nl},k})^2}{\mathrm{E}(N^{\mathrm{nl},k})\,\mathrm{E}(Y_1^{\mathrm{nl},k})^2}\\ &= \mathrm{vc}^2\big(\mathrm{E}\big(S_t^{\mathrm{CY,nl},k} \mid \Theta^k\big)\big) + \frac{\mathrm{vc}^2(Y_1^{\mathrm{nl},k}) - 1}{\mathrm{E}(N^{\mathrm{nl},k})}.\end{aligned}$$

□

Remark 4.11 The first summand in (4.13) describes the parameter risk, i.e., the volatility of those model parameters which describe the external risk Θ. Since we assume that this risk is the same for all insurers it cannot be diversified away.

The second summand in (4.13) describes the random fluctuation risk. This term is composed of the variation coefficient $\mathrm{vc}^2(Y_1^{\mathrm{nl},k})$ for individual losses and the expected number of all losses, $\mathrm{E}(N^{\mathrm{nl},k})$. Both quantities depend on the specific line of business.

FINMA estimates the variation coefficients $\mathrm{vc}(\mathrm{E}(S_t^{\mathrm{CY,nl},k} \mid \Theta))$ of the parameter risk for each line of business k and publishes these estimates for the use of the SST calculation (see Appendix 8.4.3 in [4]). FINMA also prescribes standard values for the variation coefficients $\mathrm{vc}(Y_1^{\mathrm{nl},k})$ (see Appendix 8.4.4 in [4]). Finally, FINMA prescribes the correlation coefficients $\mathrm{corr}_{k,l}$ for each line of business (see Appendix 8.4.2 in [4]).

Corollary 4.2 *Assume that the sum of normal losses over all lines of business, $S_t^{\mathrm{CY,nl}}$, is lognormally distributed. Then the distribution is uniquely determined by*

$$\mathrm{E}\big(S_t^{\mathrm{CY,nl}}\big) = \sum_k \mathrm{E}\big(S_t^{\mathrm{CY,nl},k}\big) = \sum_k \mathrm{E}\big(N^{\mathrm{nl},k}\big)\,\mathrm{E}\big(Y_1^{\mathrm{nl},k}\big)$$

and

$$\mathrm{var}\big(S_t^{\mathrm{CY,nl}}\big) = \sum_{k,l} \mathrm{corr}_{kl}\,\mathrm{vc}\big(S_t^{\mathrm{CY,nl},k}\big)\,\mathrm{vc}\big(S_t^{\mathrm{CY,nl},l}\big)\,\mathrm{E}\big(S_t^{\mathrm{CY,nl},k}\big)\,\mathrm{E}\big(S_t^{\mathrm{CY,nl},l}\big).$$

The only quantities that are not prescribed by FINMA and have to be estimated by the insurer are the expected average loss $\mathrm{E}(Y_1^{\mathrm{nl},k})$ and the expected number of losses, $\mathrm{E}(N^{\mathrm{nl},k})$.

Distribution of Major Losses Major losses consist of both individual major losses and cumulative losses. The latter could be caused by natural disasters such as hail or flooding. In addition, cumulative losses often affect more than one line of business. For instance, a hail storm may cause damage to insured properties as well as damage to cars which would be covered by car insurance.

The amount of individual major losses is modeled through a compound Poisson distribution where the losses have a Pareto distribution with a cut-off point. The amount of individual major losses for line of business k is then given by

$$S_t^{\mathrm{CY,ml},k} = \sum_{j=1}^{N^{\mathrm{ml},k}} Y_j^{\mathrm{ml},k},$$

where the random variables $Y_j^{\mathrm{ml},k}$ are independent and identically distributed, and are also independent of $N^{\mathrm{ml},k}$. The corresponding distribution has the form

$$F_{Y_1^{\mathrm{ml},k}}(x) = \begin{cases} 0 & x < \beta_k \\ 1 - (\frac{\beta_k}{x})^{\alpha_k} & \beta_k \le x < \gamma_k \\ 1 & \gamma_k \le x \end{cases}.$$

Here γ_k is the cut-off point of the Pareto distribution. The motivation of the cut-off point is that, in practice, there is usually a contractual insured maximum loss. The cut-off point is determined by each insurer individually for each line of business. However, FINMA also proposes non-binding values for γ_k which are based on industry averages. The parameter $\beta_k \in \{10^6\mathrm{CHF}, 5 \cdot 10^6\mathrm{CHF}\}$ represents the chosen split between normal losses and major losses. $\alpha_k > 0$ describes the heaviness of the tail of the Pareto distribution and is prescribed by FINMA for each choice of β_k. The number of major individual losses for each line of business k, $N^{\mathrm{ml},k} \sim \mathrm{Poisson}\left(\lambda_k^{\mathrm{ml}}\right)$, is assumed to be Poisson distributed. The Poisson parameter λ_k^{ml} is determined by each insurer individually for each line of business.

The model of cumulative losses for both hail and accident insurance is described in detail in Sects. 4.4.8.1 and 4.4.8.2 of [4]. The assumptions for the distributions are analogous to those for the distributions describing individual major losses. However, the parameters are determined differently. In a first step, a loss distribution is prescribed for the industry as a whole and in a second step, this distribution is scaled down for each insurer.

Example 4.12 We assume that the index $k = H$ corresponds to cumulative losses due to hail. The individual losses that form one cumulative loss are usually of the same order of magnitude as normal losses, they just are caused by a single hail storm. In this particular case, FINMA prescribes $\alpha_H = 1.85$ and $\gamma_H = 1.5 \times 10^9$ CHF. The cumulative loss of the whole insurance industry is given by the distribution that is parametrized by $\lambda_H^{\mathrm{ml,\ Markt}} = 0.9$ and $\beta_H^{\mathrm{Market}} = 45 \times 10^6$ CHF. The parameter $\beta_K^{\mathrm{Market}}$ is a pure normalization factor which enables FINMA to provide only one set of parameters $\alpha_H, \lambda_H, \gamma_H$. However, the split between major losses and normal losses has a significant impact on the Poisson parameter.

Assume that an insurer with market share $m_H \leq 1$ has determined major losses to be those losses which are larger than β_H. Since cumulative hail losses consist of many small individual losses, the insurer will assume the cumulative loss as a major loss if the cumulative loss of the whole market is larger than the adjusted market split $\beta_H^{\text{adj. Market}}$ given by

$$\beta_H^{\text{adj. Market}} = \frac{\beta_H}{m_H}.$$

The individual loss distribution for this adjusted market split is given by

$$F_{Y_j^{\text{ml},H,\text{adj. Market}}}(x) = \begin{cases} 0 & x < \beta_H^{\text{adj. Market}}, \\ 1 - (\frac{\beta_H^{\text{adj. Market}}}{x})^{\alpha_H} & \beta_H^{\text{adj. Market}} \leq x < \gamma_H, \\ 1 & \gamma_H \leq x. \end{cases}$$

Since this distribution does not coincide with the normalized distribution given by FINMA it is to be understood as an approximation. The adjustment of the individual loss distribution necessitates an adjustment $\lambda_H^{\text{ml, adj. Market}}$ of the Poisson parameter $\lambda_H^{\text{ml, Market}}$. We assume that the expected value for losses larger than the original market split β_H^{Market} are not altered by the adjustment of the split. Let

$$Y_{j,\text{cut below}}^{\text{ml},H,\text{adj. Market}} = 1_{\{Y_j^{\text{ml},H,\text{adj. Markt}} \geq \beta_K^{\text{Markt}}\}} Y_j^{\text{ml},H,\text{adj. Market}},$$

$$N^{\text{ml},H,\text{Markt}} \sim \text{Poisson}\left(\lambda_H^{\text{ml, Markt}}\right), \qquad N^{\text{ml},H} \sim \text{Poisson}\left(\lambda_H^{\text{ml, adj. Market}}\right),$$

and

$$S_t^{\text{CY},\text{ml},H,\text{Market}} = \sum_{j=1}^{N^{\text{ml},H,\text{Market}}} Y_j^{\text{ml},H,\text{Market}},$$

$$S_{t,\text{cut below}}^{\text{CY},\text{ml},H,\text{adj. Market}} = \sum_{j=1}^{N^{\text{ml},H}} Y_{j,\text{cut below}}^{\text{ml},H,\text{adj. Market}}.$$

Using this notation our condition can be written as

$$\mathrm{E}\big(S_t^{\text{CY},\text{ml},H,\text{Market}}\big) \stackrel{!}{=} \mathrm{E}\big(S_{t,\text{cut below}}^{\text{CY},\text{ml},H,\text{adj. Market}}\big).$$

This equation is equivalent to

$$\begin{aligned} 0 &= \mathrm{E}\big(S_t^{\text{CY},\text{ml},H,Market}\big) - \mathrm{E}\big(S_{t,\text{cut below}}^{\text{CY},\text{ml},H,\text{adj. Market}}\big) \\ &= \lambda_H^{\text{ml, Markt}} \mathrm{E}\big(Y_1^{\text{ml},H,\text{total}}\big) - \lambda_H^{\text{ml, adj. Market}} \mathrm{E}\big(Y_{1,\text{cut below}}^{\text{ml},H,\text{adj. Market}}\big) \end{aligned}$$

$$= \lambda_H^{\text{ml, Markt}} \int_{\beta_H^{\text{Markt}}}^{\gamma_H} x\alpha_H \frac{(\beta_H^{\text{Market}})^{\alpha_H}}{x^{\alpha_H+1}} \, \mathrm{d}x$$

$$- \lambda_H^{\text{ml, adj. Market}} \int_{\beta_H^{\text{Markt}}}^{\gamma_H} x\alpha_H \frac{(\beta_H^{\text{adj. Market}})^{\alpha_H}}{x^{\alpha_H+1}} \, \mathrm{d}x$$

$$= \lambda_H^{\text{ml, Market}} \frac{\alpha_H}{\alpha_H - 1} (\beta_H^{\text{Market}})^{\alpha_H} \left((\beta_H^{\text{Market}})^{-\alpha_H+1} - \gamma_H^{-\alpha_H+1} \right)$$

$$- \lambda_H^{\text{ml, adj. Market}} \frac{\alpha_H}{\alpha_H - 1} (\beta_H^{\text{adj. Market}})^{\alpha_H} \left((\beta_H^{\text{Market}})^{-\alpha_H+1} - \gamma_H^{-\alpha_H+1} \right),$$

which implies

$$\lambda_H^{\text{ml, adj. Market}} = \lambda_H^{\text{ml, Market}} \left(\frac{\beta_H^{\text{Market}}}{\beta_H^{\text{adj. Market}}} \right)^{\alpha_H}.$$

Taking into account the market share m_H of the insurer we obtain the compound Poisson distribution

$$S_t^{\text{CY,ml},H} = \sum_{j=1}^{N^{\text{ml},H}} m_H Y_j^{\text{ml},H,\text{adj. Market}} \quad \text{with } N^{\text{ml},H} \sim \text{Poisson}\left(\lambda_H^{\text{ml, adj. Market}}\right).$$

For any independent random variables S_1, S_2 with compound Poisson distribution F_1, F_2, independent individual loss distributions G_1, G_2, Poisson parameters λ_1, λ_2 the sum $S_1 + S_2$ has also a compound Poisson distribution. Its individual loss distribution is given by $G = (\lambda_1 G_1 + \lambda_2 G_2)/(\lambda_1 + \lambda_2)$ and its Poisson parameter is given by $\lambda_1 + \lambda_2$ (see for example [10, Proposition 10.9]). It follows that

$$N^{\text{ml}} \sim \text{Poisson}\left(\sum_k \lambda_k^{\text{ml}}\right), \qquad F_{Y_1^{\text{ml}}} = \frac{1}{\sum_k \lambda_k^{\text{ml}}} \sum_k \lambda_k^{\text{ml}} F_{Y_1^{\text{ml},k}}$$

represents the total distribution for major losses. The distribution $F_{Y_1^{\text{ml}}}$ for total individual losses can be calculated approximatively using the Panjer algorithm (see [10, Theorem 10.15]).

Calculation of $T_{\text{exp ins}} + T_{\text{risk ins}}$ (Part 2): The Model for the Run-off Result of PY-Losses The run-off result for PY-losses is modeled through a lognormally distributed random variable c_t^{PY} with $\mathrm{E}(c_t^{\text{PY}}) = 1$. As in the case of CY-losses, the variance stems from both parameter risk and random fluctuation risk. Volatility of PY-losses is often estimated using the Chain Ladder Method due to Mack [9]. However, Mack estimates the volatility of the final loss which includes the risks of future years whereas we are interested in the volatility of the run-off result due to risks in the current year only. Hence a direct application of Mack's method would overestimate the volatility of c_t^{PY}.

FINMA provides the following alternative for estimating the volatility of parameter risk and random fluctuation risk. Parameter risk contains both

- the risk that the industry-wide consensus regarding risk inherent in the line of business changes
- and the risk that company-specific risk estimates change.

An example of the former would be the change in the appraisal of asbestos risk during the last 50 years, while an example of the latter would be an unknown bias in data used by the specific company. As a consequence, parameter risk is hard to estimate. FINMA addresses this problem through providing a normalized variation coefficient vc_k^P for each line of business:

$$\mathrm{var}_P\Big(\sum_{\tau\leq t-1} Z^{k,\tau}\Big) = \Big(\mathrm{vc}_k^P \sum_{\tau\leq t-1} r^{\mathrm{PY},k,\tau,[t-1]}\Big)^2,$$

where:

- var_P is the variance with respect to the parameter risk,
- $Z^{k,\tau}$ are the non-discounted payments after time $t-1$ for losses that occur during the accident year $\tau \leq t-1$ in line of business k,
- $r_t^{\mathrm{PY},k,t,[t-1]}$ is the expected value of $Z^{k,\tau}$ at time $t-1$.

The variation coefficient vc_k^F of the random fluctuation risk can be determined directly from company-specific time series, provided there exist sufficient data. FINMA provides an upper bound that depends on the expected value of future loss payments and which can be used if an upper bound for the maximum loss is known [5]:

Lemma 4.5 *Let $\tau \in \mathrm{PY}$ be an accident year before t, and assume that for the line of business k there is a constant $M^{k,\tau}$ such that each loss in k during τ is bounded by $M^{k,\tau}$. Assume further that all losses stemming from this accident year are independent and that such future losses are compound Poisson distributed.*

Then the future non-discounted payments $Z^{k,\tau}$ satisfy

$$\mathrm{var}_F\big(Z^{k,\tau}\big) \leq M^{k,\tau} r^{\mathrm{PY},k,\tau,[t-1]},$$

where var_F is the variance with respect to random fluctuation risk and $r^{\mathrm{PY},k,\tau,[t-1]}$ is the expected value for all future payments.

Proof Assume that at time $t-1$ it is known that there are at least J losses which happened in accident year τ in line of business k. We denote with $X_j^{k,\tau}$ the random variable which describes the non-discounted total amount of the jth such loss ($j \in \{1,\ldots,J\}$), where we are taking into account the whole run-off of the portfolio. Let $b_j^{k,\tau} \leq X_j^{k,\tau}$ be the sum of all non-discounted payments which have already been made for this loss by time $t-1$. There may also be losses from accident year τ in line of business k which are not yet known to the company. Let N be the random variable which describes the number of these losses and $(Y_i^{k,\tau})_{i\in\{1,\ldots,N\}}$ the corresponding total loss amounts. Then the future non-discounted loss payments for losses from accident year τ in line of business k is given by

$$Z^{k,\tau} = \sum_{j=1}^{J} (X_j^{k,\tau} - b_j^{k,\tau}) + \sum_{i=1}^{N} Y_i^{k,\tau}.$$

By assumption the random variables $Y_i^{k,\tau}$ are identically distributed and N is Poisson distributed. Also by assumption, the random variables

$$(X_j^{k,\tau})_{j\in\{1,\dots,J\}}, \quad (Y_i^{k,\tau})_{i\in\{1,\dots,N\}}, \quad N$$

are independent. Hence we obtain

$$\begin{aligned}
\mathrm{var}(Z^{k,\tau}) &= \sum_{j=1}^{J} \mathrm{var}(X_j^{k,\tau} - b_j^{k,\tau}) + \mathrm{var}\left(\sum_{i=1}^{N} Y_i^{k,\tau}\right) \\
&= \sum_{j=1}^{J} \mathrm{E}((X_j^{k,\tau} - b_j^{k,\tau})^2) \overbrace{- \sum_{j=1}^{J} \mathrm{E}(X_j^{k,\tau} - b_j^{k,\tau})^2}^{\leq 0} \\
&\quad + \mathrm{E}(Y_1^{k,\tau})^2 \overbrace{\mathrm{var}(N)}^{=\mathrm{E}(N)} + \mathrm{var}(Y_1^{k,\tau})\,\mathrm{E}(N) \\
&\leq \sum_{j=1}^{J} \mathrm{E}(\overbrace{(M^{k,\tau} - b_j^{k,\tau})}^{\leq M^{k,\tau}}(X_j^{k,\tau} - b_j^{k,\tau})) + \mathrm{E}(\overbrace{(Y_1^{k,\tau})^2}^{\leq M^{k,\tau} Y_1^{k,\tau}})\,\mathrm{E}(N) \\
&\leq M^{k,\tau} \sum_{j=1}^{J} \mathrm{E}(X_j^{k,\tau} - b_j^{k,\tau}) + M^{k,\tau}\,\mathrm{E}(Y_1^{k,\tau})\,\mathrm{E}(N) = M^{k,\tau} r_t^{\mathrm{PY},k,\tau},
\end{aligned}$$

where in the last equation we have used

$$r_t^{\mathrm{PY},k,\tau,[t-1]} = \mathrm{E}(Z^{k,\tau}) = \sum_{j=1}^{J} \mathrm{E}(X_j^{k,\tau}) - \sum_{j=1}^{J} b_j^{k,\tau} + \mathrm{E}(N)\,\mathrm{E}(Y_1^{k,\tau}).$$

□

In the context of the SST insurers may use the estimate in Lemma 4.5 as an approximation,

$$\mathrm{var}_F(Z^{k,\tau}) \approx M^{k,\tau} r^{\mathrm{PY},k,\tau,[t-1]}. \tag{4.14}$$

Proposition 4.3 *Assume that random fluctuation risk, parameter risk, and losses from former accident years are all independent. Assume further that*

$\sum_k \sum_{\tau \le t-1} Z^{k,\tau}$ *is lognormally distributed. Then, using the approximation* (4.14), *the run-off result of* PY*-losses is given by*

$$\sum_k \sum_{\tau \le t-1} Z^{k,\tau} = c_t^{\mathrm{PY}} R_{\mathrm{PY}}^{[t-1]},$$

where $R_{\mathrm{PY}}^{[t-1]}$ *is the non-discounted loss reserve at time* $t-1$, c_t^{PY} *is a lognormally distributed random variable with* $\mathrm{E}(c_t^{\mathrm{PY}}) = 1$ *and*

$$\mathrm{var}(c_t^{\mathrm{PY}}) = \frac{\sum_k (\mathrm{vc}_k^P \sum_{\tau \le t-1} r^{\mathrm{PY},k,\tau,[t-1]})^2 + \sum_k \sum_{\tau \le t-1} M^{k,\tau} r_t^{\mathrm{PY},k,\tau}}{(\sum_k \sum_{\tau \le t-1} r_t^{\mathrm{PY},k,\tau})^2}.$$

Proof Clearly we have $\mathrm{E}(\sum_k \sum_{\tau \le t-1} Z^{k,\tau}) = R_{\mathrm{PY}}^{[t-1]}$, which implies $\mathrm{E}(c_t^{\mathrm{PY}}) = 1$. We also calculate

$$\begin{aligned}\mathrm{var}\big(c_t^{\mathrm{PY}} R_{\mathrm{PY}}^{[t-1]}\big) &= \sum_k \sum_{\tau \le t-1} \mathrm{var}\big(Z^{k,\tau}\big) \\ &= \sum_k \sum_{\tau \le t-1} \mathrm{var}_P\big(Z^{k,\tau}\big) + \sum_k \sum_{\tau \le t-1} \mathrm{var}_F\big(Z^{k,\tau}\big) \\ &= \sum_k \Bigg(\mathrm{vc}_k^P \sum_{\tau \le t-1} r^{\mathrm{PY},k,\tau,[t-1]}\Bigg)^2 + \sum_k \sum_{\tau \le t-1} M^{k,t} r_t^{\mathrm{PY},k,\tau}.\end{aligned}$$

Now the assertion follows from $R_{\mathrm{PY}}^{[t-1]} = \sum_k \sum_{\tau \le t-1} r_t^{\mathrm{PY},k,\tau}$. □

Calculation of $T_{\text{exp ins}} + T_{\text{risk ins}}$ (Part 3): Aggregation of Risks The insurance risk is given by

$$\begin{aligned}T_{\text{exp ins}} + T_{\text{risk ins}} &= P_t - K_t - d_{\mathrm{CY}}^{[t-1]} \mathrm{E}\big(S_t^{\mathrm{CY}}\big) \\ &\quad - d_{\mathrm{CY}}^{[t-1]}\big(S_t^{\mathrm{CY}} - \mathrm{E}\big(S_t^{\mathrm{CY}}\big)\big) - d_{\mathrm{PY}}^{[t-1]}\big(c_t^{\mathrm{PY}} - 1\big) R_{\mathrm{PY}}^{[t-1]} \\ &= P_t - K_t - d_{\mathrm{CY}}^{[t-1]} \mathrm{E}\big(S_t^{\mathrm{CY}}\big) - d_{\mathrm{CY}}^{[t-1]}\big(S_t^{\mathrm{CY,nl}} - \mathrm{E}\big(S_t^{\mathrm{CY,nl}}\big)\big) \\ &\quad - d_{\mathrm{CY}}^{[t-1]}\big(S_t^{\mathrm{CY,ml}} - \mathrm{E}\big(S_t^{\mathrm{CY,ml}}\big)\big) - d_{\mathrm{PY}}^{[t-1]}\big(c_t^{\mathrm{PY}} - 1\big) R_{\mathrm{PY}}^{[t-1]}.\end{aligned}$$

Apart from deterministic summands and a sign, the insurance risk is the sum of three random variables which have been determined in the previous sections,

- distribution of CY normal losses: $d_{\mathrm{CY}}^{[t-1]} S_t^{\mathrm{CY,nl}}$ (lognormal),
- distribution of CY mayor losses: $d_{\mathrm{CY}}^{[t-1]} S_t^{\mathrm{CY,ml}}$ (compound Poisson),
- distribution of the run-off result of PY-losses: $d_{\mathrm{PY}}^{[t-1]} c_t^{\mathrm{PY}}$ (lognormal).

It is assumed that all three distributions are independent. Then the distribution $\hat{F}_0$ of their sum can be computed through convolution.

Remark 4.12 Sect. 4.4.11 in [4] proposes as an alternative that both lognormal distributions are combined to yield an approximative lognormal distribution. This approximate approach would save one convolution.

Calculation of $T_{\text{exp invest}} + T_{\text{risk invest}}$: Market Risk and ALM Risk The sum $T_{\text{exp invest}} + T_{\text{risk invest}}$ is set to be equal to the result from Example 4.10.

Combining Insurance Risk and Investment Risk In order to obtain the distribution $\tilde{F}_0$ we need to calculate the convolution of $\hat{F}_0$ with the distribution of

$$T_{\text{exp invest}} + T_{\text{risk invest}}.$$

To do so we assume that investment risks and insurance risks are independent. We also neglect the error term T_{error}.

4.6.2 The Standard Model in Solvency II

In this section we will describe the model in the fifth Quantitative Impact Study (for short, QIS5) as a place-holder for the future Standard Model of Solvency II. In doing so we follow the Technical Document [7] of the EU Commission, which can be found online on the EU web site along with other documentation.

4.6.2.1 Fundamentals About the Risk Capital SCR

The risk capital $C^{\text{Reg}} = \text{SCR}$ (solvency capital requirement) is defined based on an economic balance sheet, where the assets are booked at market prices and the liabilities are valued with a "best estimate" plus a risk margin. In this approach, equity, the difference between assets and liabilities, has a direct economic interpretation and is termed the "net asset value" (NAV). Profits and losses are reflected by changes in the net asset value. The SCR is calibrated so that there is a high probability that an insurance company, which has enough available capital to cover the SCR, will be able to absorb all losses that may occur during the year under consideration. It is understood that account is taken of all quantifiable risks (e.g, market variations, insurance losses) to which the company is exposed. In QIS5 the risk measure, "value at risk" (VaR) has been chosen, where the safety level or confidence level is set to 99.5 %.

From the perspective of Solvency II the technical provisions are the sum of the present value of future liabilities and a market value margin (MVM). The risk-bearing capital consists of the assets that exceed these provisions, and it is composed of the Solvency Capital Requirement (SCR) and the free assets. While the SCR is the capital necessary to settle potential company losses within a 1-year horizon, the MVM serves to ensure the transferability of the insurance portfolio after the SCR

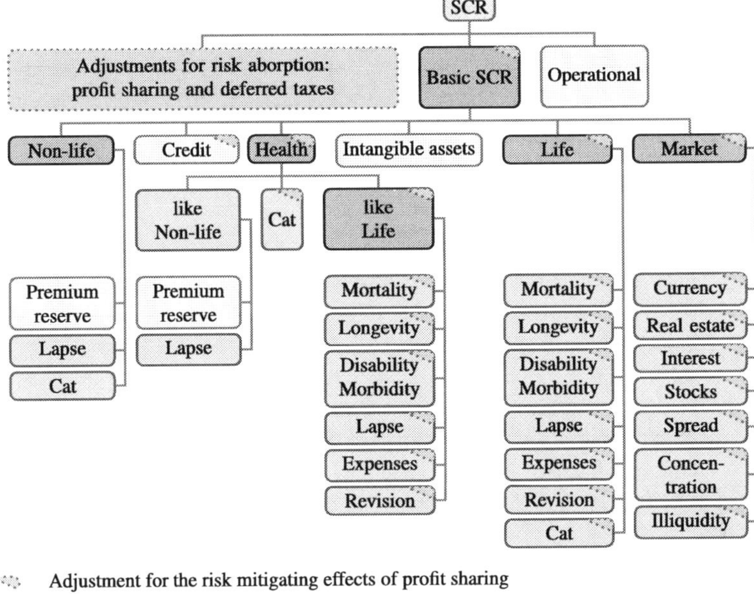

Fig. 4.8 Modular construction of the Standard Model for calculating the SCR (Source: [7])

has been used up. The MVM is determined as the risk-free discounted present value of the costs that an investor who takes over the portfolio would incur for funding the SCR during the run-off of the portfolio.

4.6.2.2 Structure of the SCR

Calculation of the SCR follows a modular construction as shown in Fig. 4.8.

The aggregation of all risk but operational risk results in the Basic SCR (BSCR). Both the SCR for operational risk (SCR_{Op}) and a (negative) adjustment Adj for the effects from deferred taxes and profit sharing are then added to the BSCR:

$$SCR = BSCR + SCR_{Op} + Adj. \tag{4.15}$$

The BSCR is divided into six modules (risk classes) SCR_{Life}, SCR_{NL}, SCR_{Health}, SCR_{Mkt} SCR_{Def}, SCR_{Intang}, most of which are composed of several submodules (risk types):

1. *Life underwriting risk* (SCR_{Life}).
   (a) Lapse risk (SCR_{Life}^{Lapse})
   (b) Expense risk (SCR_{Life}^{Exp})
   (c) Disability/morbidity risk ($SCR_{Life}^{Dis,Morb}$)
   (d) Mortality risk (SCR_{Life}^{Mort})
   (e) Longevity risk (SCR_{Life}^{Long})
   (f) Revision risk (SCR_{Life}^{Rev})
   (g) Catastrophe risk (SCR_{Life}^{Cat})
2. *Non-life underwriting risk* (SCR_{NL})
   (a) Premium and reserve risks ($SCR_{NL}^{Prem,Res}$)
   (b) Lapse risk (SCR_{NL}^{Lapse})
   (c) Catastrophe risk (SCR_{NL}^{Cat})
3. *Health underwriting risk* (SCR_{Health}). For this risk there is a choice of two hierarchies of submodules, SCR_{Health}^{SLT} for health insurance, which can be treated similarly to life insurance (e.g. in Germany and Austria), and SCR_{Health}^{NSLT} for health insurance, that can be treated similarly to non-life insurance. In addition, there is a health-specific catastrophe module SCR_{Health}^{Cat} for both types of health insurance.

   The module SCR_{Health}^{SLT} is further split into 6 submodules, analogues of the submodules of SCR_{Life}.
   (a) Lapse risk (SCR_{Health}^{Lapse})
   (b) Expense risk (SCR_{Health}^{Exp})
   (c) Disability/morbidity risk ($SCR_{Health}^{Dis,Morb}$)
   (d) Mortality risk (SCR_{Health}^{Mort})
   (e) Longevity risk (SCR_{Health}^{Long})
   (f) Revision risk (SCR_{Health}^{Rev})

   The module SCR_{Health}^{NSLT} has two submodules analogous to the submodules of SCR_{NL}.
   (a) Premium and reserve risk ($SCR_{Health}^{Prem,Res}$)
   (b) Lapse risk (SCR_{Health}^{Lapse})
4. *Default risk* (SCR_{Def}). Credit risk is split into two modules. The module SCR_{Def} measures the risk of default, for example of bonds or risk mitigating instru-

ments (reinsurance, derivatives, letters of credit, etc.). The submodule spread risk ($\mathrm{SCR}_{\mathrm{Mkt}}^{\mathrm{Spread}}$) of the market risk module $\mathrm{SCR}_{\mathrm{Mkt}}$ addresses the loss in value of fixed income instruments which may result from a change in the associated credit spreads.

5. *Market risk (*$\mathrm{SCR}_{\mathrm{Mkt}}$*).* Market risk comprehends all price risks in the capital market along with the concentration risk for the insurer's investment portfolio.

    (a) Foreign currency risk ($\mathrm{SCR}_{\mathrm{Mkt}}^{\mathrm{FX}}$)
    (b) Real estate risk ($\mathrm{SCR}_{\mathrm{Mkt}}^{\mathrm{Property}}$)
    (c) Interest rate risk ($\mathrm{SCR}_{\mathrm{Mkt}}^{\mathrm{Interest}}$)
    (d) Stock risk ($\mathrm{SCR}_{\mathrm{Mkt}}^{\mathrm{Equity}}$)
    (e) Spread risk ($\mathrm{SCR}_{\mathrm{Mkt}}^{\mathrm{Spread}}$): see also the module Default risk ($\mathrm{SCR}_{\mathrm{Def}}$)
    (f) Concentration risk ($\mathrm{SCR}_{\mathrm{Mkt}}^{\mathrm{Conc}}$)
    (g) Illiquidity risk ($\mathrm{SCR}_{\mathrm{Mkt}}^{\mathrm{Illiquidity}}$): This risk arises from the possibility that the illiquidity premium drops, which would lead to an increase in the technical provisions.

6. *Intangible asset risk (*$\mathrm{SCR}_{\mathrm{Intang}}$*).* Some intangible assets can be used as capital to cover the SCR. The $\mathrm{SCR}_{\mathrm{Intang}}$ corresponds to the risk that these intangible assets cannot be converted in times of need into material assets.

The various capital requirements are aggregated using a two-step variance/co-variance method, where the corresponding correlation matrices are predetermined. In a first step the capital requirements for each risk type in a module are aggregated. Subsequently the capital requirements per module are aggregated resulting in the Basic SCR.

The summand

$$\mathrm{Adj} = \mathrm{Adj}_{\mathrm{TP}} + \mathrm{Adj}_{\mathrm{DT}}$$

in (4.15) accounts for the risk minimizing effect of future profit sharing and deferred taxes.

The future discretionary benefit (FDB) resulting from the expected amount of future profit sharing is included in the valuation of technical provisions in Solvency II (hence the index "TP" for "technical provisions"). In case of losses due to adverse events the insurance company could reduce future profit sharing in order to cover part of these losses. The corresponding change in FDB must therefore be integrated in the calculation of the SCR. The risk mitigation of future profit sharing is captured in the SCR calculation by a parallel calculation.[18] $\mathrm{BSCR}^{\mathrm{Adj}}$ should be calculated analogously to BSCR, where, in contrast to the BSCR calculation,

[18] We are here describing the "modular method". QIS5 also describes an alternative method using so-called "equivalent scenarios".

account is taken of the fact that management can adjust profit sharing to the company's circumstances. Figure 4.8 shows the corresponding modules ("Adjustment for risk absorption: profit sharing and deferred taxes"). In this calculation the same correlation matrices are used as for BSCR, and we obtain

$$\mathrm{Adj}_{\mathrm{TP}} = -\min\bigl(\mathrm{BSCR} - \mathrm{BSCR}^{\mathrm{Adj}}, \mathrm{FDB}\bigr).$$

The adjustment for deferred taxes, $\mathrm{Adj}_{\mathrm{Tax}}$ is the change in value of deferred taxes that would result from a loss in the amount of

$$\mathrm{BSCR} + \mathrm{SCR}_{\mathrm{Op}} + \mathrm{Adj}_{\mathrm{TP}}.$$

Remark 4.13 In calculating BSCR, some risks are aggregated in several-steps using a square-root formula.

Let X^A and X^B be random variables for risks, each made up of the sum of individual risks, $X^A = \sum_{i=1}^a X_i^A$ and $X^B = \sum_{k=1}^b X_k^B$. Let SCR_A^i (resp. SCR_B^k) be the SCR for the individual risks X_i^A (resp. X_k^B) and corr^A, corr^B the corresponding correlation matrices. Then the square-root formula gives for the capital requirements of the risks X^A, X^B the values

$$\mathrm{SCR}_A = \sqrt{\sum_{i,j=1}^a \mathrm{corr}_{ij}^A \mathrm{SCR}_A^i \mathrm{SCR}_A^j}, \qquad \mathrm{SCR}_B = \sqrt{\sum_{k,l=1}^a \mathrm{corr}_{kl}^B \mathrm{SCR}_B^k \mathrm{SCR}_B^l}.$$

In order to calculate the capital requirement SCR_{A+B} of the combined risk $X^A + X^B$, a correlation coefficient corr_{AB} for X^A and X^B is provided. This leads to the double square-root formula

$$\begin{aligned}
\mathrm{SCR}_{A+B} &= \sqrt{(\mathrm{SCR}_A)^2 + 2\mathrm{corr}_{AB}\mathrm{SCR}_A\mathrm{SCR}_B + (\mathrm{SCR}_B)^2} \\
&= \Biggl[\sum_{i,j=1}^a \mathrm{corr}_{ij}^A \mathrm{SCR}_A^i \mathrm{SCR}_A^j + \sum_{k,l=1}^a \mathrm{corr}_{kl}^B \mathrm{SCR}_B^k \mathrm{SCR}_B^l \\
&\quad + 2\mathrm{corr}_{AB}\sqrt{\sum_{i,j=1}^a \mathrm{corr}_{ij}^A \mathrm{SCR}_A^i \mathrm{SCR}_A^j}\sqrt{\sum_{k,l=1}^a \mathrm{corr}_{kl}^B \mathrm{SCR}_B^k \mathrm{SCR}_B^l}\Biggr]^{\frac{1}{2}}.
\end{aligned}$$

One could alternatively do the aggregation in one step if one used a correlation matrix corr given in block-matrix form as

$$\mathrm{corr} = \begin{pmatrix} \mathrm{corr}^A & \mathrm{corr}^{AB} \\ (\mathrm{corr}^{AB})^\top & \mathrm{corr}^B \end{pmatrix}$$

with

$$\mathrm{corr}_{ik}^{AB} = \mathrm{corr}\bigl(X_i^A, X_k^B\bigr).$$

The required capital would then be

$$\widehat{\mathrm{SCR}}_{A+B} = \left[\sum_{i,j=1}^{a} \mathrm{corr}_{ij}^{A} \mathrm{SCR}_A^i \mathrm{SCR}_A^j + \sum_{k,l=1}^{a} \mathrm{corr}_{kl}^{B} \mathrm{SCR}_B^k \mathrm{SCR}_B^l \right.$$

$$\left. + 2 \sum_{i=1}^{a} \sum_{k=1}^{b} \mathrm{corr}_{ik}^{AB} \mathrm{SCR}_A^j \mathrm{SCR}_B^k \right]^{\frac{1}{2}}.$$

Naturally the matrix corr^{AB} gives more information than the previously used correlation coefficient corr_{AB}. The coefficient corr_{AB} can be understood as a sort of average correlation of the risks X_i^A and X_k^B. However, if one puts $\mathrm{corr}_{ik}^{AB} = \mathrm{corr}_{AB}$ for all i, k, it follows that

$$\mathrm{SCR}_{A+B} \leq \widehat{\mathrm{SCR}}_{A+B},$$

as long as SCR_A^i and SCR_B^k are positive and $\mathrm{corr}_{AB} > 0$ holds.

The assertion follows from the equation

$$\begin{aligned}
&\frac{(\widehat{\mathrm{SCR}}_{A+B})^2 - (\mathrm{SCR}_{A+B})^2}{2\mathrm{corr}_{AB}} \\
&= \sum_{i=1}^{a} \sum_{k=1}^{b} \mathrm{SCR}_A^j \mathrm{SCR}_B^k - \sqrt{\sum_{i,j=1}^{a} \mathrm{corr}_{ij}^{A} \mathrm{SCR}_A^i \mathrm{SCR}_A^j} \sqrt{\sum_{k,l=1}^{a} \mathrm{corr}_{kl}^{B} \mathrm{SCR}_B^k \mathrm{SCR}_B^l} \\
&\geq \sum_{i=1}^{a} \sum_{k=1}^{b} \mathrm{SCR}_A^j \mathrm{SCR}_B^k - \sqrt{\sum_{i,j=1}^{a} \mathrm{SCR}_A^i \mathrm{SCR}_A^j} \sqrt{\sum_{k,l=1}^{a} \mathrm{SCR}_B^k \mathrm{SCR}_B^l} \\
&= \sum_{i=1}^{a} \mathrm{SCR}_A^j \sum_{k=1}^{b} \mathrm{SCR}_B^k - \sqrt{\left(\sum_{i=1}^{a} \mathrm{SCR}_A^i\right)^2} \sqrt{\left(\sum_{k=1}^{a} \mathrm{SCR}_B^k\right)^2} \\
&= 0,
\end{aligned}$$

where we have used that the correlation coefficients corr_{ij}^{A} and corr_{kl}^{B} are bounded above by 1.

4.6.2.3 Scenario-Based Modules

Most actuarial risks in non-life insurance and default risk are evaluated with factor-based or analytic methods. For most market risks and actuarial risks in life insurance scenario-based calculation methods are applied.

In a scenario-based module the risk capital is estimated as the amount of the loss on the insurer's balance sheet that would follow should a specific shock occur. In other words, this loss is given by the change $\Delta\mathrm{NAV}$ in the economic equity ("net asset value"), i.e., the difference between the asset values and the technical provisions.

Example 4.13 To quantify the interest rate risk $\mathrm{SCR}_{\mathrm{Mkt}}^{\mathrm{Interest}}$ one considers two scenarios, a downward interest rate shock and an upward interest rate shock. In each scenario one obtains a resulting net asset value NAV which is compared with the corresponding best estimate. The difference $\Delta\mathrm{NAV} = \mathrm{NAV}_{\mathrm{shock}} - \mathrm{NAV}_{\mathrm{BE}}$ is calculated both with and without taking into account risk mitigation effects from future discretionary benefits (FDB). The capital requirement $\mathrm{SCR}_{\mathrm{Mkt}}^{\mathrm{Interest}}$ is then the value of $\Delta\mathrm{NAV}$ in the unfavorable interest rate scenario, where risk mitigation effects have been taken into account up to the current value of future discretionary benefits.

We consider a portfolio of pure saving products, such that the guaranteed liabilities could be replicated completely by the risk-free zero bond portfolio $\{D_t\}_{t\in\{1,\dots,8\}}$, where t denotes the duration of the bond with nominal value D_t. The assets of the company are invested in a portfolio of risk-free zero bonds $\{B_t\}_{t\in\{1,\dots,4\}}$, where B_t is the nominal value corresponding to the remaining duration t. (See Table 4.1.)

The company uses an interest rate of $i = 0.5\ \%$ for the calculation of its statutory provisions, and the guaranteed interest is also 0.5 %. Hence the statutory provisions at the *beginning* of period t are given by

$$V_{\mathrm{stat},t} = \sum_{\tau=t}^{8} \frac{D_\tau}{(1+i)^{\tau-t+1}}.$$

The company has the policy of giving a bonus to its policy holders. At the end of each year it declares a bonus rate b_t that, applied to the statutory provisions $V_{\mathrm{stat},t}$, yields the bonus

$$\mathrm{bonus}_t = b_t V_{\mathrm{stat},t}.$$

This bonus is directly paid in cash to the policy holders.[19] The bonus rate is chosen based on the actual asset management performance of the company during the year under consideration.

Let s_t be risk-free interest rate for the period $[t-1,t[$ and s_t^{BE} be its best estimate at time 0 (see Table 4.1). In order to obtain the best estimate for future bonus declarations we note that on the one hand the assets allocated to the liabilities need to cover the statutory provisions at each time t. On the other hand, the company would strive to allocate as few assets to the liabilities as possible. This implies

$$\text{value of assets at the beginning of period } [t-1,t[\ \approx V_{\mathrm{stat},t}$$

and

$$\text{return on assets during } [t-1,t[\ \approx s_t V_{\mathrm{stat},t}.$$

[19]Such a cash based bonus system simplifies our discussion but is rarely used in practice.

Table 4.1 s_t^{BE} is the best estimate of the risk free interest for the period $[t-1,t[$ based on information at time 0. The relative shocks for this interest rate curve are given by δ_{UP} and δ_{DOWN}. δ_{UP} denotes a relative interest increase and δ_{DOWN} denotes a relative decline. B_t and D_t denote nominal values corresponding to the remaining duration t

t	s_t^{BE}	B_t	D_t	δ_t^{UP}	δ_t^{DOWN}
1	0.5 %	1200	500	70 %	−75 %
2	1.0 %	500	450	70 %	−65 %
3	1.2 %	300	300	64 %	−56 %
4	1.5 %	100	250	59 %	−50 %
5	2.0 %	0	200	55 %	−46 %
6	2.2 %	0	150	52 %	−42 %
7	2.3 %	0	100	49 %	−39 %
8	2.3 %	0	50	47 %	−36 %

For simplicity we have neglected the duration structure of the asset portfolio in the latter approximation. In accordance with the company's policy for generating profits and taking into account its own costs for asset management, it will only allocate 90 % of the investment profits to its policy holders. Deducting the guaranteed interest rate, we obtain as best estimate for future bonus declarations

$$\text{E}(b_t) = \max\big(0, 90\,\%\, s_t^{\text{BE}} - i\big).$$

Observe that the guaranteed interest rate is already included in the nominal value D_t. Thus the expected value of the assets is

$$A_{\text{BE}} = \sum_{t=1}^{8} \frac{B_t}{\prod_{\tau=1}^{t}(1+s_\tau^{\text{BE}})} = 2075,$$

and the expected value of the liabilities is given by

$$V_{\text{BE}} = \sum_{t=1}^{8} \frac{D_t + \text{E}(b_t)V_{\text{stat},t}}{\prod_{\tau=1}^{t}(1+s_\tau^{\text{BE}})} = 1963.$$

As a result the "net asset value" is $\text{NAV}_{\text{BE}} = 111$.[20]

To calculate the SCR two shock scenarios are prescribed, with an instantaneous interest increase and with an instantaneous interest decrease (see Table 4.1).[21] It is an additional requirement that the downward interest shock be at least 1 %, unless

[20] While we display rounded values of intermediate results, we always use the exact values in our calculations.

[21] The values are taken from: QIS5 [7].

the shocked interest rate would then become negative. We obtain

$$s_t^{\mathrm{DOWN}} = \max\big(0, s_t^{\mathrm{BE}} + \min\big(\delta_t^{\mathrm{DOWN}} s_t^{\mathrm{BE}}, -1\ \%\big)\big),$$
$$s_t^{\mathrm{UP}} = \big(1 + \delta_t^{\mathrm{UP}}\big) s_t^{\mathrm{BE}}.$$

In our example this means that the interest decline shock is 1 % for all durations except the first. For the first two durations the resulting interest is 0 %.

If the risk mitigating effect from future discretionary bonus declarations is not accounted for, we get

$$A_{\mathrm{UP}} = \sum_{t=1}^{8} \frac{B_t}{\prod_{\tau=1}^{t}(1 + s_\tau^{\mathrm{UP}})} = 2058,$$

$$V_{\mathrm{UP}} = \sum_{t=1}^{8} \frac{D_t + \mathrm{E}(b_t) V_{\mathrm{stat},t}}{\prod_{\tau=1}^{t}(1 + s_\tau^{\mathrm{UP}})} = 1923,$$

$$A_{\mathrm{DOWN}} = \sum_{t=1}^{8} \frac{B_t}{\prod_{\tau=1}^{t}(1 + s_\tau^{\mathrm{DOWN}})} = 2099,$$

$$V_{\mathrm{DOWN}} = \sum_{t=1}^{8} \frac{D_t + \mathrm{E}(b_t) V_{\mathrm{stat},t}}{\prod_{\tau=1}^{t}(1 + s_\tau^{\mathrm{DOWN}})} = 2015.$$

From this one obtains

$$\mathrm{NAV}_{\mathrm{UP}} = A_{\mathrm{UP}} - V_{\mathrm{UP}} = 135,$$
$$\mathrm{NAV}_{\mathrm{DOWN}} = A_{\mathrm{DOWN}} - V_{\mathrm{DOWN}} = 84,$$
$$\Delta\mathrm{NAV}_{\mathrm{UP}} = -\min\big(0, (\mathrm{NAV}_{\mathrm{UP}} - \mathrm{NAV}_{\mathrm{BE}})\big) = 0,$$
$$\Delta\mathrm{NAV}_{\mathrm{DOWN}} = -\min\big(0, (\mathrm{NAV}_{\mathrm{DOWN}} - \mathrm{NAV}_{\mathrm{BE}})\big) = 28.$$

To obtain $\mathrm{BSCR}^{\mathrm{Adj}}$ the calculation is repeated, but now the risk mitigating effect from future discretionary bonus declarations is accounted for in accord with company policy. Let b_t^{UP} be the best estimate of the bonus declaration under the condition that the upward shock occurs, and let b_t^{DOWN} be the bonus declaration, if a downward shock occurs. Using the same reasoning as above, we obtain

$$b_t^{\mathrm{UP}} = \max\big(0, 90\ \%\, s_t^{\mathrm{UP}} - i\big), \qquad b_t^{\mathrm{DOWN}} = \max\big(0, 90\ \%\, s_t^{\mathrm{DOWN}} - i\big).$$

One gets

$$A_{\mathrm{UP}}^{\mathrm{Adj}} = A_{\mathrm{UP}} = 2058,$$

$$V_{\mathrm{UP}}^{\mathrm{Adj}} = \sum_{t=1}^{8} \frac{D_t + b_t^{\mathrm{UP}} V_{\mathrm{stat},t}}{\prod_{\tau=1}^{t}(1 + s_\tau^{\mathrm{UP}})} = 1958,$$

$$A_{\text{DOWN}}^{\text{Adj}} = A_{\text{DOWN}} = 2099,$$

$$V_{\text{DOWN}}^{\text{Adj}} = \sum_{t=1}^{8} \frac{D_t + b_t^{\text{DOWN}} V_{\text{stat},t}}{\prod_{\tau=1}^{t}(1 + s_\tau^{\text{DOWN}})} = 1988$$

and

$$\text{NAV}_{\text{UP}}^{\text{Adj}} = A_{\text{UP}}^{\text{Adj}} - V_{\text{UP}}^{\text{Adj}} = 99,$$

$$\text{NAV}_{\text{DOWN}}^{\text{Adj}} = A_{\text{DOWN}}^{\text{Adj}} - V_{\text{DOWN}}^{\text{Adj}} = 111,$$

$$\Delta\text{NAV}_{\text{UP}}^{\text{Adj}} = -\min\big(0, \big(\text{NAV}_{\text{UP}}^{\text{Adj}} - \text{NAV}_{\text{BE}}^{\text{Adj}}\big)\big) = 12,$$

$$\Delta\text{NAV}_{\text{DOWN}}^{\text{Adj}} = -\min\big(0, \big(\text{NAV}_{\text{DOWN}}^{\text{Adj}} - \text{NAV}_{\text{BE}}^{\text{Adj}}\big)\big) = 1.$$

These values are now used in calculating the SCR and the risk-mitigated SCR for the interest risk:

$$\text{SCR}_{\text{Mkt}}^{\text{Adj,Interest}} = \max\big(\Delta\text{NAV}_{\text{UP}}^{\text{Adj}}, \Delta\text{NAV}_{\text{DOWN}}^{\text{Adj}}\big) = 12,$$

$$\text{SCR}_{\text{Mkt}}^{\text{Interest}} = \begin{cases} \Delta\text{NAV}_{\text{UP}} & \text{if } \text{SCR}_{\text{Mkt}}^{\text{Adj,Interest}} = \Delta\text{NAV}_{\text{UP}}^{\text{Adj}} > 0 \\ \Delta\text{NAV}_{\text{DOWN}} & \text{if } \text{SCR}_{\text{Mkt}}^{\text{Adj,Interest}} = \Delta\text{NAV}_{\text{DOWN}}^{\text{Adj}} > 0 \\ 0 & \text{otherwise} \end{cases}$$

$$= 0.$$

We assume now that interest rate risk is the only risk affecting the company, and that there are no deferred taxes. Under these assumptions we have

$$\text{BSCR} = \text{SCR}_{\text{Mkt}}^{\text{Interest}}, \qquad \text{BSCR}^{\text{Adj}} = \text{SCR}_{\text{Mkt}}^{\text{Adj,Interest}}, \quad \text{and} \quad \text{SCR} = \text{BSCR} + \text{Adj},$$

where

$$\text{Adj} = -\min\big(\text{BSCR} - \text{BSCR}^{\text{Adj}}, \text{FDB}\big).$$

The future discretionary benefit FDB is the best estimate of the present value of future bonus payments,

$$\text{FDB} = \text{E}\left(\sum_{t=1}^{8} \frac{b_t V_{\text{stat},t}}{\prod_{\tau=1}^{t}(1 + s_\tau)}\right) \approx \sum_{t=1}^{8} \frac{\text{E}(b_t) V_{\text{stat},t}}{\prod_{\tau=1}^{t}(1 + s_\tau^{\text{BE}})} = 31,$$

which implies

$$\text{SCR} = 0 - \min(0 - 12, 31) = 12.$$

4.6.2.4 Non-life Underwriting Risk

Here we describe the non-life module for Solvency II in a somewhat simplified form. The non-life underwriting risk has three components:

- Premium and reserve risk,
- Lapse risk,
- Catastrophe risk.

Premium and Reserve Risk The combination premium and reserve risk includes the following risks:

- Premiums for new contracts are not sufficient to cover the risk.
- During the time horizon more and/or larger losses are sustained than expected.
- The provisions turn out to be insufficient.

The combined premium and reserve risk is $X_{\text{NL}}^{\text{Prem,Res}} = X_{\text{NL}}^{\text{Prem}} + X_{\text{NL}}^{\text{Res}}$. Here the equation

$$X_{\text{NL}}^{\text{Prem}} = \text{Benefits}^{\text{Prem}} + \text{Costs}^{\text{Prem}} - P$$

gives the benefits and costs that are not covered by the corresponding earned premium. The provisions that must be provided are already included in the benefits $\text{Benefits}^{\text{Prem}}$. For those contracts for which a reserve has already been put up, the equation

$$X_{\text{NL}}^{\text{Res}} = \text{Benefits}^{\text{Res}} + \text{Costs}^{\text{Res}} + \text{Res}^{\text{Year-end}} - \text{Res}$$

gives the portion of the sum of benefits to be paid, costs, and end-of-year provisions which exceeds the provisions made at the beginning of the year. For this we assume that the amounts are discounted correctly with the risk-free interest rate depending on when they are incurred. We put

$$\hat{X}_{\text{NL}}^{\text{Prem}} = X_{\text{NL}}^{\text{Prem}} + P \quad \text{and} \quad \hat{X}_{\text{NL}}^{\text{Res}} = X_{\text{NL}}^{\text{Res}} + \text{Res}$$

and additionally assume that

$$\hat{X}_{\text{NL}}^{\text{Prem,Res}} = \hat{X}_{\text{NL}}^{\text{Prem}} + \hat{X}_{\text{NL}}^{\text{Res}}$$

can be modeled with a lognormally distributed random variable with mean μ and standard deviation σ,

$$\ln\left(\hat{X}_{\text{NL}}^{\text{Prem,Res}}\right) \sim N\left(\ln\mu - \frac{1}{2}\ln\left(1+\text{vc}^2\right), \ln\left(1+\text{vc}^2\right)\right),$$

where $\text{vc} = \sigma/\mu$ is the coefficient of variation of $\hat{X}_{\text{NL}}^{\text{Prem,Res}}$.

Since $P + \text{Res}$ is taken to be deterministic, it holds that $\text{VaR}_{99.5\,\%}(\hat{X}_{\text{NL}}^{\text{Prem,Res}}) = P + \text{Res} + \text{VaR}_{99.5\,\%}(X_{\text{NL}}^{\text{Prem,Res}})$. By Proposition 2.1 the Solvency II risk capital is then given by

$$\begin{aligned}
\text{SCR}_{\text{NL}}^{\text{Prem,Res}} &= \text{VaR}_{99.5\,\%}\left(X_{\text{NL}}^{\text{Prem,Res}}\right) \\
&= \text{VaR}_{99.5\,\%}\left(\hat{X}_{\text{NL}}^{\text{Prem,Res}}\right) - P - \text{Res}
\end{aligned}$$

Table 4.2 The industry-wide gross coefficients of variation made available by EIOPA [7]

k	Line of business	$\text{vc}(\hat{X}_{\text{NL}}^{\text{Prem,gross},k})$	$\text{vc}(\hat{X}_{\text{NL}}^{\text{Res},k})$
1	Motor vehicle liability	10 %	9.5 %
2	Motor, other classes	7 %	10 %
3	Marine, aviation, transport (MAT)	17 %	14 %
4	Fire and other property damage	10 %	11 %
5	Third-party liability	15 %	11 %
6	Credit and suretyship	21.5 %	19 %
7	Legal expenses	6.5 %	9 %
8	Assistance	5 %	11 %
9	Miscellaneous	13 %	15 %
10	Non-proportional reinsurance—property	17.5 %	20 %
11	Non-proportional reinsurance—casualty	17 %	20 %
12	Non-proportional reinsurance—MAT	16 %	20 %

$$\begin{aligned} &= \exp\left(\ln\mu - \frac{1}{2}\ln\left(1+\text{vc}^2\right) + \sqrt{\ln\left(1+\text{vc}^2\right)}\Phi_{0,1}{}^{-1}(99.5\ \%)\right) \\ &\quad - P - \text{Res} \\ &= \frac{\mu}{\sqrt{1+\text{vc}^2}}\exp\left(\Phi_{0,1}{}^{-1}(99.5\ \%)\sqrt{\ln\left(1+\text{vc}^2\right)}\right) - P - \text{Res}. \end{aligned}$$

In determining the variable μ neither cost margins nor claim ratios play a role. EIOPA[22] [2] refers for this to Article 105 (2) of the Solvency II Directive [1] and interprets this article to the effect that expected costs and expected profits should be neglected. Therefore it follows that

$$P + \text{Res} = \text{E}\left(\hat{X}_{\text{NL}}^{\text{Prem,Res}}\right) = \mu,$$

and the formula for the Solvency II risk capital simplifies to

$$\text{SCR}_{\text{NL}}^{\text{Prem,Res}} = \mu\left(\frac{\exp(\Phi_{0,1}{}^{-1}(99.5\ \%)\sqrt{\ln(1+\text{vc}^2)})}{\sqrt{(1+\text{vc}^2)}} - 1\right).$$

The coefficient of variation vc and the expectation μ are derived as (weighted) means over the individual lines of business of the insurer.[23] The coefficient of variation vc will be derived as the volume-weighted average over the coefficients of variation for the individual lines of business.

[22]On 1 January 2011 the *Committee of European Insurance and Occupational Pensions Supervisors* (CEIOPS) was replaced by the *European Insurance and Occupational Pensions Authority* (EIOPA). In this book we will use the name EIOPA throughout, even when we are considering the period before 1 January 2011.

[23]μ will be determined later in (4.19).

To begin with, we determine the net coefficient of variation $\mathrm{vc}(\hat{X}_{\mathrm{NL}}^{\mathrm{Prem}})$ for the premium risk of line of business k. For this we use the approximation

$$\mathrm{vc}\big(\hat{X}_{\mathrm{NL}}^{\mathrm{Prem},k}\big) \approx \frac{\mathrm{vc}(\tilde{S}_k)}{\mathrm{vc}(\tilde{S}_k^{\mathrm{gross}})}\,\mathrm{vc}\big(\hat{X}_{\mathrm{NL}}^{\mathrm{Prem,gross},k}\big), \tag{4.16}$$

where $\tilde{S}_k^{\mathrm{gross}}$ denote the gross losses (including costs) and $\tilde{S}_k$ the associated net losses (including costs). This approximation is motivated by the fact that the premium risk $\hat{X}_{\mathrm{NL}}^{\mathrm{Prem},k}$ is largely determined by the loss distribution $\tilde{S}$. The gross coefficients of variation $\mathrm{vc}(\hat{X}_{\mathrm{NL}}^{\mathrm{Prem,gross},k})$ are provided by EIOPA (see Table 4.2). The factor

$$\frac{\mathrm{vc}(\tilde{S}_k)}{\mathrm{vc}(\tilde{S}_k^{\mathrm{gross}})}$$

is approximated under the assumption of a simple loss distribution and a simple reinsurance program.

Lemma 4.6 *We suppose that the gross losses $\tilde{S}_k^{\mathrm{gross}}$ of line of business k occurring in year t follow a compound Poisson distribution with the number of losses given by*

$$N_k \sim \mathrm{Poisson}\,(\lambda_k)$$

and a lognormally distributed gross loss amount Y_k^{gross} per incident. We also assume that the net losses $\tilde{S}_k$ follow a compound Poisson distribution with the same number of losses. Then we have

$$\mathrm{vc}\big(\hat{X}_{\mathrm{NL}}^{\mathrm{Prem},k}\big) \approx \mathrm{vc}\big(\hat{X}_{\mathrm{NL}}^{\mathrm{Prem,gross},k}\big)\sqrt{\frac{1+\mathrm{vc}(Y_k)^2}{1+\mathrm{vc}(Y_k^{\mathrm{gross}})^2}},$$

where Y_k denotes the net loss amount per incident and the approximation error is comparable to that of (4.16).

Proof We have

$$\mathrm{E}\big(\tilde{S}_k^{\mathrm{gross}}\big) = \lambda_k\,\mathrm{E}\big(Y_k^{\mathrm{gross}}\big)$$

and

$$\mathrm{var}\big(\tilde{S}_k^{\mathrm{gross}}\big) = \lambda_k\big(\mathrm{var}\big(Y_k^{\mathrm{gross}}\big) + \mathrm{E}\big(Y_k^{\mathrm{gross}}\big)^2\big),$$

where we have used $\mathrm{E}(N_k) = \mathrm{var}(N_k) = \lambda_k$. Analogously we obtain

$$\mathrm{E}(\tilde{S}_k) = \lambda_k\,\mathrm{E}(Y_k), \qquad \mathrm{var}(\tilde{S}_k) = \lambda_k\big(\mathrm{var}(Y_k) + \mathrm{E}(Y_k)^2\big).$$

From this follows

$$\mathrm{vc}(\tilde{S}_k) = \sqrt{\frac{\mathrm{var}(\tilde{S}_k)}{\mathrm{E}(\tilde{S}_k)^2}}$$

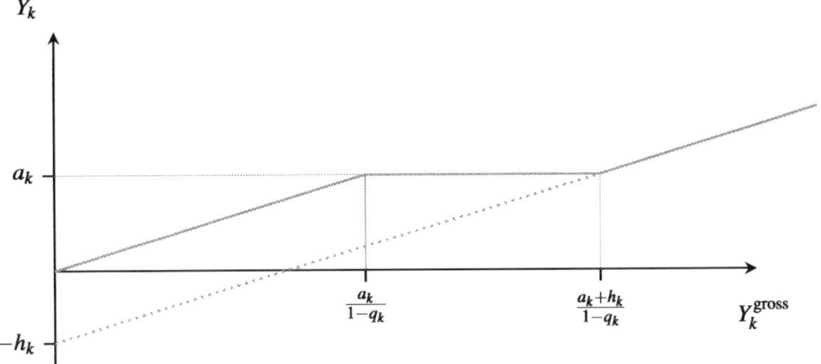

Fig. 4.9 Combination of excess-of-loss-per-event reinsurance and quota share reinsurance

$$= \mathrm{vc}\big(\tilde{S}_k^{\text{gross}}\big)\sqrt{\frac{\mathrm{E}(\tilde{S}_k^{\text{gross}})^2}{\mathrm{E}(\tilde{S}_k)^2}\frac{\mathrm{var}(\tilde{S}_k)}{\mathrm{var}(\tilde{S}_k^{\text{gross}})}}$$

$$= \mathrm{vc}\big(\tilde{S}_k^{\text{gross}}\big)\sqrt{\frac{\mathrm{E}(Y_k^{\text{gross}})^2}{\mathrm{E}(Y_k)^2}\frac{\mathrm{var}(Y_k)+\mathrm{E}(Y_k)^2}{\mathrm{var}(Y_k^{\text{gross}})+\mathrm{E}(Y_k^{\text{gross}})^2}}$$

$$= \mathrm{vc}\big(\tilde{S}_k^{\text{gross}}\big)\sqrt{\frac{1+\mathrm{vc}(Y_k)^2}{1+\mathrm{vc}(Y_k^{\text{gross}})^2}}$$

and the assertion follows from (4.16). □

Remark 4.14 There is a further implicit approximation as we have not investigated whether the assumptions of Lemma 4.6 are consistent with the assumptions on the distribution for the premium and reserve risk $X_{\text{NL}}^{\text{Prem,Res}}$.

The reinsurance program of line of business k is approximated by a pure quota share reinsurance with quota q_k and an excess-of-loss-per-event reinsurance with priority a_k and liability cap h_k (Fig. 4.9),

$$Y_k(\omega) = \begin{cases} (1-q_k)Y_k^{\text{gross}}(\omega) & \text{for } Y_k^{\text{gross}}(\omega) \le a_k \\ a_k & \text{for } a_k < (1-q_k)Y_k^{\text{gross}}(\omega) \le a_k + h_k \\ (1-q_k)Y_k^{\text{gross}}(\omega) - h_k & \text{for } a_k + h_k < (1-q_k)Y_k^{\text{gross}}(\omega). \end{cases} \quad (4.17)$$

Obviously it holds that

$$Y_k(\omega) = f\big(Y_k^{\text{gross}}(\omega)\big)$$

if we put

$$f(x) = (1-q_k)\big(x 1_{\{x: x\le \tilde{a}_k\}} + \tilde{a}_k 1_{\{x: \tilde{a}_k < x \le \tilde{a}_k + \tilde{h}_k\}} + (x-\tilde{h}_k)1_{\{x: \tilde{a}_k+\tilde{h}_k < x\}}\big)$$

with $\tilde{a}_k = a_k/(1-q_k)$, $\tilde{h}_k = h_k/(1-q_k)$. For this reinsurance structure the coefficients of variation $\mathrm{vc}(Y_k)$ and $\mathrm{vc}(Y_k^{\text{gross}})$ can be derived with the help of the following lemma.

Lemma 4.7 *Let the individual gross losses $Y_k^{\text{gross}} \sim \Phi_{0,1}(\frac{\ln()-m_k}{s_k})$ be lognormally distributed with parameters m_k, s_k, the individual net losses be given by (4.17), where we put $\tilde{a}_k = a_k/(1-q_k)$, $\tilde{h}_k = h_k/(1-q_k)$. Then for the gross distribution Y_k^{gross} holds*

$$1 + \mathrm{vc}\big(Y_k^{\text{gross}}\big)^2 = \mathrm{e}^{s_k^2}$$

and for the net distribution Y_k

$$\begin{aligned}\frac{\mathrm{E}(Y_k)}{1-q_k} &= \mathrm{e}^{m_k+s_k^2/2} - \tilde{h}_k - \tilde{a}_k\Phi_{m_k,s_k}(\ln\tilde{a}_k) + (\tilde{a}_k+\tilde{h}_k)\Phi_{m_k,s_k}\big(\ln(\tilde{a}_k+\tilde{h}_k)\big)\\ &\quad + \mathrm{e}^{m_k+s_k^2/2}\Phi_{m_k+s_k^2,s_k}(\ln\tilde{a}_k) - \mathrm{e}^{m_k+s_k^2/2}\Phi_{m_k+s_k^2,s_k}\big(\ln(\tilde{a}_k+\tilde{h}_k)\big)\end{aligned}$$

as well as

$$\begin{aligned}\frac{\mathrm{E}(Y_k)^2(1+\mathrm{vc}(Y_k)^2)}{(1-q_k)^2} &= \mathrm{e}^{2m_k+2s_k^2} - 2\tilde{h}_k\mathrm{e}^{m_k+s_k^2/2} + \tilde{h}_k^2\\ &\quad - \tilde{a}_k^2\Phi_{m_k,s_k}(\ln\tilde{a}_k) + \big(\tilde{a}_k^2 - \tilde{h}_k^2\big)\Phi_{m_k,s_k}\big(\ln(\tilde{a}_k+\tilde{h}_k)\big)\\ &\quad + 2\tilde{h}_k\mathrm{e}^{m_k+s_k^2/2}\Phi_{m_k+s_k^2,s_k}\big(\ln(\tilde{a}_k+\tilde{h}_k)\big)\\ &\quad + \mathrm{e}^{2m_k+2s_k^2}\Phi_{m_k+2s_k^2,s_k}(\ln\tilde{a}_k)\\ &\quad - \mathrm{e}^{2m_k+2s_k^2}\Phi_{m_k+2s_k^2,s_k}\big(\ln(\tilde{a}_k+\tilde{h}_k)\big).\end{aligned}$$

Proof The first claim follows directly from the properties of the lognormal distribution,

$$\begin{aligned}1 + \mathrm{vc}\big(Y_k^{\text{gross}}\big)^2 &= \frac{\mathrm{E}(Y_k^{\text{gross}})^2 + \mathrm{var}(Y_k^{\text{gross}})}{\mathrm{E}(Y_k^{\text{gross}})^2}\\ &= \big(\mathrm{e}^{2m_k+s_k^2} + \big(\mathrm{e}^{s_k^2}-1\big)\mathrm{e}^{2m_k+s_k^2}\big)\mathrm{e}^{-2m_k-s_k^2} = \mathrm{e}^{s_k^2}.\end{aligned}$$

The density of the distribution $x \mapsto \Phi_{0,1}(\frac{\ln(x)-m_k}{s_k}) = \Phi_{m_k,s_k}(\ln x)$ is

$$\phi_{m_k,s_k}^{\log}(x) = \frac{1}{x\sqrt{2\pi}\,s_k}\exp\bigg(-\frac{1}{2}\bigg(\frac{\ln x - m_k}{s_k}\bigg)^2\bigg).$$

From this we derive for any $c \geq 0$

$$\int_c^\infty x\phi_{m_k,s_k}^{\log}(x)\,\mathrm{d}x = \frac{1}{\sqrt{2\pi}\,s_k}\int_c^\infty \exp\bigg(-\frac{1}{2}\bigg(\frac{\ln x - m_k}{s_k}\bigg)^2\bigg)\,\mathrm{d}x$$

$$
\begin{aligned}
&= \frac{1}{\sqrt{2\pi} s_k} \int_{\ln c}^{\infty} \exp(y) \exp\left(-\frac{1}{2}\left(\frac{y-m_k}{s_k}\right)^2\right) \mathrm{d}y \\
&= \frac{1}{\sqrt{2\pi} s_k} \int_{\ln c}^{\infty} \exp\left(-\frac{1}{2}\left(\frac{y^2 - 2m_k y + m_k^2 - 2s_k^2 y}{s_k^2}\right)\right) \mathrm{d}y \\
&= \frac{1}{\sqrt{2\pi} s_k} \int_{\ln c}^{\infty} \exp\left(-\frac{1}{2}\left(\frac{y-(m_k+s_k^2)}{s_k}\right)^2 \right. \\
&\quad \left. + \frac{1}{2}\frac{2m_k s_k^2 + s_k^4}{s_k^2}\right) \mathrm{d}y \\
&= \exp\left(m_k + \frac{s_k^2}{2}\right) \frac{1}{\sqrt{2\pi} s_k} \int_{\ln c}^{\infty} \exp\left(-\frac{1}{2}\left(\frac{y-(m_k+s_k^2)}{s_k}\right)^2\right) \mathrm{d}y \\
&= \mathrm{e}^{m_k + s_k^2/2}\left(1 - \Phi_{m_k+s_k^2, s_k}(\ln c)\right).
\end{aligned}
$$

This implies

$$
\begin{aligned}
\frac{\mathrm{E}(Y_k)}{1-q_k} &= \frac{\mathrm{E}(f(Y_k^{\text{gross}}))}{1-q_k} \\
&= \int_0^{\tilde{a}_k} x \phi_{m_k,s_k}^{\log}(x)\,\mathrm{d}x + \tilde{a}_k \int_{\tilde{a}_k}^{\tilde{a}_k+\tilde{h}_k} \phi_{m_k,s_k}^{\log}(x)\,\mathrm{d}x \\
&\quad + \int_{\tilde{a}_k+\tilde{h}_k}^{\infty} (x - \tilde{h}_k)\phi_{m_k,s_k}^{\log}(x)\,\mathrm{d}x \\
&= \mathrm{e}^{m_k+s_k^2/2}\Phi_{m_k+s_k^2,s_k}(\ln \tilde{a}_k) + \tilde{a}_k\left(\Phi_{m_k,s_k}\left(\ln(\tilde{a}_k+\tilde{h}_k)\right) - \Phi_{m_k,s_k}(\ln \tilde{a}_k)\right) \\
&\quad + \mathrm{e}^{m_k+s_k^2/2}\left(1 - \Phi_{m_k+s_k^2,s_k}\left(\ln(\tilde{a}_k+\tilde{h}_k)\right)\right) \\
&\quad - \tilde{h}_k\left(1 - \Phi_{m_k,s_k}\left(\ln(\tilde{a}_k+\tilde{h}_k)\right)\right),
\end{aligned}
$$

which proves the second claim.

For the third claim we must find

$$
\mathrm{E}(Y_k)^2\left(1 + \mathrm{vc}(Y_k)^2\right) = \mathrm{var}(Y_k) + \mathrm{E}(Y_k)^2 = \mathrm{E}\left(Y_k^2\right).
$$

We first calculate

$$
\begin{aligned}
&\int_c^{\infty} x^2 \phi_{m_k,s_k}^{\log}(x)\,\mathrm{d}x \\
&\quad = \frac{1}{\sqrt{2\pi} s_k} \int_c^{\infty} x \exp\left(-\frac{1}{2}\left(\frac{\ln x - m_k}{s_k}\right)^2\right) \mathrm{d}x
\end{aligned}
$$

$$
\begin{aligned}
&= \frac{1}{\sqrt{2\pi}\, s_k} \int_{\ln c}^{\infty} \exp(2y) \exp\Big(-\frac{1}{2}\Big(\frac{y-m_k}{s_k}\Big)^2\Big)\,\mathrm{d}y \\
&= \frac{1}{\sqrt{2\pi}\, s_k} \int_{\ln c}^{\infty} \exp\Big(-\frac{1}{2}\Big(\Big(\frac{y-m_k-2s_k^2}{s_k}\Big)^2 - 2\big(2s_k^2+2m_k\big)\Big)\Big)\,\mathrm{d}y \\
&= \mathrm{e}^{2m_k+2s_k^2}\big(1-\Phi_{m_k+2s_k^2,s_k}(\ln c)\big).
\end{aligned}
$$

From this we get

$$
\begin{aligned}
\frac{\mathrm{E}(Y_k^2)}{(1-q_k)^2} &= \int_0^{\tilde{a}_k} x^2 \phi^{\log}_{m_k,s_k}(x)\,\mathrm{d}x + \tilde{a}_k^2 \int_{\tilde{a}_k}^{\tilde{a}_k+\tilde{h}_k} \phi^{\log}_{m_k,s_k}(x)\,\mathrm{d}x \\
&\quad + \int_{\tilde{a}_k+\tilde{h}_k}^{\infty} \big(x^2 - 2x\tilde{h}_k + \tilde{h}_k^2\big)\phi^{\log}_{m_k,s_k}(x)\,\mathrm{d}x \\
&= \mathrm{e}^{2m_k+2s_k^2}\Phi_{m_k+2s_k^2,s_k}(\ln \tilde{a}_k) + \tilde{a}_k^2\big(\Phi_{m_k,s_k}\big(\ln(\tilde{a}_k+\tilde{h}_k)\big) - \Phi_{m_k,s_k}(\ln \tilde{a}_k)\big) \\
&\quad + \mathrm{e}^{2m_k+2s_k^2}\big(1-\Phi_{m_k+2s_k^2,s_k}\big(\ln(\tilde{a}_k+\tilde{h}_k)\big)\big) \\
&\quad - 2\tilde{h}_k \mathrm{e}^{m_k+s_k^2/2}\big(1-\Phi_{m_k+s_k^2,s_k}\big(\ln(\tilde{a}_k+\tilde{h}_k)\big)\big) \\
&\quad + \tilde{h}_k^2\big(1-\Phi_{m_k,s_k}\big(\ln(\tilde{a}_k+\tilde{h}_k)\big)\big),
\end{aligned}
$$

whence the third claim follows. □

From the Lemmas 4.6 and 4.7 we can now approximately calculate the coefficients of variation.

$$
\begin{aligned}
\mathrm{vc}\big(\hat{X}_{\mathrm{NL}}^{\mathrm{Prem},k}\big) &\approx \mathrm{vc}\big(\hat{X}_{\mathrm{NL}}^{\mathrm{Prem,gross},k}\big)\sqrt{\frac{1+\mathrm{vc}(Y_k)^2}{1+\mathrm{vc}(Y_k^{\mathrm{gross}})^2}} \\
&= \frac{\mathrm{vc}(\hat{X}_{\mathrm{NL}}^{\mathrm{Prem,gross},k})}{\sqrt{1+\mathrm{vc}(Y_k^{\mathrm{gross}})^2}}\sqrt{\frac{\mathrm{E}(Y_k)^2(1+\mathrm{vc}(Y_k)^2)}{(1-q_k)^2}\Big(\frac{\mathrm{E}(Y_k)}{1-q_k}\Big)^{-2}} \\
&= \mathrm{vc}\big(\hat{X}_{\mathrm{NL}}^{\mathrm{Prem,gross},k}\big)\mathrm{e}^{-s_k^2/2} \\
&\quad \times \Big(\mathrm{e}^{2m_k+2s_k^2} - 2\tilde{h}_k\mathrm{e}^{m_k+s_k^2/2} + \tilde{h}_k^2 \\
&\quad - \tilde{a}_k^2\Phi_{m_k,s_k}(\ln\tilde{a}_k) + \big(\tilde{a}_k^2-\tilde{h}_k^2\big)\Phi_{m_k,s_k}\big(\ln(\tilde{a}_k+\tilde{h}_k)\big) \\
&\quad + 2\tilde{h}_k\mathrm{e}^{m_k+s_k^2/2}\Phi_{m_k+s_k^2,s_k}\big(\ln(\tilde{a}_k+\tilde{h}_k)\big) \\
&\quad + \mathrm{e}^{2m_k+2s_k^2}\Phi_{m_k+2s_k^2,s_k}(\ln\tilde{a}_k) \\
&\quad - \mathrm{e}^{2m_k+2s_k^2}\Phi_{m_k+2s_k^2,s_k}\big(\ln(\tilde{a}_k+\tilde{h}_k)\big)\Big)
\end{aligned}
$$

$$\times \left(\mathrm{e}^{m_k+s_k^2/2} - \tilde{h}_k - \tilde{a}_k \Phi_{m_k,s_k}(\ln \tilde{a}_k)\right.$$
$$+ (\tilde{a}_k + \tilde{h}_k)\Phi_{m_k,s_k}\big(\ln(\tilde{a}_k + \tilde{h}_k)\big) + \mathrm{e}^{m_k+s_k^2/2}\Phi_{m_k+s_k^2,s_k}(\ln \tilde{a}_k)$$
$$\left.- \mathrm{e}^{m_k+s_k^2/2}\Phi_{m_k+s_k^2,s_k}\big(\ln(\tilde{a}_k + \tilde{h}_k)\big)\right)^{-1}. \tag{4.18}$$

The formula is a bit long, but it is explicit and so easily manageable.

For the reserve risk $\hat{X}_{\mathrm{NL}}^{\mathrm{Res},k}$ of line of business k the effect of reinsurance is not determined on a company-specific basis but EIOPA prescribes directly the net coefficients of variation $\mathrm{vc}(\hat{X}_{\mathrm{NL}}^{\mathrm{Res},k})$ (see Table 4.2).

Let $P_t^{\mathrm{written},k}$ be the premium written in year t, and P_t^k the premium earned in year t, by the line of business k. We denote by upr_t^k the unearned premium reserve from the previous year $t-1$ at the start of year t. Then we have

$$P_t^k = P_t^{\mathrm{written},k} + \mathrm{upr}_t^k - \mathrm{upr}_{t+1}^k.$$

Consider those contracts which are already in existence at time t and let $P_t^{\mathrm{future},k}$ denote the present value of net premiums associated with them which are expected to be earned after the time $t+1$. Therefore $P_t^{\mathrm{future},k}$ is part of the premium reserve. This term vanishes if all the insurance contracts have a term of at most one year (with no option of extension).

As a volume measure for the premium risk we choose

$$\begin{aligned} V_t^{\mathrm{Prem},k} &= \max\big(P_{t-1}^{\mathrm{written},k}, P_t^{\mathrm{written},k}, P_t^k\big) + P_t^{\mathrm{future},k} \\ &= \max\big(P_{t-1}^k - \mathrm{upr}_{t-1}^k + \mathrm{upr}_t^k, P_t^k - \mathrm{upr}_t^k + \mathrm{upr}_{t+1}^k, P_t^k\big) + P_t^{\mathrm{future},k}, \end{aligned}$$

and assume $\mathrm{E}(\hat{X}_{\mathrm{NL}}^{\mathrm{Prem},k}) = V_t^{\mathrm{Prem},k}$. These assumptions include a financial cushion because of the maximization in the definition of $V_t^{\mathrm{Prem},k}$.

Just as in Sect. 4.6.1.5, let R^k be the deterministic best estimate of the non-discounted loss provisions at time 0 for claims outstanding which have occurred in previous years. Let the deterministic payment patterns for these losses be denoted by $(\beta_t^k)_{t\geq 1}$. These satisfy the normalization condition $\sum_{t=1}^{\infty} \beta_t^k = 1$. The volume measure for the reserve risk is the best estimate of the necessary loss provisions, discounted at the risk-free interest rate, and so given by

$$V_t^{\mathrm{Res},k} = R^k \sum_{t=1}^{\infty} \frac{\beta_t^k}{\prod_{\tau=1}^{t}(1+s_\tau)}.$$

It is assumed that $\mathrm{E}(\hat{X}_{\mathrm{NL}}^{\mathrm{Res},k}) = V_t^{\mathrm{Res},k}$.

Clearly then it holds for the premium & reserve risk that

$$\mu = \mathrm{E}(\hat{X}_{\mathrm{NL}}^{\mathrm{Prem,Res}}) = \sum_{k=1}^{12} \mathrm{E}\big(\hat{X}_{\mathrm{NL}}^{\mathrm{Prem},k}\big) + \sum_{k=1}^{12} \mathrm{E}\big(\hat{X}_{\mathrm{NL}}^{\mathrm{Res},k}\big)$$

$$= \sum_{k=1}^{12} \left(V_t^{\mathrm{Prem},k} + V_t^{\mathrm{Res},k} \right). \tag{4.19}$$

For the correlation between the premium risk and the reserve risk, EIOPA prescribes the estimate $\mathrm{corr}^k_{\mathrm{Prem,Res}} = 0.5$. The coefficient of variation of the combined premium and reserve risk is then given by

$$\mathrm{vc}\left(\hat{X}_{\mathrm{NL}}^{\mathrm{Prem,Res},k}\right) = \frac{\sqrt{(\sigma_t^{\mathrm{Prem},k})^2 + 2\mathrm{corr}^k_{\mathrm{Prem,Res}}\sigma_t^{\mathrm{Prem},k}\sigma_t^{\mathrm{Res},k} + (\sigma_t^{\mathrm{Res},k})^2}}{V_t^{\mathrm{Prem},k} + V_t^{\mathrm{Res},k}}, \tag{4.20}$$

where we have set $\sigma_t^{\mathrm{Prem},k} = V_t^{\mathrm{Prem},k}\,\mathrm{vc}(\hat{X}_{\mathrm{NL}}^{\mathrm{Prem},k})$ and $\sigma_t^{\mathrm{Res},k} = V_t^{\mathrm{Res},k}\,\mathrm{vc}(\hat{X}_{\mathrm{NL}}^{\mathrm{Res},k})$.

To calculate $\mathrm{vc}(\hat{X}_{\mathrm{NL}}^{\mathrm{Prem,Res}})$ we employ a correlation matrix $(\mathrm{corr}_{kl})_{k,l\in\{1,\ldots,12\}}$ for the combined premium and reserve risks of the individual lines of business,

$$\mathrm{vc}\left(\hat{X}_{\mathrm{NL}}^{\mathrm{Prem,Res}}\right) = \frac{\sqrt{\sum_{k,l=1}^{12} \mathrm{corr}_{kl}\,\sigma^{\mathrm{Prem,Res},k}\sigma^{\mathrm{Prem,Res},l}}}{\mu},$$

where we put $\sigma^{\mathrm{Prem,Res},k} = (V_t^{\mathrm{Prem},k} + V_t^{\mathrm{Res},k})\,\mathrm{vc}(\hat{X}_{\mathrm{NL}}^{\mathrm{Prem,Res},k})$,[24] and the correlation matrix $(\mathrm{corr}_{kl})_{k,l\in\{1,\ldots,12\}}$ reads

$$\begin{pmatrix}
1 & 0.50 & 0.50 & 0.25 & 0.50 & 0.25 & 0.50 & 0.25 & 0.50 & 0.25 & 0.25 & 0.25 \\
 & 1 & 0.25 & 0.25 & 0.25 & 0.25 & 0.50 & 0.50 & 0.50 & 0.25 & 0.25 & 0.25 \\
 & & 1 & 0.25 & 0.25 & 0.25 & 0.25 & 0.50 & 0.50 & 0.25 & 0.25 & 0.50 \\
 & & & 1 & 0.25 & 0.25 & 0.25 & 0.50 & 0.50 & 0.50 & 0.25 & 0.50 \\
 & & & & 1 & 0.50 & 0.50 & 0.25 & 0.50 & 0.25 & 0.50 & 0.25 \\
 & & & & & 1 & 0.50 & 0.25 & 0.50 & 0.25 & 0.50 & 0.25 \\
 & & & & & & 1 & 0.25 & 0.50 & 0.25 & 0.50 & 0.25 \\
 & & & & & & & 1 & 0.50 & 0.50 & 0.25 & 0.25 \\
 & & & & & & & & 1 & 0.25 & 0.25 & 0.50 \\
 & & & & & & & & & 1 & 0.25 & 0.25 \\
 & & & & & & & & & & 1 & 0.25 \\
 & & & & & & & & & & & 1
\end{pmatrix}.$$

Remark 4.15 The costs of and provisions for reinsurance do not enter directly into calculating the SCR, since they are already accounted for in determining the value of the assets.

Lapse Risk Lapse risk is only relevant if calculation of the reserves proceeds from the assumption that a certain portion of the insured will also pay premiums in the

[24]For geographically diversified portfolios this formula is adjusted in the Standard Model.

future (that is after the year under consideration). The lapse risk is determined with scenarios. To do this the insurance business is projected forward under the following assumptions:

(i) 50 % higher lapse rate than expected in each following year, where a total cancellation rate should not exceed 100 %,
(ii) the expected lapse rate in each following year,
(iii) 50 % lower lapse rate than expected in each following year.

We denote the net results (net value of Assets – Liabilities) by $\mathrm{NAV}^{\mathrm{Lapse}}_{50\,\%}$, $\mathrm{NAV}^{\mathrm{Lapse}}_{0\,\%}$, $\mathrm{NAV}^{\mathrm{Lapse}}_{-50\,\%}$. Then the Solvency II risk capital for non-life lapse risk is given by

$$\mathrm{SCR}^{\mathrm{Lapse}}_{\mathrm{NL}} = \max\big(\mathrm{E}\big(\mathrm{NAV}^{\mathrm{Lapse}}_{0\,\%}\big) - \mathrm{E}\big(\mathrm{NAV}^{\mathrm{Lapse}}_{50\,\%}\big), \mathrm{E}(\mathrm{NAV}_{0\,\%}) - \mathrm{E}\big(\mathrm{NAV}^{\mathrm{Lapse}}_{-50\,\%}\big)\big).$$

These expected values are not calculated stochastically but in approximation as a result of a calculation with expected parameters and correspondingly adjusted lapse rates.

Catastrophe Risk Catastrophe risk is captured by the analysis of standardized catastrophe scenarios or by an alternative factor-based formula. In general the scenario-based methods are preferred. Factor-based methods are employed

- if the standardized scenarios are not applicable,
- if partial internal models are not appropriate,
- for the line of business "Miscellaneous" ($k = 9$).

Standardized Catastrophe Scenarios Catastrophe scenarios are differentiated into natural catastrophes,

- storm,
- earthquake,
- flood,
- hail,

and catastrophes that result from human actions,

- motor vehicles (e.g., a car accident causes a train crash, fire in a tunnel, a bus accident in which an entire football team of the national professional soccer league dies)
- seafaring disasters (e.g. collision of a passenger ship with an oil tanker),
- credit (very much higher defaults and recession),
- aviation,
- indemnity,
- terrorism.

It is assumed that the risks of catastrophes caused by humans are independent. It would be beyond the scope of this book to describe the standardized catastrophe scenarios individually. Natural catastrophes involve both geographical dependencies and dependencies amongst lines of business.

Factor-Based Catastrophe Models In the factor-based catastrophe model individual catastrophic events are modeled by multiplying the gross written premium by a factor. Some of these events affect more than one line of business, and some lines of business are affected by more than one individual event. It is assumed that each insurance contract is affected by at most one event. To avoid double payments, the premium for each line of business k is apportioned amongst the events i affecting the line of business.

Using the factor-based formula, the capital for the non-life catastrophe risk is

$$\mathrm{SCR}_{\mathrm{NL}}^{\mathrm{Cat}} = \Bigg[\Bigg(\sqrt{\sum_{i\in\{1,2,3,5\}} \left(c_i P_{t,i}^{\mathrm{gross,written}}\right)^2} + c_{11} P_{t,11}^{\mathrm{gross,written}}\Bigg)^2 + \sum_{i\in\{4,7,8,9,10,12\}} \left(c_i P_{t,i}^{\mathrm{gross,written}}\right)^2 + \left(c_6 P_{t,6}^{\mathrm{gross,written}} + c_{13} P_{t,13}^{\mathrm{gross,written}}\right)^2\Bigg]^{\frac{1}{2}},$$

where

$$P_{t,i}^{\mathrm{gross,written}} = \sum_{k\in A_i} \lambda_{ik} P_t^{\mathrm{gross,written},k}$$

and the factors c_i and the sets A_i are defined in Table 4.3. The coefficients λ_{ik} model the apportionment of the gross written premium for line of business k over the events affecting this line of business, and thus must satisfy

$$\sum_{i\in\{j:\, k\in A_j\}} \lambda_{ik} = 1.$$

These coefficients are fixed by the company.

Total Risk Premium and reserve risk, lapse risk, and catastrophe risk are aggregated together using a correlation coefficient $\mathrm{corr}_{\mathrm{PremResLapse,Cat}}^{\mathrm{NL}}$:

$$\mathrm{SCR}_{\mathrm{NL}} = \Bigg[\left(\mathrm{SCR}_{\mathrm{NL}}^{\mathrm{Prem,Res}}\right)^2 + \left(\mathrm{SCR}_{\mathrm{NL}}^{\mathrm{Lapse}}\right)^2 + \left(\mathrm{SCR}_{\mathrm{NL}}^{\mathrm{Cat}}\right)^2 + 2\mathrm{corr}_{\mathrm{PremResLapse,Cat}}^{\mathrm{NL}} \mathrm{SCR}_{\mathrm{NL}}^{\mathrm{Cat}} \sqrt{\left(\mathrm{SCR}_{\mathrm{NL}}^{\mathrm{Prem,Res}}\right)^2 + \left(\mathrm{SCR}_{\mathrm{NL}}^{\mathrm{Lapse}}\right)^2}\Bigg]^{\frac{1}{2}},$$

where the value $\mathrm{corr}_{\mathrm{PremResLapse,Cat}}^{\mathrm{NL}} = 0.25$ is prescribed.

Example 4.14 The company XYZ Inc has 3 lines of business, fire F, liability L, and theft T. XYZ Inc has reinsured each of these lines of business with a simple quota

Table 4.3 Catastrophe factors and their assignment to lines of business

i	Event	A_i	c_i
1	Storm	$\{2,4\}$	175 %
2	Flood	$\{2,4\}$	113 %
3	Earthquake	$\{2,4\}$	120 %
4	Hail	$\{2\}$	30 %
5	Major fires, explosions	$\{4\}$	175 %
6	Major MAT disaster	$\{3\}$	100 %
7	Major motor vehicle liability disasters	$\{1\}$	40 %
8	Major third party liability disaster	$\{5\}$	85 %
9	Credit	$\{6\}$	139 %
10	Miscellaneous	$\{9\}$	40 %
11	Non-proportional reinsurance – property	$\{10\}$	250 %
12	Non-proportional reinsurance – MAT	$\{11\}$	250 %
13	Non-proportional reinsurance – casualty	$\{12\}$	250 %

Table 4.4 Company-specific inputs for the calculation of premium and reserve risk

	PY $= t-1$			CY $= t$			NY $= t+1$		
	F	L	T	F	L	T	F	L	T
$P_t^{\text{gross, written},k}$	500	200	50	600	300	100	–	–	–
upr_t^k	50	20	5	70	40	12	75	45	15
$V_t^{\text{Res},k}$	–	–	–	200	150	20	–	–	–
q_k	25 %	20 %	20 %	25 %	20 %	20 %	25 %	20 %	20 %
a_k	–	–	–	0.2	1	1	–	–	–
h_k	–	–	–	0.5	0	0	–	–	–
$\text{E}(Y_k^{\text{gross}})$	–	–	–	0.1	0.1	0.1	–	–	–
$\text{vc}(Y_k^{\text{gross}})$	–	–	–	50 %	60 %	70 %	–	–	–

share, and the line of business fire, over and above that, has an excess-of-loss-per-event reinsurance contract with the parameters $a_F = 0.2$, $h_F = 0.5$. All insurance contracts have terms of one year, and there is no option of automatic extension.

Loss quotas, costs and reinsurance provisions are not needed in the calculation. Table 4.4 provides the company-specific quantities needed for the calculation.

From this we obtain for the earned premiums

$$P_{\text{PY}} = \begin{pmatrix} 355 \\ 180 \\ 33 \end{pmatrix}, \qquad P_{\text{CY}} = \begin{pmatrix} 445 \\ 235 \\ 77 \end{pmatrix}$$

and thus

Table 4.5 Interest rate curve and run-off pattern of the reserves

t	s_t	β_t^F	β_t^L	β_t^T
1	3 %	$\frac{10}{17}$	$\frac{10}{28}$	$\frac{10}{14}$
2	3.1 %	$\frac{6}{17}$	$\frac{9}{28}$	$\frac{4}{14}$
3	3.15 %	$\frac{1}{17}$	$\frac{6}{28}$	0
4	3.2 %	0	$\frac{2}{28}$	0
5	3.2 %	0	$\frac{1}{28}$	0
6	3.2 %	0	0	0

$$V_t^{\text{Prem}} = \begin{pmatrix} 450 \\ 240 \\ 80 \end{pmatrix}.$$

Table 4.5 shows the interest rate curve and the run-off pattern for the reserves found by XYZ Inc. From this we see

$$V_t^{\text{Res}} = \begin{pmatrix} 191.43 \\ 140.83 \\ 19.25 \end{pmatrix} \quad \text{and} \quad \mu = 1121.51.$$

To determine the coefficients of variation $\text{vc}(X_{\text{NL}}^{\text{Prem},k})$ XYZ Inc first calculates the net individual loss distributions,

$$m = \begin{pmatrix} -2.41 \\ -2.46 \\ -2.50 \end{pmatrix}, \qquad s = \begin{pmatrix} 0.47 \\ 0.55 \\ 0.63 \end{pmatrix},$$

$$\text{E}(Y) = \begin{pmatrix} 0.075 \\ 0.080 \\ 0.080 \end{pmatrix}, \qquad \text{vc}(Y) = \begin{pmatrix} 0.48 \\ 0.60 \\ 0.70 \end{pmatrix}.$$

XYZ Inc assigns theft insurance to the line of business "Miscellaneous" in the Table 4.2. The provided gross coefficients of variance for the premium risk then are

$$\left(\text{vc}\left(X_{\text{NL}}^{\text{Prem,gross},k}\right)\right)_{k\in\{4,5,9\}} = \begin{pmatrix} 10.0\,\% \\ 15.0\,\% \\ 13.0\,\% \end{pmatrix},$$

and Eq. (4.18) gives us

$$\text{vc}\left(X_{\text{NL}}^{\text{Prem},k}\right) = \begin{pmatrix} 0.099 \\ 0.150 \\ 0.130 \end{pmatrix}.$$

Since the prescribed coefficients of variation for the reserve risk are

$$\left(\mathrm{vc}\left(X_{\mathrm{NL}}^{\mathrm{Res},k}\right)\right)_{k\in\{4,5,9\}} = \begin{pmatrix} 11.0\,\% \\ 11.0\,\% \\ 15.0\,\% \end{pmatrix},$$

we obtain

$$\left(\mathrm{vc}\left(\hat{X}_{\mathrm{NL}}^{\mathrm{Prem,Res},k}\right)\right)_{k\in\{4,5,9\}} = \begin{pmatrix} 0.091 \\ 0.120 \\ 0.122 \end{pmatrix}.$$

Using the relevant piece of the correlation matrix,

$$(\mathrm{corr}_{kl})_{k,l\in\{4,5,9\}} = \begin{pmatrix} 1 & 0.25 & 0.50 \\ 0.25 & 1 & 0.50 \\ 0.50 & 0.50 & 1 \end{pmatrix},$$

XYZ Inc calculates

$$\mathrm{vc}\left(\hat{X}_{\mathrm{NL}}^{\mathrm{Prem,Res}}\right) = 0.0807.$$

The SCR for the premium and reserve risk is thus

$$\mathrm{SCR}_{\mathrm{NL}}^{\mathrm{Prem,Res}} = 1376.$$

XYZ Inc uses an internal model to calculate the net results in the scenarios where the lapse rate is as expected, 50 % higher than expected, and 50 % lower than expected. The company obtains

$$\mathrm{NAV}_{0\,\%} = 3000, \qquad \mathrm{NAV}_{50\,\%} = 2800, \qquad \mathrm{NAV}_{-50\,\%} = 3100.$$

From this we find as SCR for the lapse risk

$$\mathrm{SCR}_{\mathrm{NL}}^{\mathrm{Lapse}} = \max(3000 - 2800, 3000 - 3100) = 200.$$

The SCR for the combined risk is therefore

$$\mathrm{SCR}_{\mathrm{NL}}^{\mathrm{Prem,Res,Lapse}} = \sqrt{1376^2 + 200^2} = 1390.$$

In respect of the catastrophe risk XYZ Inc argues that the standard scenarios are not appropriate for the insurance niche which XYZ Inc serves. So XYZ Inc uses the factor-based catastrophe model. The lines of business fire, liability and theft correspond to the indices $F = 4$, $L = 5$, $T = 9$ in the Table 4.3. Since the sets A_i with $\{4, 5, 9\} \cap A_i \neq \emptyset$ are given by

$$A_1 = A_2 = A_3 = \{2, 4\}, \qquad A_5 = \{4\}, \qquad A_8 = \{5\}, \qquad A_{10} = \{9\},$$

it follows that

$$P_{t,i}^{\mathrm{gross,written}} = \lambda_{iF}\, P_t^{\mathrm{gross,written},F} \qquad \text{for } i \in \{1, 2, 3, 5\},$$

Table 4.6 Apportionments λ_{ik} of the gross premium of each line of business for catastrophic events

Event	Line of business		
	F	L	T
1	10 %	–	–
2	20 %	–	–
3	30 %	–	–
5	40 %	–	–
8	–	100 %	–
10	–	–	100 %

$$P_{t,8}^{\text{gross, written}} = P_t^{\text{gross, written}, L},$$

$$P_{t,10}^{\text{gross, written}} = P_t^{\text{gross, written}, T},$$

$$P_{t,i}^{\text{gross, written}} = 0 \quad \text{for } i \in \{4, 6, 7, 9, 11, 12, 13\}.$$

The factors λ_{ik} found by the company are shown in Table 4.6. They deliver

$$\begin{aligned}
\mathrm{SCR}_{\mathrm{NL}}^{\mathrm{Cat}} &= \Big[\Big(c_1\lambda_{1F}P_t^{\text{gross, written}, F}\Big)^2 + \Big(c_2\lambda_{2F}P_t^{\text{gross, written}, F}\Big)^2 \\
&\quad + \Big(c_3\lambda_{3F}P_t^{\text{gross, written}, F}\Big)^2 + \Big(c_5\lambda_{5F}P_t^{\text{gross, written}, F}\Big)^2 \\
&\quad + \Big(c_8 P_t^{\text{gross, written}, L}\Big)^2 + \Big(c_{10} P_t^{\text{gross, written}, T}\Big)^2\Big]^{\frac{1}{2}} \\
&= \Big[(1.75 \times 60)^2 + (1.13 \times 120)^2 + (1.2 \times 180)^2 + (1.75 \times 240)^2 \\
&\quad + (0.85 \times 300)^2 + (0.4 \times 100)^2\Big]^{\frac{1}{2}} \\
&= 564.9.
\end{aligned}$$

Overall XYZ Inc obtains

$$\mathrm{SCR}_{\mathrm{NL}} = \sqrt{1390^2 + 2 \times 0.25 \times 1390 \times 564.9 + 564.9^2} = 1626.$$

References

1. Commission of the European Communities, Directive of the European Parliament and the Council on the taking-up and pursuit of the business of Insurance and Reinsurance, Solvency II (recast), November 2009. PE-CONS 3643/6/09 REV 6, approved by the European Parliament on 2009-04-22 (p. 181)

2. Committee of European Insurance and Occupational Pensions Supervisors (CEIOPS), CEIOPS' Advice for Level 2 Implementing Measures on Solvency II: SCR Standard Formula, Article 111, Non-life Underwriting Risk, October 2009. CEIOPS-DOC-41/09, CP48. (p. 181)
3. Eidgenössische Finanzmarktaufsicht (FINMA), Technical document on the Swiss Solvency Test, October 2006. Version 2 (p. 134)
4. Eidgenössische Finanzmarktaufsicht (FINMA), Technisches Dokument zum Swiss Solvency Test, http://www.finma.ch/d/beaufsichtigte/versicherungen/schweizer-solvenztest/Seiten/default.aspx (2006) (pp. 118, 125, 139, 145, 146, 147, 148, 163, 164, 170)
5. Eidgenössische Finanzmarktaufsicht (FINMA), Wegleitung für Versicherungsunternehmen betreffend die Abschätzung des Zufallsrisikos der Abwicklung von Schadenrückstellungen in der Nichtleben-Versicherung, http://www.finma.ch/d/beaufsichtigte/versicherungen/schweizer-solvenztest/Seiten/default.aspx, February 2010 (p. 167)
6. Eidgenössische Finanzmarktaufsicht (FINMA), Wegleitung zum SST-Marktrisiko-Standardmodell, http://www.finma.ch/d/beaufsichtigte/versicherungen/schweizer-solvenztest/Seiten/default.aspx, October 2011 (p. 134)
7. European Commission, QIS5 Technical specification. Annex to Call for Advice from CEIOPS on QIS5, July 2010 (pp. 122, 170, 171, 177, 181)
8. M. Kriele, J. Wolf, On market value margins and cost of capital. Blätter DGVFM **28**(2), 195–219 (2007) (p. 117)
9. T. Mack, *Schadenversicherungsmathematik* (Verlag Versicherungswirtschaft E.V., Karlsruhe, 1997) (p. 166)
10. A. McNeil, R. Frey, P. Embrechts, *Quantitative Risk Management. Concepts, Techniques, Tools* (Princeton University Press, Princeton, 2005) (p. 166)

Chapter 5
Allocation of Capital

5.1 Introduction

If the random variable X results from more than one driver of risk, then the question arises how these drivers of risk contribute to the risk as a whole. The answer would make it possible to evaluate individual risks while taking account of the diversification effects of the whole portfolio. We consider m risky portfolios, whose risks are described by random variables $X_1, \ldots, X_m$ on the probability space $(\Omega, \mathscr{F}, \mathbf{P})$. The risk of the portfolio as a whole is then modeled by the random variable $X = X_1 + \cdots + X_m$. Reciprocally we can begin with the whole company and split it into business areas:

Definition 5.1 Let $X \in \mathscr{M}(\Omega, \mathbb{R})$ be a random variable describing the losses of a company. An *segmentation into business areas* (or for short, a *segmentation*) is a random vector

$$(X_1, \ldots, X_m)$$

with $\sum_{i=1}^{m} X_i = X$. We denote the set of segmentations by $\mathscr{M}_X(\Omega)$.

If $(X_1, \ldots, X_m)$ is a segmentation into business areas, then X_i is the loss that should be assigned to the *business area* i.

The notion of a "business areas" can be taken as very broad. A segmentation into business areas could be done, for example, based on

- product groups,
- profit centers,
- lines of business,
- divisions,
- the business associated with various distribution channels,
- risks that somehow touch the whole business.

For each of these cases, the contribution of each business area to the whole can be of interest from a management standpoint.

M. Kriele, J. Wolf, *Value-Oriented Risk Management of Insurance Companies*, EAA Series, DOI 10.1007/978-1-4471-6305-3_5,

The summation condition in Definition 5.1 simply says that any loss to the whole business must come from a loss in some business area. There are naturally risks that implicate several business areas without it's being clear to which area the risk (and the associated losses) can be ascribed. Examples for this would be operational risks that affect the whole company, such as, say, a terrorist attack on head office or a reputational loss resulting from management errors, or, perhaps, the market risk in life insurance, since there is often no natural correspondence of investments to insurance products. In these cases the assignment of company-wide risks to individual business areas is a matter of modeling, and is subject to a certain arbitrariness that should be considered in interpreting the results. All the same, it is also reasonable to define a business area "Corporate" of its own, which takes up the risks that can only be assigned with difficulty to single business areas.

The remainder of this chapter is devoted to the allocation of risk capital to business areas. The allocation of risk capital is not absolutely necessary for good risk management. One could equally well consider sensitivities in a committee in order to come to an optimal solution for the whole company. In practice, however, it often shows itself to be more effective to give individual areas broad autonomy. It is then expected that each area will maximize its own benefit, without necessarily making a priority of the benefit of the company as a whole. In order to harmonize the interests of the areas with the interests of the company a good system of incentives must be put in place. With the help of the allocation of risk capital it is possible to construct an incentive system that will be viewed as fair.

After choosing a risk measure ρ the total risk capital $\rho(X)$ takes into account the diversification between the various business areas, while $\rho(X_i)$ represents the risk capital that the business area i would need if it were independent. If the company simply held the capital $\rho(X_i)$ for each business area i separately, then it would hold more capital than needed, since the existence of diversification effects would imply

$$\rho(X) < \sum_{i=1}^{m} \rho(X_i).$$

The saving in capital $\sum_{i=1}^{m} \rho(X_i) - \rho(X)$ arises from the interaction of risks, so that the contribution of the individual risks to the savings is usually varied. Since the diversification effect only occurs at the level of the company as a whole from the interaction of risks, it is not possible to determine objectively the part played by single business areas in the overall diversification effect. So there is no optimal method of risk capital allocation. The real problem with risk capital is to take account of this effect "as fairly as possible". That a risk capital allocation be viewed as fair is an absolute necessity for value-oriented management of a company that compares the performance of each business area with the risk contribution from that business area. The definition of "fair" is thus part of the problem of allocation of risk capital.

Remark 5.1 Performance data, which are defined as expected values, can easily be allocated to business areas because of the linearity of expected values. If other performance indicators are chosen (such as, say, the median) techniques that are described in this chapter must be employed.

Definition 5.2 Let X be the random variable that describes the loss of the whole company and let $(X_1, \dots, X_m) \in \mathscr{M}_X(\Omega)$ be a segmentation. A *capital allocation* is a vector $\bar{\Lambda} \in \mathbb{R}^m$, whose i-th component $\bar{\Lambda}_i$ represents the *capital allocated* to business area i.

In this generality the definition of a capital allocation is of little significance. We will formulate two axioms below that describe economicly motivated properties. The difference between an axiom and a definition is that an axiom has a direct bearing on an area outside mathematics. Both our axioms are intended to be evident, so that any reasonable capital allocation can be required to fulfill them. That is why they are formulated as axioms and not as definitions.[1]

A natural requirement connects the capital allocation with a risk measure:

$$\bar{\Lambda}_1 + \dots + \bar{\Lambda}_m \geq \rho(X),$$

since otherwise, in total, less risk capital would be allocated to the business areas than necessary for the whole company. If $\bar{\Lambda}_1 + \dots + \bar{\Lambda}_m > \rho(X)$, then unnecessary *excess capital* is being allocated to business areas, which could be returned to the owners of the business.[2] So we start here from the ideal situation of equality:

Axiom 5.1 (No excess capital) Let $(X_1, \dots, X_m) \in \mathscr{M}_X(\Omega)$ be the segmentation into business areas. Then all the capital must be allocated to business areas:

$$\rho(X) = \sum_{i=1}^{m} \bar{\Lambda}_i.$$

We also want to take for granted that no business area is assigned more capital than the risk capital that it would require on its own. If this postulate were violated then there would be an anti-diversification effect from the business area's point of view. The economic inference for that area of business would be to split off from the company.

Axiom 5.2 (Diversification) Let $(X_1, \dots, X_m) \in \mathscr{M}_X(\Omega)$ be the segmentation into business areas. The capital allocated to any business area does not exceed its individual risk capital, that is, it is true that

$$\bar{\Lambda}_i \leq \rho(X_i)$$

for each $i = 1, \dots, n$.

---

[1] A good example for a fruitful axiom system is that for Euclidean geometry, which represents a (just approximately valid) physical theory of space.

[2] In practice, there are perfectly sound reasons to hold excess capital. This will be discussed in Sect. 7.1.1 under the rubrics "risk appetite" and "risk tolerance". It is however difficult to motivate that the excess capital should be divided among business areas, at least when the business areas are not independent businesses.

Definition 5.3 A capital allocation $\bar{\Lambda}$ is called an *allotment* for the segmentation

$$(X_1, \ldots, X_m) \in \mathscr{M}_X(\Omega),$$

if the Axioms 5.1 and 5.2 are fulfilled.

5.2 Examples

To illustrate the following examples, we examine an insurance company that has three lines of business: fire insurance with a loss function X_F, water damage insurance with loss function X_W, computer theft insurance with loss function X_C. To simplify the calculations we assume here (unrealistically) that these risks are multi-normally distributed. Since water damage occurs in putting out a fire, it is plausible that X_F and X_W are correlated with each other. We suppose that the correlation is 50 %. In addition, we assume that X_C is independent of X_F and X_W. Overall we obtain for the distribution $\bar{X} = (X_F, X_W, X_C)$ the correlation matrix

$$\text{corr} = \begin{pmatrix} 1 & 50\,\% & 0 \\ 50\,\% & 1 & 0 \\ 0 & 0 & 1 \end{pmatrix}.$$

Let the expected profit of the lines of business and the standard deviation be

$$-\mathrm{E}(\bar{X}^\top) = \begin{pmatrix} 10 \\ 5 \\ 5 \end{pmatrix}, \qquad \sigma(\bar{X}^\top) = \begin{pmatrix} 12 \\ 2.5 \\ 7.5 \end{pmatrix}$$

(for the sign see our definition of a loss function given in Sect. 2.2). The covariance matrix is then given by

$$\text{cov} = \begin{pmatrix} (\sigma_F)^2 & \text{cov}_{FW} & 0 \\ \text{cov}_{FW} & (\sigma_W)^2 & 0 \\ 0 & 0 & (\sigma_C)^2 \end{pmatrix} = \begin{pmatrix} 144 & 15 & 0 \\ 15 & 6.25 & 0 \\ 0 & 0 & 56.25 \end{pmatrix}.$$

For our example, we choose as our risk measure the expected shortfall $\rho = \mathrm{ES}_\alpha$ at confidence level $\alpha = 99.5\,\%$.

In further calculations, for simplicity of notation, we use

$$\beta = \frac{\phi_{0,1}(\Phi_{0,1}^{-1}(\alpha))}{1-\alpha} = 2.9, \qquad \mu = \mathrm{E}(X), \qquad \mu_i = \mathrm{E}(X_i) \quad \text{for } i \in \{F, W, C\}$$

We report interim results with rounding, but we always carry the unrounded values through to the final result.

Since $(X_F, X_W, X_C)^\top$ is multinormally distributed and

$$X = X_F + X_W + X_C = \overbrace{\begin{pmatrix}1 & 1 & 1\end{pmatrix}}^{=:A} \begin{pmatrix} X_F \\ X_W \\ X_C \end{pmatrix}$$

X is similarly normally distributed and has the covariance matrix

$$\begin{aligned} A \operatorname{cov} A^\top &= \begin{pmatrix}1 & 1 & 1\end{pmatrix} \begin{pmatrix} (\sigma_F)^2 & \operatorname{cov}_{FW} & 0 \\ \operatorname{cov}_{FW} & (\sigma_W)^2 & 0 \\ 0 & 0 & (\sigma_C)^2 \end{pmatrix} \begin{pmatrix} 1 \\ 1 \\ 1 \end{pmatrix} \\ &= \begin{pmatrix}1 & 1 & 1\end{pmatrix} \begin{pmatrix} (\sigma_F)^2 + \operatorname{cov}_{FW} \\ \operatorname{cov}_{FW} + (\sigma_W)^2 \\ (\sigma_C)^2 \end{pmatrix} = (\sigma_F)^2 + (\sigma_W)^2 + (\sigma_C)^2 + 2\operatorname{cov}_{FW}. \end{aligned}$$

We deduce from Proposition 2.5

$$\begin{aligned} \sigma(X) &= \sqrt{(\sigma_F)^2 + (\sigma_W)^2 + (\sigma_C)^2 + 2\operatorname{cov}_{FW}} \\ &= \sqrt{144 + 6.25 + 56.25 + 2 \times 15} \\ &= 15.4 \end{aligned} \tag{5.1}$$

and therefore

$$\rho(X) = \mathrm{E}(X) + \beta\sigma(X) = -20 + 2.9 \times 15.4 = 24.5.$$

For the expected shortfall of the marginal distributions we get analogously

$$\begin{aligned} \rho(X_F) &= \mu_F + \beta\sigma_F = -10 + 2.9 \times 12 = 24.7, \\ \rho(X_W) &= \mu_W + \beta\sigma_W = -5 + 2.9 \times 2.5 = 2.2, \\ \rho(X_C) &= \mu_C + \beta\sigma_C = -5 + 2.9 \times 7.5 = 16.7. \end{aligned}$$

This section is inspired by [6], where further developments can also be found.

5.2.1 Proportional Capital Allocation

Definition 5.4 The *proportional capital allocation* for the segmentation

$$(X_1, \ldots, X_m) \in \mathscr{M}_X(\Omega)$$

and the business area i is defined by

$$\Lambda_i^{\text{proportional}} = \frac{\rho(X_i)\rho(X)}{\sum_{j=1}^m \rho(X_j)}.$$

This approach to allocation has the advantage that it is simple to calculate, but the dependence structure of the segmentation is not taken into account by this capital allocation. This is in contradiction to the goal of a "fair" allotment of the diversification effect.

Proposition 5.1 *For positive, subadditive risk measures proportional capital allocation is an allotment.*

Proof Axiom 5.1 follows from

$$\sum_{i=1}^{m} \bar{\Lambda}_i^{\text{proportional}} = \frac{\sum_{i=1}^{m} \rho(X_i)\rho(X)}{\sum_{j=1}^{m} \rho(X_j)} = \rho(X).$$

Since ρ is subadditive, we have

$$\rho(X) = \rho\left(\sum_{i=j}^{m} X_j\right) \leq \sum_{i=j}^{m} \rho(X_j)$$

and therefore $\bar{\Lambda}_i^{\text{proportional}} \leq \rho(X_i)$. Thus Axiom 5.2 is also satisfied. □

Example 5.1 Since in our example $\rho(X_i) = \mu_i + \beta\sigma_i$ $(i \in \{F, W, C\})$, we deduce

$$\frac{\bar{\Lambda}_F^{\text{proportional}}}{\rho(X)} = \frac{\mu_F + \beta\sigma_F}{\mu + \beta(\sigma_F + \sigma_W + \sigma_C)} = \frac{24.7}{43.6} = 56.6\ \%,$$

$$\frac{\bar{\Lambda}_W^{\text{proportional}}}{\rho(X)} = \frac{\mu_W + \beta\sigma_W}{\mu + \beta(\sigma_F + \sigma_W + \sigma_C)} = \frac{2.2}{43.6} = 5.1\ \%,$$

$$\frac{\bar{\Lambda}_C^{\text{proportional}}}{\rho(X)} = \frac{\mu_C + \beta\sigma_C}{\mu + \beta(\sigma_F + \sigma_W + \sigma_C)} = \frac{16.7}{43.6} = 38.3\ \%.$$

5.2.2 *Marginal Principles*

5.2.2.1 Discrete Marginal Principle (Merton and Perold)

To take diversification into account one can, in principle, replace the proportional allocation of the risk capital $\rho(X_i)$ by the contribution of business area i to the total risk. This amount can be seen as the difference between the risk capitals that one would obtain with and without X_i.

Definition 5.5 The *discrete, marginal capital allocation* for the segmentation

$$(X_1, \ldots, X_m) \in \mathcal{M}_X(\Omega)$$

and business area i, is defined to be

$$\bar{\Lambda}_i^{\textit{discr. marginal}} = \frac{(\rho(X) - \rho(X - X_i))\rho(X)}{\sum_{j=1}^{m}(\rho(X) - \rho(X - X_j))}.$$

Discrete, marginal capital allocation accounts for all the interdependence of the business areas i in the current structure of the company, but does not take account of the effect of a broadening of a business area. One is primarily interested in allocating capital as fairly as possible for a given segmentation to business areas.

Remark 5.2 The discrete, marginal capital allocation is in general not an allotment, even if we require subadditivity of the risk measure. To see this, let (X_1, X_2) be a segmentation. We write $\rho(X_i) = \rho_i$ and $\varepsilon = \rho_1 + \rho_2 - \rho(X)$. Then we have

$$\begin{aligned}\bar{\Lambda}_1^{\text{discr. marginal}} &= \frac{(\rho_1 + \rho_2 - \varepsilon - \rho_2)(\rho_1 + \rho_2 - \varepsilon)}{2\rho_1 + 2\rho_2 - 2\varepsilon - \rho_2 - \rho_1} = \frac{(\rho_1 - \varepsilon)(\rho_1 + \rho_2 - \varepsilon)}{\rho_1 + \rho_2 - 2\varepsilon} \\ &= \rho_1 - \varepsilon + \frac{\varepsilon(\rho_1 - \varepsilon)}{\rho_1 + \rho_2 - 2\varepsilon} = \rho_1 - \varepsilon \frac{\rho_1 + \rho_2 - 2\varepsilon - \rho_1 + \varepsilon}{\rho_1 + \rho_2 - 2\varepsilon} \\ &= \rho_1 - \varepsilon \frac{\rho_2 - \varepsilon}{\rho_1 + \rho_2 - 2\varepsilon}.\end{aligned}$$

If $\rho_2 < \varepsilon$ and $\rho_1 + \rho_2 - 2\varepsilon > 0$, then Axiom 5.2 is unsatisfied. This axiom is also not obeyed in our standard example.

Example 5.2 Analogously to the calculation of $\rho(X)$, we obtain

$$\sigma(X - X_F) = \sqrt{(\sigma_W)^2 + (\sigma_C)^2} = \sqrt{62.5} = 7.9, \tag{5.2}$$

$$\sigma(X - X_W) = \sqrt{(\sigma_F)^2 + (\sigma_C)^2} = \sqrt{200.25} = 14.2, \tag{5.3}$$

$$\sigma(X - X_C) = \sqrt{(\sigma_F)^2 + (\sigma_W)^2 + 2\,\mathrm{cov}_{FW}} = \sqrt{180.25} = 13.4 \tag{5.4}$$

and from this

$$\begin{aligned}\rho(X) - \rho(X - X_F) &= \mu + \beta\sigma - \bigl(\mu_W + \mu_C + \beta\sigma(X - X_F)\bigr) \\ &= \mu_F + \beta\bigl(\sigma - \sigma(X - X_F)\bigr) \\ &= -10 + 2.9 \times (15.4 - 7.9) = 11.6, \\ \rho(X) - \rho(X - X_W) &= \mu + \beta\sigma - \bigl(\mu_C + \mu_F + \beta\sigma(X - X_W)\bigr) \\ &= \mu_W + \beta\bigl(\sigma - \sigma(X - X_W)\bigr) \\ &= -5 + 2.9 \times (15.4 - 14.2) = -1.4, \\ \rho(X) - \rho(X - X_C) &= \mu + \beta\sigma - \bigl(\mu_F + \mu_W + \beta\sigma(X - X_C)\bigr) \\ &= \mu_C + \beta\bigl(\sigma - \sigma(X - X_C)\bigr) \\ &= -5 + 2.9 \times (15.4 - 13.4) = 0.6.\end{aligned}$$

The relative capital allocation comes out with

$$Z = \sum_j \big(\rho(X) - \rho(X - X_j)\big) = 11.6 - 1.4 + 0.6 = 10.8$$

as

$$\frac{\bar{\Lambda}_F^{\text{discr. marginal}}}{\rho(X)} = \frac{1}{Z}\big(\rho(X) - \rho(X - X_F)\big) = \frac{11.6}{10.8} = 107\ \%,$$

$$\frac{\bar{\Lambda}_W^{\text{discr. marginal}}}{\rho(X)} = \frac{1}{Z}\big(\rho(X) - \rho(X - X_W)\big) = \frac{-1.4}{10.8} = -13\ \%,$$

$$\frac{\bar{\Lambda}_C^{\text{discr. marginal}}}{\rho(X)} = \frac{1}{Z}\big(\rho(X) - \rho(X - X_C)\big) = \frac{0.6}{10.8} = 6\ \%.$$

For the absolute capital allocation we obtain

$$\bar{\Lambda}_F^{\text{discr. marginal}} = 107\ \% \times 24.5 = 26.3 > \rho(X_F),$$

$$\bar{\Lambda}_W^{\text{discr. marginal}} = -13\ \% \times 24.5 = -3.3 < 0,$$

$$\bar{\Lambda}_C^{\text{discr. marginal}} = 6\ \% \times 24.5 = 1.5.$$

Remark 5.3 Example 5.2 exposes an essential weakness in the discrete marginal principle. The “fire” business area is allocated more capital under the discrete marginal principle than when diversification is not taken into account. Such a result can hardly be explained to the management of the business area “fire”.

5.2.2.2 Continuous Marginal Principle (Myers and Read)

An alternative to the discrete marginal principle which emphasized notions of fairness, would be a capital allocation principle for which adding an infinitesimal extra bit of capital to the allocation would be handled fairly. This would be sensible if one wished to use the capital allocation to find out which business areas should be encouraged to grow. For such a measure, management considerations are primary.

Definition 5.6 Let $(X_1, \ldots, X_m)$ be a segmentation for X and ρ a risk measure. Then

$$\tilde{X} : \xi \mapsto \sum_{i=1}^{m} \xi_i X_i$$

is the *volume parametrization induced by* the segmentation of X, if $\xi \mapsto \rho(\tilde{X}(\xi))$ is differentiable in a neighborhood of $\xi = (1, \ldots, 1)$.

The idea behind Definition 5.6 is that the i-th component of the quantity $\xi = (\xi_1, \ldots, \xi_m)$ has to do with the i-th business area and represents a growth factor. Differentiability is necessary so that one can consider infinitesimal contributions to the total risk capital.

In Definition 5.6 it is implicitly assumed that the growth of one business area exerts no influence on other business areas. This is seldom the case in reality. The following definition is a generalization of Definition 5.6 which offers the freedom of describing mutual interactions of business areas.

Definition 5.7 A *volume parametrization* of a loss distribution X for a risk measure ρ is a mapping

$$\tilde{X} : U \subset \mathbb{R}^m \to \mathscr{M}(\Omega, \mathbb{R})$$
$$\xi \mapsto \tilde{X}(\xi),$$

where

(i) U is an open neighborhood of $(1, \ldots, 1) \in \mathbb{R}^m$,
(ii) $\xi \mapsto \rho(\tilde{X}(\xi))$ is differentiable on U,
(iii) $\tilde{X}(1, \ldots, 1) = X$.

In Definition 5.7 one does not explicitly start with a segmentation. But we continue to identify the i-th component of the volume parameter ξ with the growth of the i-th business area, and, in practical, concrete examples of Definition 5.7, as a rule, a segmentation is the basis.

Definition 5.8 Let X be a loss distribution and $\tilde{X}$ a volume parametrization of X for the risk measure ρ. Then the *continuous, marginal capital allocation* is given by

$$\bar{\Lambda}_i^{\text{cont. marginal}} = \frac{\frac{\partial \rho \circ \tilde{X}(\xi)}{\partial \xi_i} \rho(X)}{\sum_{j=1}^m \frac{\partial \rho \circ \tilde{X}(\xi)}{\partial \xi_j}}. \tag{5.5}$$

Remark 5.4 Obviously Axiom 5.1 is fulfilled. On the other hand, it makes no sense to speak of fulfilling Axiom 5.2 since there is no natural segmentation associated with a continuous, marginal capital allocation.

We would now like to specialize continuous, marginal capital allocation to the case of a volume parametrization induced by a segmentation $(X_1, \ldots, X_m)$ and a homogeneous risk measure. Since for a homogeneous risk measure ρ the homogeneity condition for $t \in \mathbb{R}$ is

$$t\rho\left(\sum_{i=1}^m \xi_i X_i\right) = \rho\left(\sum_{i=1}^m t\xi_i X_i\right)$$

it follows, on taking the derivative at $t = 1$, that

$$\rho\left(\sum_{i=1}^{m} \xi_i X_i\right) = \sum_{i=1}^{m} \xi_i \frac{\partial \rho(\sum_{j=1}^{m} \xi_j X_j)}{\partial \xi_i}.$$

This derivative rule for homogeneous functions is called Euler's Principle. With it we get a simplification of the marginal capital allocation described by (5.5) as follows:

Definition 5.9 Let ρ be a homogeneous risk measure. The *Euler capital allocation* for the segmentation $(X_1, \ldots, X_m) \in \mathscr{M}_X(\Omega)$ and the business area i is given by

$$\bar{\Lambda}_i^{\text{Euler}} = \left.\frac{\partial \rho(\sum_{j=1}^{m} \xi_j X_j)}{\partial \xi_i}\right|_{\xi=(1,\ldots,1)}.$$

Proposition 5.2 *The Euler capital allocation is, for a homogeneous subadditive risk measure, actually an allotment.*

Proof In Remark 5.4 we have already found that Axiom 5.1 is satisfied. From

$$\begin{aligned}
\bar{\Lambda}_i^{\text{Euler}} &= \left.\frac{\partial \rho(\sum_{j=1}^{m} \xi_j X_j)}{\partial \xi_i}\right|_{\xi=(1,\ldots,1)} \\
&= \lim_{\varepsilon\to 0} \frac{\rho((1+\varepsilon)X_i + \sum_{j=1, j\neq i}^{m} X_j) - \rho(X_i + \sum_{j=1, j\neq i}^{m} X_j)}{\varepsilon} \\
&\leq \lim_{\varepsilon\to 0} \frac{\rho(\varepsilon X_i) + \rho(\sum_{j=1}^{m} X_j) - \rho(\sum_{j=1}^{m} X_j)}{\varepsilon} \\
&= \lim_{\varepsilon\to 0} \frac{\rho(\varepsilon X_i)}{\varepsilon} = \lim_{\varepsilon\to 0} \varepsilon \frac{\rho(X_i)}{\varepsilon} = \rho(X_i)
\end{aligned}$$

follows that Axiom 5.2 holds. □

Remark 5.5 Let ρ be a homogeneous risk measure and $\tilde{X}(\xi) = \sum_{j=1}^{m} \xi_j X_j$ be the volume parametrization induced by the segmentation $(X_1, \ldots, X_m)$ of X. Then it holds that

$$\frac{\partial}{\partial \xi_i} \frac{\mathrm{E}(\tilde{X}(\xi))}{\rho(\tilde{X}(\xi))} = \rho\big(\tilde{X}(\xi)\big)^{-2}\left(\mathrm{E}(X_i)\rho\big(\tilde{X}(\xi)\big) - \mathrm{E}\big(\tilde{X}(\xi)\big)\frac{\partial \rho(\tilde{X}(\xi))}{\partial \xi_i}\right).$$

Under the natural assumption that $\bar{\Lambda}_i^{\text{Euler}} > 0$ there follows

$$\left(\frac{\partial}{\partial \xi_i} \frac{\mathrm{E}(\tilde{X}(\xi))}{\rho(\tilde{X}(\xi))}\right)_{\xi=(1,\ldots,1)} > 0 \quad \Leftrightarrow \quad \frac{\mathrm{E}(X_i)}{\bar{\Lambda}_i^{\text{Euler}}} > \frac{\mathrm{E}(X)}{\rho(X)}.$$

This means that the return to risk ratio of the i-th business area will be estimated by the Euler principle to be above average exactly when the expansion of that business area improves the return to risk ratio of the whole business.

We now specialize the Euler capital allocation to the case when $\rho(X) = a\,\mathrm{E}(X) + \beta\sigma(X)$ is a linear combination of the expected value and the standard deviation. Obviously this risk measure is homogeneous. We have

$$\begin{aligned}
\frac{\partial\rho\circ\tilde{X}(\xi)}{\partial\xi_i} &= a\frac{\partial}{\partial\xi_i}\mathrm{E}\left(\sum_{j=1}^{m}\xi_j X_j\right) + \beta\frac{\partial}{\partial\xi_i}\sqrt{\mathrm{cov}\left(\sum_{j=1}^{m}\xi_j X_j, \sum_{k=1}^{m}\xi_k X_k\right)} \\
&= a\,\mathrm{E}(X_i) + \frac{\beta}{2}\frac{1}{\sigma}\frac{\partial}{\partial\xi_i}\mathrm{cov}\left(\sum_{i=1}^{m}\xi_j X_j, \sum_{k=1}^{m}\xi_k X_k\right) \\
&= a\,\mathrm{E}(X_i) + \frac{\beta}{2}\frac{1}{\sigma}2\sum_{j=1}^{m}\xi_j\,\mathrm{cov}(X_i, X_j) \\
&= a\,\mathrm{E}(X_i) + \beta\frac{\mathrm{cov}(X_i, \tilde{X})}{\sigma}.
\end{aligned}$$

Because of the normalization $\tilde{X}(1, \ldots, 1) = X$ we obtain the following capital allocation.

Definition 5.10 The *covariance capital allocation* for the business area i and the segmentation $(X_1, \ldots, X_m) \in \mathcal{M}_X(\Omega)$ is given by

$$\bar{\Lambda}_i^{\text{Covariance}} = a\,\mathrm{E}(X_i) + \beta\frac{\mathrm{cov}(X_i, X)}{\sigma}.$$

Proposition 5.3 *The covariance capital allocation is an allotment.*

Proof This follows from Proposition 5.2 and the subadditivity of the risk measure

$$X \mapsto a\,\mathrm{E}(X) + \beta\sigma(X). \qquad \square$$

Example 5.3 In our example, let us set $\tilde{X}(\xi) = \sum_{i=1}^{m}\xi_i X_i$. Then the continuous, marginal capital allocation reduces to the covariance capital allocation. Since

$$\begin{aligned}
\mathrm{cov}(X_i, X) &= \mathrm{cov}(X_i, X_F) + \mathrm{cov}(X_i, X_W) + \mathrm{cov}(X_i, X_C) \\
&= \mathrm{cov}_{iF} + \mathrm{cov}_{iW} + \mathrm{cov}_{iC}
\end{aligned}$$

and $a = 1$, we obtain

$$\begin{aligned}
\bar{\Lambda}_F^{\text{Covariance}} &= \mu_F + \frac{\beta}{\sigma}\big((\sigma_F)^2 + \text{cov}_{FW}\big) \\
&= -10 + \frac{2.9}{15.4} \times (144 + 15) = 19.9, \\
\bar{\Lambda}_W^{\text{Covariance}} &= \mu_W + \frac{\beta}{\sigma}\big((\sigma_W)^2 + \text{cov}_{FW}\big) \\
&= -5 + \frac{2.9}{15.4} \times (6.25 + 15) = -1.0, \\
\bar{\Lambda}_C^{\text{Covariance}} &= \mu_C + \frac{\beta}{\sigma}(\sigma_C)^2 \\
&= -5 + \frac{2.9}{15.4} \times 56.25 = 5.6
\end{aligned}$$

respectively,

$$\begin{aligned}
\frac{\bar{\Lambda}_F^{\text{Covariance}}}{\rho(X)} &= \frac{19.9}{24.5} = 81\ \%, \\
\frac{\bar{\Lambda}_W^{\text{Covariance}}}{\rho(X)} &= \frac{-1.0}{24.5} = -4\ \%, \\
\frac{\bar{\Lambda}_C^{\text{Covariance}}}{\rho(X)} &= \frac{5.6}{24.5} = 23\ \%.
\end{aligned}$$

Remark 5.6 We get a negative capital allocation for the business area W, since the expected yield $-\mu_W$ is higher than the risk assigned to W. In this case the risk can be absorbed by the expected yield.

It is also possible to construct examples in which the diversification effect caused by the business areas is higher than the associated risk. This can lead to a negative allocation that is not offset by the expected yield. If, for example $\text{corr}_{FW} = -50\ \%$ is assumed instead of $\text{corr}_{FW} = 50\ \%$, we have

$$\begin{aligned}
\bar{\Lambda}_F^{\text{Covariance}} &= 18.1, \\
\bar{\Lambda}_W^{\text{Covariance}} &= -6.9, \\
\bar{\Lambda}_C^{\text{Covariance}} &= 7.2.
\end{aligned}$$

Therefore in this case $-\mu_W + \bar{\Lambda}_W^{\text{Covariance}} < 0$. Such a negative capital allocation can, however, in some cases be entirely appropriate, if it reflects that the business area in question works as an internal hedge for other business areas. Think, for example, of an insurer of annuities that also introduces term insurance.

For a negative capital allocation one cannot use the RORAC as a risk-adjusted performance measure since a negative RORAC cannot be interpreted.

5.2.3 Game-Theoretic Capital Allocation Principles

There cannot ever really be a fair allotment. It is, however, possible to identify allotment schemes which are generally considered unfair and that one thus wishes to exclude. For example, it would be unfair if the capital allotment $\bar{\Lambda}$ of a group B of business areas were to be higher than the capital needed by them. These business areas would then be providing a cross-subsidy for remaining areas.

Let $\bar{X} = (X_1, \ldots, X_m)$ be a segmentation into business areas. If ρ is the risk measure, then the capital needed for the group $B \subseteq \{1, \ldots, m\}$ of business areas is given by

$$\zeta_{\bar{X}}(B) = \rho\left(\sum_{i \in B} X_i\right).$$

If ρ is subadditive, then for all $B, C \subseteq \{1, \ldots, m\}$ the subadditivity condition $\zeta_{\bar{X}}(B \cup C) \leq \zeta_{\bar{X}}(B) + \zeta_{\bar{X}}(C)$ holds. With this we can reduce our risk measures, in our discussion, to simpler maps with analogous properties, but defined on finite sets. This motivates the following definition.

Definition 5.11 A mapping $\zeta : \mathscr{P}(\{1, \ldots, m\}) \to \mathbb{R}$, $B \mapsto \zeta(B)$ is a *discrete risk measure*, if $\zeta(\emptyset) = 0$ and $\zeta(B) \geq 0$ for all $B \in \mathscr{P}(\{1, \ldots, m\})$.

A discrete risk measure is called *subadditive*, if $\zeta(B \cup C) \leq \zeta(B) + \zeta(C)$ holds for all $B, C \in \mathscr{P}(\{1, \ldots, m\})$.

Just as in the case of the concept of risk measure, the concept of an allotment can be formulated "discretely":

Definition 5.12 A *discrete allotment* for a discrete risk measure ζ is a vector $\bar{\lambda}$ with

(i) $\zeta(\{1, \ldots, m\}) = \sum_{i=1}^{m} \bar{\lambda}_i$
(ii) $\bar{\lambda}_i \leq \zeta(\{i\})$ for all $i \in \{1, \ldots, m\}$.

Clearly a segmentation, a subadditive risk measure and an allotment induce a discrete risk measure ζ and a discrete allotment for ζ. Therefore discrete allotments are relevant for capital allocation.

The notion of an unfair allotment can be formulated through the notion of the *dominant, discrete allotment*:

Definition 5.13 Let ζ be a discrete risk measure. A discrete allotment $\bar{\lambda}'$ *dominates* the discrete allotment $\bar{\lambda}$ with respect to B, if there holds

(i) $\bar{\lambda}'_i < \bar{\lambda}_i \ \forall i \in B$
(ii) $\zeta(B) \leq \sum_{i \in B} \bar{\lambda}'_i$.

The dominant discrete allotment $\bar{\lambda}'$ assigns all business areas in B less risk capital than the dominated discrete allotment $\bar{\lambda}$ would. In addition, this group B of business areas can obtain an allotment more advantageous to them by splitting off as a separate enterprise from the whole company.

The following theorem gives a criterion for this type of unfairness not to occur.

Theorem 5.1 *Let ζ be a subadditive, discrete risk measure. There is no discrete allotment that dominates the discrete allotment $\bar{\lambda}$ for any subset of business areas, if and only if for all $B \subset \{1, \dots, m\}$ it holds that:* $\sum_{i \in B} \bar{\lambda}_i \leq \zeta(B)$.

Proof If $\bar{\lambda}'$ is a discrete allotment that dominates $\bar{\lambda}$ with respect to B, then

$$\sum_{i \in B} \bar{\lambda}_i > \sum_{i \in B} \bar{\lambda}'_i \geq \zeta(B).$$

Conversely, let $\bar{\lambda}$ be a discrete allotment that is not dominated by an allotment with respect to B. Since by the definition of a discrete allotment

$$\zeta\big(\{1, \dots, m\}\big) = \sum_{i=1}^{m} \bar{\lambda}_i \quad \text{and} \quad \zeta\big(\{i\}\big) \geq \bar{\lambda}_i$$

for all $i \in \{1, \dots, m\}$, let us assume that there is a subset B with $1 < \#B < m$ and

$$\zeta(B) < \sum_{i \in B} \bar{\lambda}_i.$$

We choose such a set B with maximal cardinality $\#B$. We find an $\varepsilon > 0$, so that $\sum_{i \in B} \bar{\lambda}_i - \varepsilon \#B > \zeta(B)$. Because

$$\begin{aligned} \sum_{i \notin B} \zeta\big(\{i\}\big) &\geq \zeta\big(\{1, \dots, m\}\big) - \zeta(B) = \sum_{i=1}^{m} \bar{\lambda}_i - \zeta(B) \\ &= \sum_{i \in B} \bar{\lambda}_i - \varepsilon \#B - \zeta(B) + \sum_{i \notin B} \bar{\lambda}_i + \varepsilon \#B \\ &> \sum_{i \notin B} \bar{\lambda}_i + \varepsilon \#B, \end{aligned}$$

$\bar{\lambda}_i \leq \zeta(\{i\})$ and the choice of $\#B$, we can select $\varepsilon_i > 0, i \notin B$, so that

$$\bar{\lambda}'_i := \bar{\lambda}_i + \varepsilon_i \leq \zeta\big(\{i\}\big), \quad i \notin B,$$

and $\sum_{i \notin B} \varepsilon_i = \varepsilon \#B$. By putting

$$\bar{\lambda}'_i := \bar{\lambda}_i - \varepsilon, \quad i \in B,$$

we have found a discrete allotment $\bar{\lambda}' = (\bar{\lambda}'_1, \dots, \bar{\lambda}'_m)$ that dominates $\bar{\lambda}$ wrt. B, in contradiction to our assumption. Thus for all subsets B we must have

$$\zeta(B) \geq \sum_{i \in B} \bar{\lambda}_i.$$

□

A subgoal of the game-theoretic method is to find an allotment algorithm whose result is not dominated by any other allotment with respect to an arbitrary group of business areas.

5.2.3.1 The Shapley Algorithm

Definition 5.14 Let $\zeta: \mathscr{P}(\{1,\ldots,m\}) \to \mathbb{R}$ be a discrete risk measure and $B \subset \{1,\ldots,m\}$ a portion of the business areas. The *capital contribution* of the business area $i \in \{1,\ldots,m\}$ to the business areas B is given by $\Delta_i(\zeta, B) = \zeta(B \cup \{i\}) - \zeta(B)$.

For $i \in B$, it obviously holds that $\Delta_i(\zeta, B) = 0$.

At best, one would have a method of constructing a discrete allotment that cannot be dominated by any other discrete allotment. Unfortunately no general procedure for this is known. The next best method is to define a suggestive axiom system that largely determines an allotment. This is the path we take below.

Definition 5.15 Let R be a subset of the discrete risk measures

$$\zeta: \mathscr{P}\big(\{1,\ldots,m\}\big) \to \mathbb{R}.$$

A mapping $\lambda\colon R \to \mathbb{R}^m$ is called a *Shapley algorithm*, if for all $\zeta, \zeta_1, \zeta_2 \in R$ the following properties hold:

(i) For each pair of business areas i, j, with $\Delta_i(\zeta, B) = \Delta_j(\zeta, B)$ for all B with $i, j \notin B$, it always holds that $\lambda_i(\zeta) = \lambda_j(\zeta)$.

(ii) For business areas i, which, for all B with $i \notin B$ provide the capital contribution

$$\Delta_i(\zeta, B) = \zeta\big(\{i\}\big)$$

we have $\lambda_i(\zeta) = \zeta(\{i\})$.

(iii) $\lambda(\zeta_1 + \zeta_2) = \lambda(\zeta_1) + \lambda(\zeta_2)$.

In general $\lambda(\zeta)$ is not a discrete allotment. However the Conditions (i), (ii), (iii) are motivated by their use for discrete allotments:

Condition (i) is a uniqueness condition: two business areas that always bring the same capital contribution to any other subset of business areas should not be distinguished in capital allocation.

Condition (ii) says that business areas that bring no contribution to diversification get their full individual risk capital allocated.

Condition (iii) is a linearity condition, whose content is hard to motivate since the addition of two discrete risk measures has no direct operational meaning. It does lead to the mathematically fruitful subject of linear operators, which greatly simplifies the study of Shapley algorithms. From the point of view of relevance to applications, however, this condition deserves careful scrutiny.

Theorem 5.2 *There is a Shapley algorithm on the space of discrete risk measures and it is unique. It is given by*

$$\bar{\lambda}_i^{\text{Shapley}}(\zeta) = \sum_{B \subseteq \{1,\dots,m\}\setminus\{i\}} \frac{\#B!(m-1-\#B)!}{m!} \Delta_i(\zeta, B).$$

If ζ is a subadditive discrete risk measure, then $\bar{\lambda}^{\text{Shapley}}(\zeta)$ is a discrete allotment.

Remark 5.7 Theorem 5.2 does not assert that the Shapley Algorithm is uniquely determined on the smaller space of subadditive discrete risk measures.

In preparation for the proof of Theorem 5.2 we set up a lemma on the boundary values of a subset of business areas.

Definition 5.16 The *boundary values* of a subset B of business areas for a discrete risk measure ζ are recursively defined as follows:

$$r_{\{i\}}(\zeta) := \zeta(\{i\}) \quad \text{for all } i = 1, \dots, m,$$

$$r_B(\zeta) := \zeta(B) - \sum_{L \subset B} r_L(\zeta) \quad \text{for all } B \subseteq \{1, \dots, m\} \text{ with } \#B > 1.$$

Lemma 5.1 *For the boundary value $r_B(\zeta)$ of the subset B there holds*

$$r_B(\zeta) = \sum_{C \subseteq B} (-1)^{\#B-\#C} \zeta(C).$$

Proof For $\#B = 1$ the assertion is obvious. For $\#B > 1$ we see inductively that

$$\begin{aligned} r_B(\rho) &= \zeta(B) - \sum_{C \subset B} r_C(\zeta) \\ &= \zeta(B) - \sum_{C \subset B} \sum_{D \subseteq C} (-1)^{\#C-\#D} \zeta(D) \\ &= \zeta(B) - \sum_{C \subset B} \sum_{i=0}^{\#B-\#C-1} \binom{\#B-\#C}{i} (-1)^i \zeta(C) \\ &= \zeta(B) - \sum_{C \subset B} (-1)^{\#B-\#C-1} \zeta(C) \\ &= \sum_{C \subseteq B} (-1)^{\#B-\#C} \rho(C), \end{aligned}$$

where we have used $\sum_{i=0}^{n} \binom{n+1}{i} (-1)^i = (-1+1)^{n+1} - (-1)^{n+1} = (-1)^n$ by the general binomial formula. □

Proof of Theorem 5.2 By assumption, R is the set of all discrete risk measures.

1. For $B \subseteq \{1, \ldots, m\}$ and $k \geq 0$, let the discrete risk measure $\zeta_{B,k}$ be given by

$$\zeta_{B,k}(C) = \begin{cases} k, & \text{if } B \subseteq C \text{ and } B \neq \emptyset, C \neq \emptyset \\ 0, & \text{if } B \nsubseteq C \text{ or } B = \emptyset \text{ or } C = \emptyset. \end{cases}$$

We show that for any Shapley algorithm λ on R

$$\lambda_i(\zeta_{B,k}) = \begin{cases} \frac{k}{\#B}, & i \in B \\ 0, & i \notin B. \end{cases} \tag{5.6}$$

holds.
For $i \notin B$, we have $\zeta_{B,k}(\{i\}) = 0$ and thus

$$\Delta_i(\zeta_{B,k}, C) = 0 \text{ for all } C \subseteq \{1, \ldots, m\}.$$

The Shapley Condition (ii) therefore implies $\lambda_i(\zeta_{B,k}) = 0$.
If $i, j \in B$ with $i \neq j$, then

$$\Delta_i(\zeta_{B,k}, C) = \Delta_j(\zeta_{B,k}, C) = 0 \text{ for all } C \subseteq \{1, \ldots, m\} \setminus \{i, j\}.$$

From the Shapley Condition (i) follows $\lambda_i(\zeta_{B,k}) = \lambda_j(\zeta_{B,k})$, which, taking note of

$$\sum_{i \in B} \lambda_i(\zeta_{B,k}) = \sum_{i=1}^{m} \lambda_i(\zeta_{B,k}) = \zeta_{B,k}\big(\{1, \ldots, m\}\big) = k,$$

implies the relationship $\lambda_i(\zeta_{B,k}) = \frac{k}{\#B}$ for $i \in B$.

2. There exist subsets $\mathscr{D} \subset \mathscr{P}\{1, \ldots, m\}$ with the property that an arbitrary discrete risk measure ζ can be represented uniquely as a sum

$$\zeta = \sum_{\emptyset \neq B \in \mathscr{D}} \zeta_{B,k_B} - \sum_{\emptyset \neq B' \in \mathscr{P}\{1,\ldots,m\} \setminus \mathscr{D}} \zeta_{B',k'_{B'}} \tag{5.7}$$

We have $\mathscr{D} = \{B \in \mathscr{P}\{1, \ldots, m\} : r_B(\zeta) \geq 0\}$. The values $k_B, k'_{B'}$ are given by the boundary values $k_B = r_B(\zeta)$ and $k'_{B'} = -r_{B'}(\zeta)$.
To simplify the notation in the proof of this assertion, we define, for $B \in \mathscr{P}\{1, \ldots, m\}$ and $k < 0$, the map $\zeta_{B,k} = -\zeta_{B,-k}$. We can then write (5.7) more simply as

$$\zeta = \sum_{\emptyset \neq B} \zeta_{B,k_B} \tag{5.8}$$

where, however, in general not all summands are risk measures. It is clearly enough to show the existence and uniqueness of the representation (5.8).

First we demonstrate the existence of the representation. Let $C \subseteq \{1, \dots, m\}$. Using Lemma 5.1 and the definition of $\zeta_{B,k}$ we transform the right-hand side:

$$\begin{aligned}
\sum_{B\subseteq\{1,\dots,m\}} \zeta_{B,r_B(\zeta)}(C) &= \sum_{B\subseteq C} r_B(\zeta) \\
&= \sum_{B\subseteq C}\sum_{D\subseteq B} (-1)^{\#B-\#D}\zeta(D) \\
&= \sum_{D\subseteq C}\sum_{B:D\subseteq B\subseteq C} (-1)^{\#B-\#D}\zeta(D) \\
&= \sum_{D\subseteq C}\left(\sum_{b=\#D}^{\#C}\sum_{\substack{B:D\subseteq B\subseteq C\\ \#B=b}} (-1)^{b-\#D}\right)\zeta(D).
\end{aligned}$$

Since a set B with b elements that contains a set D can be chosen from a set C in $\binom{\#C-\#D}{b-\#D}$ different ways, we deduce by applying the binomial formula that

$$\begin{aligned}
\sum_{B\subseteq\{1,\dots,m\}} \zeta_{B,r_B(\zeta)}(C) &= \sum_{D\subseteq C}\left(\sum_{b=\#D}^{\#C}\binom{\#C-\#D}{b-\#D}(-1)^{b-\#D}\right)\zeta(D)) \\
&= \sum_{D\subseteq C}\left(\sum_{b=0}^{\#C-\#D}\binom{\#C-\#D}{b}(-1)^{b}\right)\zeta(D) \\
&= \sum_{D\subseteq C} 1_{\{0\}}(\#C-\#D)\zeta(D) \\
&= \zeta(C).
\end{aligned}$$

Now we prove the uniqueness of the parameter k_B. If we have two different representations

$$\zeta = \sum_{B\subseteq\{1,\dots,m\}} \zeta_{B,k_B} = \sum_{B\subseteq\{1,\dots,m\}} \zeta_{B,\tilde{k}_B}$$

then we show the equality of the parameters k_B and $\tilde{k}_B$ by applying them to $C \subseteq \{1, \dots, m\}$ using induction on $\#C$. First we get for $C = \{i\}$ the equality $k_{\{i\}} = \tilde{k}_{\{i\}}$ from

$$\zeta(\{i\}) = \sum_{B\subseteq\{i\}} k_B = k_{\{i\}}.$$

Now suppose $k_B = \tilde{k}_B$ to already be shown for all B with $\#B < \#C$. From

$$\zeta(C) = \sum_{B\subseteq\{1,\dots,m\}} \zeta_{B,\tilde{k}_B}(C) = \sum_{B\subseteq C} \tilde{k}_B = \sum_{B\subseteq C} k_B = \sum_{B\subset C} \overbrace{k_B}^{=\tilde{k}_B} + k_C$$

follows $k_C = \tilde{k}_C$. So we see the representation is unique.

3. We have

$$\lambda_i(\zeta) = \sum_{C\subseteq\{1,\dots,m\}\setminus\{i\}} \frac{(m-\#C-1)!\#C!}{m!}\Delta_i(\zeta, C). \tag{5.9}$$

Let $\mathscr{D} \subset \mathscr{P}\{1,\dots,m\}$ be the set of all subsets B with $r_B(\zeta) \geq 0$. Then $\zeta_{B,r_B(\zeta)}$ and $\zeta_{B',-r_{B'}(\zeta)}$ are discrete risk measures for all $B \in \mathscr{D}$ and $B' \in \mathscr{P}\{1,\dots,m\} \setminus \mathscr{D}$. The Shapley Condition (iii) and the expression (5.7) thus imply

$$\begin{aligned}
\lambda(\zeta) + &\sum_{B'\in\mathscr{P}\{1,\dots,m\}\setminus\mathscr{D}} \lambda(\zeta_{B',-r_{B'}(\zeta)}) \\
&= \lambda\Bigg(\zeta + \sum_{B'\in\mathscr{P}\{1,\dots,m\}\setminus\mathscr{D}} \zeta_{B',-r_{B'}(\zeta)}\Bigg) \\
&= \lambda\Bigg(\sum_{B\in\mathscr{D}} \zeta_{B,r_B(\zeta)}\Bigg) \\
&= \sum_{B\in\mathscr{D}} \lambda(\zeta_{B,r_B(\zeta)}).
\end{aligned}$$

From (5.6) it now follows that

$$\lambda_i(\zeta) = \sum_{\substack{B\in\mathscr{D}\\ i\in B}} \frac{r_B(\zeta)}{\#B} - \sum_{\substack{B'\in\mathscr{P}\{1,\dots,m\}\setminus\mathscr{D}\\ i\in B'}} \frac{-r_{B'}(\zeta)}{\#B'} = \sum_{B:\, i\in B} \frac{r_B(\zeta)}{\#B}.$$

From Lemma 5.1 we obtain

$$\begin{aligned}
\lambda_i(\zeta) &= \sum_{B:\, i\in B} \frac{1}{\#B} \sum_{C\subseteq B} (-1)^{\#B-\#C} \zeta(C) \\
&= \sum_{C} \Bigg(\sum_{B:\, C\cup\{i\}\subseteq B} \frac{1}{\#B}(-1)^{\#B-\#C} \Bigg) \zeta(C).
\end{aligned} \tag{5.10}$$

We can rearrange the outer sums by combining the terms for C and $C \cup \{i\}$ for each C with $i \notin C$. Then we obtain

$$\begin{aligned}
\lambda_i(\zeta) &= \sum_{C\subseteq\{1,\dots,m\}\setminus\{i\}} \Bigg(\sum_{B:\, C\cup\{i\}\subseteq B} \frac{1}{\#B}(-1)^{\#B-\#C} \Bigg) \big(\zeta(C) - \zeta\big(C\cup\{i\}\big)\big) \\
&= -\sum_{C\subseteq\{1,\dots,m\}\setminus\{i\}} \Bigg(\sum_{B:\, C\cup\{i\}\subseteq B} \frac{1}{\#B}(-1)^{\#B-\#C} \Bigg) \Delta_i(\zeta, C).
\end{aligned}$$

With the help of the binomial formula we now calculate

$$-\sum_{B:\,C\cup\{i\}\subseteq B}\frac{(-1)^{\#B-\#C}}{\#B}=\sum_{b=\#C+1}^{m}\frac{1}{b}(-1)^{b-(\#C+1)}\binom{m-(\#C+1)}{b-(\#C+1)}$$

$$=\sum_{b=\#C+1}^{m}\int_0^1 x^{b-1}\,\mathrm{d}x(-1)^{b-\#C-1}\binom{m-\#C-1}{b-\#C-1}$$

$$=\int_0^1\sum_{b=\#C+1}^{m}\binom{m-\#C-1}{b-\#C-1}$$

$$\times x^{b-\#C-1}(-1)^{b-\#C-1}x^{\#C}\,\mathrm{d}x$$

$$=\int_0^1(1-x)^{m-\#C-1}x^{\#C}\,\mathrm{d}x$$

$$=\frac{(m-\#C-1)!(\#C)!}{m!},$$

where we applied the definition and properties of the Beta function in the last step. Altogether we have demonstrated that an allotment algorithm that fulfills the properties required in Definition 5.15 has the form (5.9).

4. The mapping λ defined by (5.9) is a Shapley algorithm.
Assume given $i,j\in\{1,\dots,m\}$ so that $\Delta_i(\zeta,B)=\Delta_j(\zeta,B)$ holds for all $B\subseteq\{1,\dots,m\}\setminus\{i,j\}$. Then it also holds that

$$\zeta\big(B\cup\{i\}\big)=\Delta_i(\zeta,B)+\zeta(B)=\Delta_j(\zeta,B)+\zeta(B)=\zeta\big(B\cup\{j\}\big)$$

and further that

$$\Delta_i\big(\zeta,B\cup\{j\}\big)=\zeta\big(B\cup\{j\}\cup\{i\}\big)-\overbrace{\zeta\big(B\cup\{j\}\big)}^{=\zeta(B\cup\{i\})}=\Delta_j\big(\zeta,B\cup\{i\}\big).$$

It follows that

$$\lambda_i(\zeta)=\sum_{C\subseteq\{1,\dots,m\}\setminus\{i\}}\frac{(m-\#C-1)!\#C!}{m!}\Delta_i(\zeta,C)$$

$$=\sum_{B\subseteq\{1,\dots,m\}\setminus(\{i\}\cup\{j\})}\frac{(m-\#B-1)!\#B!}{m!}\overbrace{\Delta_i(\zeta,B)}^{=\Delta_j(\zeta,B)}$$

$$+\sum_{B\subseteq\{1,\dots,m\}\setminus(\{i\}\cup\{j\})}\frac{(m-(\#B+1)-1)!(\#B+1)!}{m!}$$

$$\times\sum_{j\in\{1,\dots,m\}\setminus\{i\}}\overbrace{\Delta_i\big(\zeta,B\cup\{j\}\big)}^{\Delta_j(\zeta,B\cup\{i\})}$$

$$=\lambda_j(\zeta),$$

where in the last equation we used $\Delta_i(\zeta, B \cup \{i\}) = \Delta_j(\zeta, B \cup \{j\}) = 0$. Therefore, Shapley Condition (i) is satisfied.
If i is a business area with $\Delta_i(\zeta, B) = \zeta(\{i\})$ for all $B \subset \{1, \ldots, m\}$, then

$$\begin{aligned}\lambda_i(\zeta) &= \sum_{B \subseteq \{1,\ldots,m\} \setminus \{i\}} \frac{(m - \#B - 1)!\#B!}{m!} \Delta_i(\zeta, B) \\ &= \zeta\big(\{i\}\big) \sum_{B \subseteq \{1,\ldots,m\} \setminus \{i\}} \frac{(m - \#B - 1)!\#B!}{m!}.\end{aligned}$$

Shapley Condition (ii) is therefore satisfied because

$$\begin{aligned}\sum_{B \subseteq \{1,\ldots,m\} \setminus \{i\}} \frac{(m - \#B - 1)!(\#B)!}{m!} &= \sum_{b=0}^{m-1} \binom{m-1}{b} \frac{(m-b-1)!\,b!}{m!} \\ &= \frac{1}{m} \sum_{b=0}^{m-1} 1 \\ &= 1. \end{aligned} \tag{5.11}$$

Finally the validity of Condition (iii) follows directly from

$$\begin{aligned}\Delta_i(\zeta_1 + \zeta_2, B) &= \zeta_1\big(B \cup \{i\}\big) + \zeta_2\big(B \cup \{i\}\big) - \big(\zeta_1(B) + \zeta_2(B)\big) \\ &= \Delta_i(\zeta_1, B) + \Delta_i(\zeta_2, B).\end{aligned}$$

5. It remains to show that for a subadditive, discrete risk measure ζ the vector $\lambda(\zeta)$ really is a discrete allotment.
By subadditivity of ζ we have $\Delta_i(\zeta, B) \le \zeta(\{i\})$ for all i, B. Equations (5.9) and (5.11) thus imply $\lambda_i(\zeta) \le \zeta(\{i\})$.
Finally we need to prove that $\sum_{i=1}^m \lambda_i(\zeta) = \zeta(\{1, \ldots, m\})$. We have

$$\begin{aligned}\sum_{i=1}^m \lambda_i(\zeta) &= \sum_{i=1}^m \sum_{B \subseteq \{1,\ldots,m\} \setminus \{i\}} \frac{(m - \#B - 1)!\#B!}{m!} \big(\zeta\big(B \cup \{i\}\big) - \zeta(B)\big) \\ &= \sum_{i=1}^m \sum_{B \subseteq \{1,\ldots,m\} \setminus \{i\}} \frac{(m - (\#B + 1))!((\#B + 1) - 1)!}{m!} \zeta\big(B \cup \{i\}\big) \\ &\quad - \sum_{i=1}^m \sum_{B \subseteq \{1,\ldots,m\} \setminus \{i\}} \frac{(m - \#B - 1)!\#B!}{m!} \zeta(B).\end{aligned}$$

If $A \subset \{1, \ldots, m\}$, then for each $i \in A$ one can find exactly one $B \subset \{1, \ldots, m\} \setminus \{i\}$ with $A = B \cup \{i\}$. Hence

$$\sum_{i=1}^{m} \sum_{B\subseteq\{1,\dots,m\}\setminus\{i\}} \frac{(m-(\#B+1))!((\#B+1)-1)!}{m!}\zeta\big(B\cup\{i\}\big)$$
$$= \sum_{A\subseteq\{1,\dots,m\}} \#A\frac{(m-\#A)!(\#A-1)!}{m!}\zeta(A)$$
$$= \sum_{A\subseteq\{1,\dots,m\}} \frac{(m-\#A)!\#A!}{m!}\zeta(A).$$

Since there are $m-\#A$ business areas $i \notin A$, we obtain

$$\sum_{i=1}^{m} \sum_{B\subseteq\{1,\dots,m\}\setminus\{i\}} \frac{(m-\#B-1)!\#B!}{m!}\zeta(B)$$
$$= \sum_{A\subset\{1,\dots,m\}} (m-\#A)\frac{(m-\#A-1)!\#A!}{m!}\zeta(A)$$
$$= \sum_{A\subset\{1,\dots,m\}} \frac{(m-\#A)!\#A!}{m!}\zeta(A).$$

Therefore

$$\sum_{i=1}^{m}\lambda_i(\zeta) = \sum_{A\subseteq\{1,\dots,m\}} \frac{(m-\#A)!\#A!}{m!}\zeta(A) - \sum_{A\subset\{1,\dots,m\}} \frac{(m-\#A)!\#A!}{m!}\zeta(A)$$
$$= \frac{(m-m)!m!}{m!}\zeta\big(\{1,\dots,m\}\big)$$
$$= \zeta\big(\{1,\dots,m\}\big)$$

□

Remark 5.8 The formula given in Theorem 5.2 allows a direct interpretation. Since there are exactly $\binom{m-1}{\#B}$ different subsets B of $\{1,\dots,m\}\setminus\{i\}$ with $\#B$ elements, from

$$\bar{\lambda}_i^{\text{Shapley}}(\zeta) = \sum_{B\subseteq\{1,\dots,m\}\setminus\{i\}} \frac{\#B!(m-1-\#B)!}{m!}\Delta_i(\zeta,B)$$
$$= \frac{1}{m}\sum_{B\subseteq\{1,\dots,m\}\setminus\{i\}} \frac{1}{\binom{m-1}{\#B}}\Delta_i(\zeta,B),$$

it follows that, up to a proportionality factor $1/m$, the Shapley allotment $\bar{\lambda}_i^{\text{Shapley}}(\zeta)$ is precisely the average capital contribution of business area i to all subsets of other business areas. The proportionality factor ensures that the overall allotment to all business areas just makes up the total risk capital.

In the Shapley algorithm, business areas are seen as wholes, so that just as in the discrete marginal principle management measures are secondary.

Example 5.4 We have $m = 3$ business areas and have to look at the following subsets, doing so using the results of Example 5.2:

$$B_1 = \{F, W, C\}: \quad \zeta\big(\{F, W, C\}\big) = \rho\bigg(\sum_{j \in B_1} X_j\bigg) = \mu_F + \mu_W + \mu_C + \beta\sigma(X)$$

$$B_2 = \{F, W\}: \quad \zeta\big(\{F, W\}\big) = \rho\bigg(\sum_{j \in B_2} X_j\bigg) = \mu_F + \mu_W + \beta\sigma(X - X_C)$$

$$B_3 = \{F, C\}: \quad \zeta\big(\{F, C\}\big) = \rho\bigg(\sum_{j \in B_3} X_j\bigg) = \mu_F + \mu_C + \beta\sigma(X - X_W)$$

$$B_4 = \{W, C\}: \quad \zeta\big(\{W, C\}\big) = \rho\bigg(\sum_{j \in B_4} X_j\bigg) = \mu_W + \mu_C + \beta\sigma(X - X_F)$$

$$B_5 = \{F\}: \quad \zeta\big(\{F\}\big) = \rho\bigg(\sum_{j \in B_5} X_j\bigg) = \mu_F + \beta\sigma_F$$

$$B_6 = \{W\}: \quad \zeta\big(\{W\}\big) = \rho\bigg(\sum_{j \in B_6} X_j\bigg) = \mu_W + \beta\sigma_W$$

$$B_7 = \{C\}: \quad \zeta\big(\{C\}\big) = \rho\bigg(\sum_{j \in B_7} X_j\bigg) = \mu_C + \beta\sigma_C$$

$$B_8 = \emptyset: \quad \zeta(\emptyset) = \rho\bigg(\sum_{j \in B_8} X_j\bigg) = 0.$$

From this we obtain for the non-vanishing capital contributions of the business area F

$$\begin{aligned}
\Delta_F(\zeta, B_4) &= \mu_F + \beta\big(\sigma(X) - \sigma(X - X_F)\big), \\
\Delta_F(\zeta, B_6) &= \mu_F + \beta\big(\sigma(X - X_C) - \sigma_W\big), \\
\Delta_F(\zeta, B_7) &= \mu_F + \beta\big(\sigma(X - X_W) - \sigma_C\big), \\
\Delta_F(\zeta, B_8) &= \mu_F + \beta\sigma_F
\end{aligned}$$

and so

$$\bar{\lambda}_F^{\text{Shapley}}(\zeta) = \sum_{B \subseteq \{1,\dots,m\} \setminus \{F\}} \frac{\#B!(2 - \#B)!}{6} \Delta_F(\zeta, B)$$

$$
\begin{aligned}
&= \frac{2!(2-2)!}{6}\big(\mu_F + \beta\big(\sigma(X) - \sigma(X - X_F)\big)\big) \\
&\quad + \frac{1!(2-1)!}{6}\big(\mu_F + \beta\big(\sigma(X - X_C) - \sigma_W\big)\big) \\
&\quad + \frac{1!(2-1)!}{6}\big(\mu_F + \beta\big(\sigma(X - X_W) - \sigma_C\big)\big) \\
&\quad + \frac{0!(2-0)!}{6}(\mu_F + \beta\sigma_F) \\
&= \mu_F + \frac{\beta}{3}\sigma \\
&\quad + \frac{\beta}{6}\big(-2\sigma(X - X_F) + \sigma(X - X_W) + \sigma(X - X_C)\big) \\
&\quad + \frac{\beta}{6}(2\sigma_F - \sigma_W - \sigma_C).
\end{aligned}
$$

For the business areas W and C, we obtain analogously

$$
\begin{aligned}
\bar{\lambda}_W^{\text{Shapley}}(\zeta) &= \mu_W + \frac{\beta}{3}\sigma \\
&\quad + \frac{\beta}{6}\big(\sigma(X - X_F) - 2\sigma(X - X_W) + \sigma(X - X_C)\big) \\
&\quad + \frac{\beta}{6}(-\sigma_F + 2\sigma_W - \sigma_C), \\
\bar{\lambda}_C^{\text{Shapley}}(\zeta) &= \mu_C + \frac{\beta}{3}\sigma \\
&\quad + \frac{\beta}{6}\big(\sigma(X - X_F) + \sigma(X - X_W) - 2\sigma(X - X_C)\big) \\
&\quad + \frac{\beta}{6}(-\sigma_F - \sigma_W + 2\sigma_C).
\end{aligned}
$$

As a check one easily sees that

$$
\bar{\lambda}_F^{\text{Shapley}}(\zeta) + \bar{\lambda}_W^{\text{Shapley}}(\zeta) + \bar{\lambda}_C^{\text{Shapley}}(\zeta) = \rho(X)
$$

since in the sum, except for the term $\mu_i + \beta\sigma/3$, all other terms cancel.

We obtain

$$
\begin{aligned}
\bar{\lambda}_F^{\text{Shapley}}(\zeta) &= -10 + \frac{2.9}{3} \times 15.4 + \frac{2.9}{6}(-2 \times 7.9 + 14.2 + 13.4) \\
&\quad + \frac{2.9}{6}(2 \times 12.0 - 2.5 - 7.5) \\
&= -10 + 14.8 + 5.7 + 6.7 = 17.2,
\end{aligned}
$$

$$\bar{\lambda}_W^{\text{Shapley}}(\zeta) = -5 + \frac{2.9}{3} \times 15.4 + \frac{2.9}{6}(7.9 - 2 \times 14.2 + 13.4)$$
$$+ \frac{2.9}{6}(-12.0 + 2 \times 2.5 - 7.5)$$
$$= -5 + 14.8 - 3.4 - 7.0 = -0.5,$$
$$\bar{\lambda}_C^{\text{Shapley}}(\zeta) = -5 + \frac{2.9}{3} \times 15.4 + \frac{2.9}{6}(7.9 + 14.2 - 2 \times 13.4)$$
$$+ \frac{2.9}{6}(-12.0 - 2.5 + 2 \times 7.5)$$
$$= -5 + 14.8 - 2.3 + 0.2 = 7.8.$$

5.2.3.2 Aumann-Shapley Algorithm

The Aumann-Shapley algorithm is an infinitesimal version of the Shapley algorithm.

We use the property mentioned in Remark 5.8 that a business area's Shapley allotment $\bar{\lambda}_i^{\text{Shapley}}(\zeta)$ is just the average capital contribution of the business area i to all other business areas. The infinitesimal analog of the capital contribution $\Delta_i(\zeta, B) = \zeta(B \cup \{i\}) - \zeta(B)$ is just the derivative of $\rho \circ \tilde{X}$ with respect to the volume parameter for the i-th business area, $\frac{\partial \rho \circ \tilde{X}}{\xi_i}$. To get the average capital contribution we integrate the infinitesimal capital contributions. As a result we have the following definition.

Definition 5.17 Let X be a random variable describing loss and $\tilde{X} : U \subset \mathbb{R}^m \to \mathcal{M}(\Omega, \mathbb{R})$ a volume-parametrization for X with the risk measure ρ that has the following properties:

(i) U contains a neighborhood of the line from $(0, \dots, 0)$ to $(1, \dots, 1)$,
(ii) $\tilde{X}(0, \dots, 0) = 0$.

The *Aumann-Shapley* capital allocation is given by

$$\bar{\Lambda}_i^{\text{Aumann-Shapley}}(\rho) = \int_0^1 \frac{\partial \rho \circ X(t(1, \dots, 1))}{\partial \xi_i} \, dt.$$

In contrast to the Shapley algorithm, here management considerations are primary.

Proposition 5.4 *The Aumann-Shapley capital allocation fulfills Axiom* 5.1.

Proof The assertion follows directly from

$$\rho\big(X(1, \dots, 1)\big) = \rho\big(X(1, \dots, 1)\big) - \rho\big(X(0, \dots, 0)\big)$$

$$
\begin{aligned}
&= \int_0^1 \frac{\mathrm{d}\rho(t(1,\ldots,1))}{\mathrm{d}t}\mathrm{d}t \\
&= \sum_{i=1}^m \int_0^1 \left(\frac{\partial\rho(t(\xi_1,\ldots,\xi_m))}{\partial\xi_i}\xi_i \right)_{|\xi=(1,\ldots,1)} \mathrm{d}t \\
&= \sum_{i=1}^m \bar{\Lambda}_i^{\text{Aumann-Shapley}}(\rho).
\end{aligned}
$$

□

Axiom 5.2 may not be satisfied, if the Aumann-Shapley capital allocation is not based on a segmentation.

Proposition 5.5 *Suppose the volume parametrization* $\tilde{X}$ *is induced by a segmentation and let* ρ *be a homogeneous, subadditive risk measure. Then the Aumann-Shapley capital allocation is an allotment.*

Proof From

$$
\begin{aligned}
\rho\big(X(\tilde{\xi})\big) &= \rho\big(X(\tilde{\xi})\big) - \rho\big(X(0)\big) \\
&= \int_0^1 \frac{\mathrm{d}\rho \circ X(t\tilde{\xi})}{\mathrm{d}t}\mathrm{d}t = \sum_{i=1}^m \tilde{\xi}_i \int_0^1 \frac{\partial\rho \circ X(t\tilde{\xi})}{\partial\xi_i}\mathrm{d}t
\end{aligned}
$$

follows the validity of Axiom 5.1.

Analogously to the proof of Proposition 5.2 we calculate

$$
\begin{aligned}
&\bar{\Lambda}_i^{\text{Aumann-Shapley}}(\rho) \\
&\quad= \int_0^1 \left(\frac{\partial\rho(\sum_{j=1}^m \xi_j X_j)}{\partial\xi_i} \right)_{|t(1,\ldots,1)} \mathrm{d}t \\
&\quad= \int_0^1 \lim_{\varepsilon\to 0} \frac{\rho((t+\varepsilon)X_i + t\sum_{j=1,j\neq i}^m X_j) - \rho(tX_i + t\sum_{j=1,j\neq i}^m X_j)}{\varepsilon}\mathrm{d}t \\
&\quad\leq \int_0^1 \rho(X_i)\,\mathrm{d}t = \rho(X_i).
\end{aligned}
$$

So Axiom 5.2 is also satisfied. □

Example 5.5 We take it, as in Example 5.3, that $\tilde{X}$ is given by

$$
\tilde{X}(\xi) = \xi_F X_F + \xi_W X_W + \xi_C X_C.
$$

It holds that

$$
\begin{aligned}
\rho \circ \tilde{X}(t\tilde{\xi}) &= t(\tilde{\xi}_F\mu_F + \tilde{\xi}_W\mu_W + \tilde{\xi}_C\mu_C) \\
&\quad + \beta\sigma(t\tilde{\xi}_F X_F + t\tilde{\xi}_W X_W + t\tilde{\xi}_C X_C)
\end{aligned}
$$

$$= t(\tilde{\xi}_F \mu_F + \tilde{\xi}_W \mu_W + \tilde{\xi}_C \mu_C)$$

$$+ \beta\sigma \left(\begin{pmatrix} t\tilde{\xi}_F & t\tilde{\xi}_W & t\tilde{\xi}_C \end{pmatrix} \begin{pmatrix} X_F \\ X_W \\ X_C \end{pmatrix} \right)$$

$$= t(\tilde{\xi}_F \mu_F + \tilde{\xi}_W \mu_W + \tilde{\xi}_C \mu_C)$$

$$+ \beta \sqrt{ \begin{pmatrix} t\tilde{\xi}_F & t\tilde{\xi}_W & t\tilde{\xi}_C \end{pmatrix} \operatorname{cov} \begin{pmatrix} t\tilde{\xi}_F \\ t\tilde{\xi}_W \\ t\tilde{\xi}_C \end{pmatrix} }$$

$$= t(\tilde{\xi}_F \mu_F + \tilde{\xi}_W \mu_W + \tilde{\xi}_C \mu_C)$$

$$+ \beta \sqrt{(t\tilde{\xi}_F \sigma_F)^2 + (t\tilde{\xi}_W \sigma_W)^2 + (t\tilde{\xi}_C \sigma_C)^2 + 2t\tilde{\xi}_F t\tilde{\xi}_W \operatorname{cov}_{FW}}.$$

Thus we obtain at $\tilde{\xi} = (1, \ldots, 1)$

$$\frac{\partial \rho \circ \tilde{X}(t\tilde{\xi})}{\partial \xi_F} = \mu_F + \frac{\beta t (2(\sigma_F)^2 + 2\operatorname{cov}_{FW})}{2t\sqrt{(\sigma_F)^2 + (\sigma_W)^2 + (\sigma_C)^2 + 2\operatorname{cov}_{FW}}}$$

$$= \mu_F + \beta \frac{(\sigma_F)^2 + \operatorname{cov}_{FW}}{\sigma(X)}$$

$$\frac{\partial \rho \circ \tilde{X}(t\tilde{\xi})}{\partial \xi_W} = \mu_W + \frac{\beta t (2(\sigma_W)^2 + 2\operatorname{cov}_{FW})}{2t\sqrt{(\sigma_F)^2 + (\sigma_W)^2 + (\sigma_C)^2 + 2\operatorname{cov}_{FW}}}$$

$$= \mu_W + \beta \frac{(\sigma_W)^2 + \operatorname{cov}_{FW}}{\sigma(X)}$$

$$\frac{\partial \rho \circ \tilde{X}(t\tilde{\xi})}{\partial \xi_F} = \mu_C + \frac{\beta t 2(\sigma_C)^2}{2t\sqrt{(\sigma_F)^2 + (\sigma_W)^2 + (\sigma_C)^2 + 2\operatorname{cov}_{FW}}}$$

$$= \mu_C + \beta \frac{(\sigma_C)^2}{\sigma(X)}.$$

Since the integrand no longer depends on t, it follows that

$$\bar{\Lambda}_F^{\text{Aumann-Shapley}}(\rho) = \mu_F + \frac{\beta((\sigma_F)^2 + \operatorname{cov}_{FW})}{\sigma}$$

$$= -10 + \frac{2.9 \times (144 + 15)}{15.4} = 19.9,$$

$$\bar{\Lambda}_W^{\text{Aumann-Shapley}}(\rho) = \mu_W + \frac{\beta((\sigma_W)^2 + \operatorname{cov}_{FW})}{\sigma}$$

$$= -5 + \frac{2.9 \times (6.25 + 15)}{15.4} = -1.0,$$

$$\begin{aligned}\bar{\Lambda}_C^{\text{Aumann-Shapley}}(\rho) &= \mu_C + \frac{\beta(\sigma_C)^2}{\sigma} \\ &= -5 + \frac{2.9 \times 56.25}{15.4} = 5.6\end{aligned}$$

and also

$$\frac{\bar{\Lambda}_F^{\text{Aumann-Shapley}}(\rho)}{\rho(X)} = \frac{19.9}{24.5} = 81\ \%,$$

$$\frac{\bar{\Lambda}_W^{\text{Aumann-Shapley}}(\rho)}{\rho(X)} = \frac{-1.0}{24.5} = -4\ \%,$$

$$\frac{\bar{\Lambda}_C^{\text{Aumann-Shapley}}(\rho)}{\rho(X)} = \frac{5.6}{24.5} = 23\ \%.$$

So for our example we get the same capital allocation as in the case of the covariance capital allocation.

5.2.4 Kalkbrener's Axioms

In the previous section we set out and considered different capital allocations and examined whether they satisfy Axioms 5.1 and 5.2. In this section we will proceed in the opposite direction in that we—as far as possible—derive a capital allocation from an axiom system. This section is based on the work of Kalkbrener [4].

While we have up to this point on each occasion considered a loss function X and a segmentation into business areas, we will formulate axioms in this section that concern the space of possible loss functions and arbitrary business areas. We are shifting, in a sense, from a point consideration of a concrete company situation to the mathematical vector space of all possible loss functions. In addition, we will extrapolate some financial properties to an extent that is not backed any more by a direct financial interpretation. As a result our axiom system is not as well motivated as Axioms 5.1 and 5.2, but consequently does offer a much richer structure, which will allow us to come to more far-reaching conclusions.

Since a capital allocation for a segmentation into business areas is supposed to encode how much each business area has contributed to the diversification effect within the company as a whole, the capital allocation must consider all of these business areas simultaneously. On the other hand, capital allocation should not depend on an indiscriminate segmentation of the company into business areas. If, for example, a company has the lines of business "Fire", "Hail", "Storm", "Earthquake", "Indemnity", and "Theft", it is of no significance for the capital allocation to the line of business "Theft" whether the company views the other lines of business individually, or, perhaps, puts "Hail", "Storm", and "Earthquake" together in a line of business "Natural Phenomena".

This means it should be possible to give the capital allocation for a business area whose losses are described by a random variable U, as a function of U and the random variable X describing the losses of the whole company. So we will express the capital allocation as a bivariate mapping $\Lambda(U, X)$, where $\Lambda(U, X)$ gives the capital amount to be allotted to the subportfolio U of the entire portfolio X. As a typical application one chooses $U = X_i$ if there is a prescribed segmentation $(X_1, \ldots, X_m)$.

Definition 5.18 A *global capital allocation* is a mapping

$$\Lambda\colon \mathscr{M}(\Omega, \mathbb{R}) \times \mathscr{M}(\Omega, \mathbb{R}) \to \mathbb{R}$$
$$(U, Y) \mapsto \Lambda(U, Y).$$

If $0 \leq U \leq X$, then a business area can be constructed so that U is the random variable that describes the losses of that business area. $\Lambda(U, X)$ will then be taken as the capital allocated to this business area. If it is not the case that $0 \leq U \leq X$, then, in contrast, we have no financial interpretation of the value $\Lambda(U, X)$.

The name "global capital allocation" is motivated by the fact that we are now considering the space of all random variables instead of a single chosen random variable X.

In this section we consider a fixed risk measure ρ.

Lemma 5.2 *If for a global capital allocation Λ and any segmentation Axiom* 5.1 *holds, then the equation $\Lambda(X, X) = \rho(X)$ follows.*

Proof This follows by applying Axiom 5.1 to he trivial segmentation (X) into a single business area. □

This motivates our first Axiom:

Axiom 5.3 (Total allocation) For each $X \in \mathscr{M}(\Omega, \mathbb{R})$ it holds that $\Lambda(X, X) = \rho(X)$.

By analogy with the definition of coherent risk measures, we would like to identify some intuitive axioms that a capital allocation map should fulfill.

Axiom 5.4 (Linearity) For all $U, V, X \in \mathscr{M}(\Omega, \mathbb{R})$ and $a, b \in \mathbb{R}$, it holds that

$$\Lambda(aU + bV, X) = a\Lambda(U, X) + b\Lambda(V, X).$$

Kalkbrener's Linearity Axiom 5.4 is a strengthening of Axiom 5.1. To justify linearity we consider a segmentation $(X_1, \ldots, X_m)$ of X. The equation $\sum_{i=1}^{m} \Lambda(X_i, X) = \rho(X) = \Lambda(X, X)$ follows directly from Axiom 5.1, which says there can be no excess capital, and Axiom 5.3. Since X and the segmentation $(X_1, \ldots, X_m)$ were arbitrary, additivity, $\Lambda(U + V, X) = \Lambda(U, X) + \Lambda(V, X)$, follows if $0 \leq U, V$ and $U + V = X$. If $U + V < X$, then, upon setting $W = X - U - V$, we have the equation

$$\Lambda(U+V+W,X)=\Lambda(U,X)+\Lambda(V,X)+\Lambda(W,X)$$

and similarly

$$\Lambda(U+V+W,X)=\Lambda(U+V,X)+\Lambda(W,X).$$

From this follows additivity $\Lambda(U+V,X)=\Lambda(U,X)+\Lambda(V,X)$ for any $0\leq U,V$ with $U+V\leq X$. To motivate the multiplicative part of linearity, first assume $k,l\in\mathbb{N}$. Then the business areas that are described by U can be divided into k identical business areas, each described by $V=U/k$. While $lU\leq X$ holds, we can consider the sum of l of these business areas, $W=lU/k$. We derive from additivity

$$\begin{aligned}\frac{k}{l}\Lambda(W,X)&=\frac{1}{l}\overbrace{\big(\Lambda(W,X)+\cdots+\Lambda(W,X)\big)}^{k\text{ summands}}=\frac{1}{l}\overbrace{\big(\Lambda(V,X)+\cdots+\Lambda(V,X)\big)}^{lk\text{ summands}}\\&=\overbrace{\Lambda(V,X)+\cdots+\Lambda(V,X)}^{k\text{ summands}}=\Lambda\big(\overbrace{V+\cdots+V}^{k\text{ summands}},X\big)\\&=\Lambda(kV,X)=\Lambda\left(\frac{k}{l}W,X\right).\end{aligned}$$

So it has been shown that, for business areas with $U\geq 0$ and positive rational numbers $\frac{k}{l}$ with $lU\leq X$, multiplicativity follows from additivity. Assume furthermore that additivity holds also for negative random variables, and for random variables that for which $U+V\leq X$ does not hold. This is a mathematically well-motivated extrapolation, which however cannot be clearly motivated by the description of business reality. With this additional assumption, multiplicativity for all rational numbers follows. Since any real number can be approximately as closely as desired by a rational number, it suggests itself to postulate multiplicativity for arbitrary real numbers. To motivate the requirement of continuity, it should be remarked that discontinuity in the description of reality can almost always be explained as an artifact of a simplifying model.

Lemma 5.3 *If for a global capital allocation Λ and any segmentation Axiom* 5.2 *holds, then for $U,X\in\mathcal{M}(\Omega,\mathbb{R})$ with $0\leq U\leq X$ we have the relationship $\Lambda(U,X)\leq\Lambda(U,U)$.*

Proof This follows by applying Axiom 5.2 to the segmentation $(U,X-U)$. □

The next Axiom strengthens Axiom 5.2

Axiom 5.5 (Stong Diversification) If $U,X\in\mathcal{M}(\Omega,\mathbb{R})$, then $\Lambda(U,X)\leq\Lambda(U,U)$.

For $0\leq U\leq X$, assuming the diversification property of Axiom 5.2, Axiom 5.5 follows directly from Lemma 5.3. If $\{\omega\in\Omega:U(\omega)\geq X(\omega)\text{ or }U(\omega)<0\}$ is not a null set, there is no direct interpretation. As we did for the linearity of Axiom 5.4, we motivate Axiom 5.5 as a strengthening of Axiom 5.2 through mathematical extrapolation to all random variables U, X.

Definition 5.19 Let ρ be a risk measure. A *global allotment* with respect to ρ is a global capital allocation Λ, for which the Axioms 5.3, 5.4, and 5.5 hold.

For each global allotment, any segmentation $(X_1, \ldots, X_m)$ of X induces an allotment. But the theorem below shows that, in contrast to allotments in general, global allotments are only possible for a special class of risk measures.

Theorem 5.3 *Let ρ be a risk measure and Λ a global allotment. Then ρ is positively homogeneous and subadditive.*

Proof Let $a \geq 0$. Then by Axioms 5.4 and 5.5

$$a\Lambda(U,U) \overset{\text{linearity}}{\overbrace{=}} \underbrace{\Lambda(aU,U) \leq \Lambda(aU,aU)}_{\text{strong diversification}} \overset{\text{linearity}}{\overbrace{=}} \underbrace{a\Lambda(U,aU) \leq a\Lambda(U,U)}_{\text{strong diversification}}$$

holds.

It follows on taking into consideration Axiom 5.5 that

$$a\rho(U) = a\Lambda(U,U) = \Lambda(aU,aU) = \rho(aU),$$

so ρ is positively homogeneous. Subadditivity follows from

$$\begin{aligned}\rho(U+V) &= \Lambda(U+V,U+V) = \Lambda(U,U+V) + \Lambda(V,U+V) \\ &\leq \Lambda(U,U) + \Lambda(V,V) \\ &= \rho(U) + \rho(Y).\end{aligned}$$

□

In particular, it is not possible for general loss distributions to have a capital allocation for the risk measure VaR_α satisfying the axioms. Even if some of our axioms are more demanding than one might wish, this result is another indication that the expected shortfall can be more robustly motivated for use in company management than the value at risk.

We will now show that for each positively homogeneous subadditive risk measure there does exist a global allotment.

Lemma 5.4 (Hahn-Banach) *Let E be a real vector space and $\rho\colon E \to \mathbb{R}$ a convex map. Let $F \subseteq E$ be a subspace and $f\colon F \to \mathbb{R}$ a linear map with $f(U) \leq \rho(U)$ for all $U \in F$. Then there exists a linear map $h\colon E \to \mathbb{R}$ with $h(V) \leq \rho(V)$ and $h(U) = f(U)$ for all $V \in E$ and $U \in F$.*

Proof See [5, Theorem III.5]. □

Lemma 5.5 *Let $\rho\colon \mathscr{M}(\Omega,\mathbb{R}) \to \mathbb{R}$ be a positively homogeneous, subadditive map and $V \in \mathscr{M}(\Omega,\mathbb{R})$. then there exists a linear map $h_V\colon \mathscr{M}(\Omega,\mathbb{R}) \to \mathbb{R}$ with*

(i) $h_V(V) = \rho(V)$,
(ii) $h_V(U) \leq \rho(U)$ *for all* $U \in \mathscr{M}(\Omega,\mathbb{R})$.

Proof We want to apply Lemma 5.4. By its positive homogeneity and subadditivity the map ρ satisfies the convexity condition $\rho(tU + (1-t)V) \leq t\rho(U) + (1-t)\rho(V)$ for all $U, V \in \mathscr{M}(\Omega,\mathbb{R})$ and $t \in [0,1]$. Consider the vector space $F = \{cV : c \in \mathbb{R}\} \subseteq \mathscr{M}(\Omega,\mathbb{R})$ and the functional $f\colon cV \mapsto c\rho(V)$. Then f satisfies $f(V) = \rho(V)$, $f(cV) = \rho(cV)$ for all $c \geq 0$. From

$$\begin{aligned}0 &= \rho(0) = \rho(-cV + cV)\\ &\leq \rho(-cV) + \rho(cV) = \rho(-cV) - f(-cV)\end{aligned}$$

for all $c \geq 0$ follows $f(-cV) \leq \rho(-cV)$ for all $c \geq 0$, and therefore $f(U) \leq \rho(U)$ for all $U \in F$. Lemma 5.4 thus gives the existence of the map h_V. □

Theorem 5.4 (Existence) *For each positively homogeneous, subadditive risk measure ρ there is a global allotment Λ.*

Proof Lemma 5.5 guarantees that for each $V \in \mathscr{M}(\Omega,\mathbb{R})$ there is a linear map $h_V\colon \mathscr{M}(\Omega,\mathbb{R}) \to \mathbb{R}$ with $h_V(V) = \rho(V)$ and $h_V(U) \leq \rho(U)$ for all $U \in \mathscr{M}(\Omega,\mathbb{R})$. Let

$$\begin{aligned}\Lambda\colon \mathscr{M}(\Omega,\mathbb{R}) \times \mathscr{M}(\Omega,\mathbb{R}) &\to \mathbb{R}\\ (U,V) &\mapsto \Lambda(U,V) = h_V(U).\end{aligned}$$

By construction of Λ we have $\Lambda(U,U) = h_U(U) = \rho(U)$ for all U, so that Λ satisfies Axiom 5.3. The linearity (Axiom 5.4) is obvious, since $h_V(\cdot)$ is linear. From

$$h_V(U) \leq \rho(U) = h_U(U)$$

follows the diversification condition, Axiom 5.5. □

The global allotment Λ guaranteed by Theorem 5.4 is not necessarily unique. Uniqueness does follow under an additional continuity assumption.

Definition 5.20 (Continuity) Let $X \in \mathscr{M}(\Omega,\mathbb{R})$. A global capital allocation Λ is called *continuous* in X, if for each $U \in \mathscr{M}(\Omega,\mathbb{R})$ it holds that

$$\lim_{\varepsilon \to 0} \Lambda(U, X + \varepsilon U) = \Lambda(U, X).$$

In practice X will be a random variable corresponding to the total loss for the company. If the global capital allocation were not continuous in X then an infinitesimal change in the total portfolio could lead to a noticeable change in the capital allocation, which is not plausible. If one requires continuity not only for the total loss but also for all $X \in \mathscr{M}(\Omega, \mathbb{R})$, then the space of those risk measures, which are compatible with the axioms, will be too restrictive from an economic point of view.[3]

Theorem 5.5 *Let $X \in \mathscr{M}(\Omega, \mathbb{R})$, ρ be a risk measure and Λ a global allotment. If Λ is continuous in X, then*

$$\Lambda(U, X) = \lim_{\varepsilon \to 0} \frac{\rho(X + \varepsilon U) - \rho(X)}{\varepsilon}$$

for all $U \in \mathscr{M}(\Omega, \mathbb{R})$.

Proof Let $\varepsilon, \eta \in \mathbb{R}$. There follows

$$\begin{aligned}\rho(X + \eta U) &= \Lambda(X + \eta U, X + \eta U) \\ &\geq \Lambda(X + \eta U, X + \varepsilon U) \\ &= \Lambda\big(X + \varepsilon U + (\eta - \varepsilon)U, X + \varepsilon U\big) \\ &= \Lambda(X + \varepsilon U, X + \varepsilon U) + (\eta - \varepsilon)\Lambda(U, X + \varepsilon U) \\ &= \rho(X + \varepsilon U) + (\eta - \varepsilon)\Lambda(U, X + \varepsilon U)\end{aligned}$$

and from this

$$\rho(X + \eta U) - \rho(X + \varepsilon U) \geq (\eta - \varepsilon)\Lambda(U, X + \varepsilon U).$$

Analogously we obtain

$$\rho(X + \varepsilon U) - \rho(X + \eta U) \geq (\varepsilon - \eta)\Lambda(U, X + \eta U),$$

which is equivalent to

$$\rho(X + \eta U) - \rho(X + \varepsilon U) \leq (\eta - \varepsilon)\Lambda(U, X + \eta U).$$

If $\eta > \varepsilon$, it follows that

$$\Lambda(U, X + \varepsilon U) \leq \frac{\rho(X + \eta U) - \rho(X + \varepsilon U)}{\eta - \varepsilon} \leq \Lambda(U, X + \eta U).$$

Since Λ in the second argument is continuous in X, the limit $(\varepsilon, \eta) \to (0, 0)$ exists, and the assertion of the Theorem follows. □

From Theorem 5.5 results, in particular, the following uniqueness statement about capital allocation for a given overall loss amount X.

[3]Sebastian Maass has made the observation that, assuming continuity for $X = 0$ and homogeneity of the risk measure, it follows that $\Lambda(U, 0) = \rho(U)$ for all $U \in \mathscr{M}(\Omega, \mathbb{R})$. In particular, $U \mapsto \rho(U)$ is then linear, which is not the case for sensible risk measures ρ. His remark follows immediately from the representation for Λ given in Theorem 5.5 below, if one puts $X = 0$.

Corollary 5.1 *If a global allotment Λ for a risk measure ρ is continuous at X, then $\Lambda(U,X)$ for all $U \in \mathcal{M}(\Omega,\mathbb{R})$ is uniquely determined by ρ.*

Corollary 5.2 *Let Λ be a global allotment for the risk measure ρ. If Λ is continuous in X and $(X_1,\ldots,X_m)$ is a segmentation, then $\Lambda(X_i,X)$ is the Euler capital allocation for business area i.*

Proof The assertion follows directly from Theorem 5.5,

$$\begin{aligned}\Lambda(X_i,X) &= \lim_{\varepsilon\to 0}\frac{\rho(X+\varepsilon X_i)-\rho(X)}{\varepsilon} = \left(\frac{\partial\rho(X+\varepsilon X_i)}{\partial\varepsilon}\right)_{|\varepsilon=0}\\ &= \left(\frac{\partial\rho(\sum_{j=1,j\neq i}^{m} X_j+\xi_i X_i)}{\partial\xi_i}\right)_{|\xi_i=1}\\ &= \bar{\Lambda}_i^{\text{Euler}}.\end{aligned}$$

□

Theorem 5.6 (Uniqueness) *Assume $U,V\in\mathcal{M}(\Omega,\mathbb{R})$ and that ρ is a positively homogeneous, subadditive risk measure. If the directional derivative*

$$\lim_{\varepsilon\to 0}\frac{\rho(V+\varepsilon U)-\rho(V)}{\varepsilon},$$

exists then the global allotment in (U,V) is unique and given by this directional derivative.

Proof By Theorem 5.4 there does exist a global allotment Λ. Since

$$\begin{aligned}\rho(V+\varepsilon U)-\rho(V) &= \Lambda(V+\varepsilon U, V+\varepsilon U)-\Lambda(V,V)\\ &\geq \Lambda(V+\varepsilon U, V)-\Lambda(V,V)\\ &= \varepsilon\Lambda(U,V)\end{aligned}$$

holds for each $\varepsilon\in\mathbb{R}$, it follows that for $\varepsilon<0$

$$\frac{\rho(V+\varepsilon U)-\rho(V)}{\varepsilon}\leq\Lambda(U,V)$$

and for $\varepsilon>0$

$$\frac{\rho(V+\varepsilon U)-\rho(V)}{\varepsilon}\geq\Lambda(U,V).$$

In both cases the left-hand side converges to the directional derivative. Therefore it must coincide with $\Lambda(U,V)$. □

Example 5.6 The risk measure $\rho(\cdot)=\mathrm{E}(\cdot)+\beta\sigma(\cdot)$ is homogeneous and subadditive. In addition, the directional derivative $\Lambda_\rho(U,X)$ exists for $X\neq 0$. From Corollary 5.2 and the special form of our risk measure it follows that the global allotment for our example is just the covariance capital allocation (see Example 5.3).

The most important special case is the expected shortfall for confidence level α. Since this risk measure is positively homogeneous and subadditive there is a global allotment.

Proposition 5.6 *For the risk measure* ES_α *there exists a global allotment given by*

$$\Lambda_{\mathrm{ES}_\alpha}(U,X)=\frac{1}{1-\alpha}\mathrm{E}(U 1_{X,\mathrm{VaR}_\alpha(X),\alpha}), \tag{5.12}$$

where we used the notation

$$1_{V,x,\alpha}=1_{\{V>x\}}+\beta_{V,\alpha}(x)1_{\{V=x\}}$$

$$\beta_{V,\alpha}(x)=\begin{cases}\frac{\mathbf{P}(V\leq x)-\alpha}{\mathbf{P}(V=x)} & \textit{if } \mathbf{P}(V=x)>0\\ 0 & \textit{otherwise}.\end{cases}$$

Proof We have to show that the mapping given by (5.12) satisfies the Axioms 5.3, 5.4, and 5.5.

The validity of Axiom 5.4 follows immediately from the linearity of the expectation value, and Axiom 5.3 is satisfied by Lemma 2.5(iii).

It remains to show $\Lambda_{\mathrm{ES}_\alpha}(U,X)\leq \Lambda_{\mathrm{ES}_\alpha}(U,U)$ in order to verify Axiom 5.5.

For $V\in\mathcal{M}(\Omega,\mathbb{R})$ set

$$\begin{aligned}A_V&=\{\omega\in\Omega:\ V(\omega)>\mathrm{VaR}_\alpha(V)\},\\ B_V&=\{\omega\in\Omega:\ V(\omega)=\mathrm{VaR}_\alpha(V)\},\\ C_V&=\Omega\setminus(A_V\cup B_V).\end{aligned}$$

These sets are disjoint and measurable, and $A_V\cup B_V\cup C_V=\Omega$. With the abbreviation $\beta_V=\beta_{V,\mathrm{VaR}_\alpha(V)}$ there follows from Lemma 2.5(ii)

$$1-\alpha=\mathbf{P}(A_V)+\beta_V\mathbf{P}(B_V). \tag{5.13}$$

For $X,U\in\mathcal{M}(\Omega,\mathbb{R})$ the sets

$$\begin{matrix}A_X\cap A_U, & A_X\cap B_U, & A_X\cap C_U,\\ B_X\cap A_U, & B_X\cap B_U, & B_X\cap C_U,\\ C_X\cap A_U, & C_X\cap B_U, & C_X\cap C_U,\end{matrix}$$

are also a disjoint and measurable partition of Ω. We obtain

$$\begin{aligned}(1-\alpha)\Lambda_{\mathrm{ES}_\alpha}(U,X)&=\int_{A_X}U\,\mathrm{d}\mathbf{P}+\beta_X\int_{B_X}U\,\mathrm{d}\mathbf{P}\\ &=\int_{A_X\cap A_U}U\,\mathrm{d}\mathbf{P}+\int_{A_X\cap B_U}U\,\mathrm{d}\mathbf{P}+\int_{A_X\cap C_U}U\,\mathrm{d}\mathbf{P}\\ &\quad+\beta_X\int_{B_X\cap A_U}U\,\mathrm{d}\mathbf{P}+\beta_X\int_{B_X\cap B_U}U\,\mathrm{d}\mathbf{P}+\beta_x\int_{B_X\cap C_U}U\,\mathrm{d}\mathbf{P}.\end{aligned}$$

Since $U \leq \text{VaR}_\alpha(U)$ holds on $B_U \cup C_U$, it follows that

$$\begin{aligned}(1-\alpha)\Lambda_{\text{ES}_\alpha}(U,X) &\leq \int_{A_X\cap A_U} U\,\mathrm{d}\mathbf{P} + \beta_x \int_{B_X\cap A_U} U\,\mathrm{d}\mathbf{P} \\ &\quad + \text{VaR}_\alpha(U)(\mathbf{P}(A_X\cap B_U) + \beta_X \mathbf{P}(B_X\cap B_U) \\ &\quad + \mathbf{P}(A_X\cap C_U) + \beta_X \mathbf{P}(B_X\cap C_U)) \\ &= \int_{A_X\cap A_U} U\,\mathrm{d}\mathbf{P} + \beta_X \int_{B_X\cap A_U} U\,\mathrm{d}\mathbf{P} \\ &\quad + \text{VaR}_\alpha(U)\big(1-\alpha-\mathbf{P}(A_X\cap A_U) - \beta_X\mathbf{P}(B_X\cap A_U)\big) \\ &= \int_{A_X\cap A_U} U\,\mathrm{d}\mathbf{P} \\ &\quad + \beta_X \int_{B_X\cap A_U} U\,\mathrm{d}\mathbf{P} + (1-\beta_X)\,\text{VaR}_\alpha(U)\mathbf{P}(B_X\cap A_U) \\ &\quad + \text{VaR}_\alpha(U)\big(1-\alpha-\mathbf{P}(A_X\cap A_U) - \mathbf{P}(B_X\cap A_U)\big) \\ &\leq \int_{(A_X\cup B_X)\cap A_U} U\,\mathrm{d}\mathbf{P} \\ &\quad + \text{VaR}_\alpha(U)\big(1-\alpha-\mathbf{P}(A_X\cap A_U) - \mathbf{P}(B_X\cap A_U)\big) \\ &\leq \int_{A_U} U\,\mathrm{d}\mathbf{P} - \text{VaR}_\alpha(U)\mathbf{P}(C_X\cap A_U) \\ &\quad + \text{VaR}_\alpha(U)\big(1-\alpha-\mathbf{P}(A_X\cap A_U) - \mathbf{P}(B_X\cap A_U)\big) \\ &= \int_{A_U} U\,\mathrm{d}\mathbf{P} - \text{VaR}_\alpha(U)\beta_U\mathbf{P}(B_U) \\ &= (1-\alpha)\Lambda_{\text{ES}_\alpha}(U,U),\end{aligned}$$

where we used (5.13) in last step but one. □

The global allotment for the expected shortfall can be described vividly to the effect that each business area is allocated exactly the portion of the risk capital which corresponds to the losses that business area could generate. After all $\Lambda_{\text{ES}_\alpha}(U,X)$ is just the expected value of the losses caused by U that can result from events with total losses greater than $\text{VaR}_\alpha(X)$. Thus this method corresponds very well to our intuitive notion of fairness.

We now want to demonstrate how we can determine this global allotment for a continuous overall loss distribution function F_X in the framework of a Monte Carlo simulation.

1. For a given Monte Carlo simulation, which considers several different business areas at the same time, it is possible for each $p \in [0,1]$ to ascertain those risks

that contribute to VaR_p. For this the Monte Carlo run is examined that shows as the loss just $X = \mathrm{VaR}_p(X)$.

If, for example, 10000 simulations are carried out and $p = 99.5\ \%$ is chosen, then first all runs are sorted by increasing loss, and then from this list number 9950 is selected.

2. The expected shortfall $\mathrm{ES}_\alpha(X)$ for $\alpha = 99\ \%$ would be, for 10000 simulations, simply the arithmetic mean of $\mathrm{VaR}_p(X)$ for the 100 runs that correspond to $p > \alpha$.
3. To determine the capital allocation for business areas, the contribution to the total risk from each business area is determined. If we want to calculate, in our example, the allocation for a business area U we would find the loss amount $V(i, \alpha, X, U)$ for area U in each of the 100 runs i with which $\mathrm{ES}_\alpha(X)$ was estimated, and thereafter take the average

$$\frac{\sum_{i=9901}^{10000} V(i, \alpha, X, U)}{100}.$$

This is just the discretized version of the global allotment for the expected shortfall, where we have used that the second integral in (5.12) vanishes because F_X is continuous.

Remark 5.9 It is clear that, for these discretized methods also, the sum of capital allocations over all business areas gives the total risk capital.

Remark 5.10 The method can only be modeled if the overall distribution of all risks can be modeled. This is a practical reason why simple copulas are to be preferred to linear correlations.

Remark 5.11 The contribution of a business area U found by this method can be markedly different from the expected shortfall for this business area. It is conceivable, for example, that large losses for U occur just in those simulations that lead to profits for the other business areas, and thus do not enter into the calculation of the expected shortfall for the total risk.

Remark 5.12 This method can readily be generalized to spectral measures, but is less well suited to the value at risk: With the one run that determines $\mathrm{VaR}_\alpha(X)$ one cannot also estimate the contributions of individual business areas well. This is so because a larger loss by business area A and a smaller loss by business area B have the same effect on the overall loss as a smaller loss for A and a larger one for B, but lead to quite different allocations. If the parameters are adjusted a little, or a different generator of randomness is chosen, there will result a different composition of individual losses. In our example, run 9900 is employed in the determination of $\mathrm{VaR}_\alpha(\mathrm{X})$. Even if we allow for an error of $\pm 0.1\ \%$, and therefore allow ourselves to work with 20 runs that perhaps approximate $\mathrm{VaR}_\alpha(X)$ adequately, that few runs will still not be enough for an allocation to several business areas. In contrast, the method functions, as a rule, well for general spectral measures, since, because there is an integration, one averages over many more runs.

The main advantage of allocation based on the total distribution compared with indicator-based allocation is that this method can be better communicated, and so leads to higher acceptance levels. If there are many business areas, then the computational effort required is also less than for the Shapley algorithm.

5.3 Capital Allocation for Groups

Insurers are operating more and more as groups, whether as an insurance group or as a financial group. Capital requirements for solvency can thus be defined for several levels within the group — from the business unit up to the level of the group. Diversification effects arise from aggregating and balancing risks within the group.

Diversification is fundamental for both risk management within insurance groups and increasingly for solvency requirements from insurance groups. Group diversification effects are best quantified by comparison of summed stand-alone solvency capital requirements and aggregated group target capital.

The balancing of risks within a group demands, however, mobility of capital. This could be hampered, if the business units are in different legal jurisdictions.

Furthermore, that a business unit belongs to a group brings additional group risks. For example, a business unit could be sold by the group management and be passed to run-off. In addition, damage to its image or other negative events for a business unit can have an adverse effect on all business units.

The effects of restricted capital mobility and group risks on the risk capital and the effectively available capital are hard to model. There are different methods conceivable and under discussion (for example [1–3]).

In general, both informal and legally binding instruments for transferring capital and risk in a group can be used, to balance risks between business units. However, for regulatory purposes it is important to minimize any moral risks. For this reason, the Swiss Group Solvency Test [2] only allows to take legally binding instruments into account. However, this does include sale of business units at their market prices, defined as risk-bearing capital less the market value margin for the run-off of the asset-liability portfolio.[4] From the point of view of solvency therefore a group consists in its business units (parent company and subsidiaries) and a network of legally binding agreements which transfer capital and risks and result in clearly defined cash flows under clearly defined conditions. The group target capital is defined as the sum of all target capital requirements for the individual business units. The network of risk and capital transfer agreements can, naturally, be optimized with a view to the group solvency capital requirements (see [3]). The diversification effect can then finally be measured by comparison of the thus defined group solvency capital requirements with and without use of risk and capital transfer agreements.

[4] Recall that in the Swiss Solvency Test, the market value margin is counted as part of the capital requirement rather than as a part of the liabilities

Diversification effects can obviously also be determined at the subgroup level, in that one may, for example, group business units according to their regional locations. It holds quite generally that the diversification effects for subgroups are smaller the deeper the level at which one puts together subgroups (see [3]). That, however, does **not** happen from the point of view of the individual business units! The example below serves as an illustration:

Example 5.7 We consider a Swiss group consisting of three business units distributed over two geographical regions, A and B, where the first two are in A and the third is in B. For the sake of simplicity we assume complete mobility of capital. Theoretically this corresponds to a perfect net of risk and capital transfer agreements which can generate the necessary cash flow for any state of the world. Let $\Delta\mathrm{RTK}_i$ be the change in the risk-bearing capital of business unit i without risk and capital transfer agreements. We assume that $\Delta\mathrm{RTK}_1, \Delta\mathrm{RTK}_2, \Delta\mathrm{RTK}_3$ are multivariately normally distributed with mean zero, variances $\mathrm{var}(\Delta\mathrm{RTK}_i) = 100$, and correlation matrix

$$\mathrm{corr}(\Delta\mathrm{RTK}) = \begin{pmatrix} 1 & 0 & 0.75 \\ 0 & 1 & 0 \\ 0.75 & 0 & 1 \end{pmatrix}.$$

Let the risk measure be the 99 % expected shortfall. Then we get from Proposition 2.1 the stand-alone capital requirements

$$\mathrm{TC}_i = \mathrm{ES}_{99\,\%}[-\Delta\mathrm{RTK}_i] = 2.6652 \times \sqrt{\mathrm{var}(-\Delta\mathrm{RTK}_i)} = 26.652.$$

Thanks to complete capital mobility, we can simply add the cash flows $\Delta\mathrm{RTK}_i$ at the group level, and obtain as the group target capital

$$\begin{aligned} \mathrm{TC}_G &= \mathrm{ES}_{99\,\%}\left[-\sum_{i=1}^{3} \Delta\mathrm{RTK}_i\right] \\ &= 2.6652 \times \sqrt{100 + 100 + 100 + 2 \times 0.75 \times 100} \\ &= 56.54. \end{aligned}$$

We allocate this group target capital with the help of the covariance principle (see Sect. 5.2.2.2) to the business units and obtain the diversified target capital requirements at group level

$$\begin{aligned} \mathrm{TC}_1^G = \mathrm{TC}_3^G &= 2.6652 \times \frac{\mathrm{cov}(\Delta\mathrm{RTK}_1, \sum_{i=1}^{3} \Delta\mathrm{RTK}_i)}{\sqrt{\mathrm{var}(\sum_{i=1}^{3} \Delta\mathrm{RTK}_i)}} \\ &= 2.6652 \times \frac{100 + 0.75 \times 100}{\sqrt{100 + 100 + 100 + 2 \times 0.75 \times 100}} = 21.99, \\ \mathrm{TC}_2^G &= 2.6652 \times \frac{100}{\sqrt{100 + 100 + 100 + 2 \times 0.75 \times 100}} = 12.56. \end{aligned}$$

As is the case for any compatible capital allocation method, we have $\mathrm{TC}_1^G + \mathrm{TC}_2^G + \mathrm{TC}_3^G = \mathrm{TC}_G$.

Analogously the solvency capital requirement for the whole group in the region A is

$$\mathrm{TC}_A = \mathrm{ES}_{99\,\%}\left[-\sum_{i=1}^{2} \Delta\mathrm{RTK}_i\right] = 2.6652 \times \sqrt{100+100} = 37.7.$$

Using covariance methods we obtain, with diversification at the regional level, the target capital requirements

$$\mathrm{TC}_1^A = \mathrm{TC}_2^A = 2.6652 \times \frac{\mathrm{cov}(\Delta\mathrm{RTK}_1, \sum_{i=1}^2 \Delta\mathrm{RTK}_i)}{\sqrt{\mathrm{var}(\sum_{i=1}^2 \Delta\mathrm{RTK}_i)}}$$

$$= 2.6652 \times \frac{100}{\sqrt{100+100}} = 18.85.$$

Obviously there results a smaller diversification benefit for business unit one at group level than at the regional level: $\mathrm{TC}_1^G > \mathrm{TC}_1^A$! This is to be traced back to the high correlation of 0.75 with business unit three. Nonetheless it does hold, as remarked above, that region A, as a subgroup, profits from being part of the whole group: $\mathrm{TC}_1^G + \mathrm{TC}_2^G < \mathrm{TC}_A$.

References

1. Committee of European Insurance and Occupational Pensions Supervisors (CEIOPS). Advice on sub-group supervision, diversification effects, cooperation with third countries and issues related to the MCR and the SCR in a group context, November 2006. Document CEIOPS-DOC-05/06 (p. 234)
2. Eidgenössische Finanzmarktaufsicht (FINMA), *Draft: Modelling of Groups and Group Effects* (2006) (p. 234)
3. D. Filipović, M. Kupper, Optimal capital and risk transfers for group diversification. Math. Finance **18**(1), 55–76 (2008) (pp. 234, 235)
4. M. Kalkbrener, An axiomatic approach to capital allocation. Math. Finance **15**(3), 425–438 (2005) (p. 224)
5. M. Reed, B. Simon, *Methods of Modern Mathematical Physics I: Functional Analysis* (Academic Press, New York, 1980) (p. 227)
6. M. Urban, Allokation von Risikokapital auf Versicherungsportfolios. Master thesis, TU München (2002) (p. 201)

Chapter 6
Performance Measurement

To judge the performance of a company it is not enough to determine its profits. Beyond that one should also take into account the risk linked to its profits. In order to do so one can use cost of capital (Sect. 4.1.2), but there are also other possibilities (Sect. 6.6.4). Both absolute and relative measurements lend themselves to the task of performance measurement.

We denote by C_t the necessary risk capital, by s_t the risk-free interest rate and by k_t the spread (or additional interest) that is supposed to compensate for the risk taken on. If we do not take into account the feedback effect from investing the risk capital itself, then we are assuming that risk capital is invested without risk and we mark the corresponding quantities with a tilde $\tilde{}$.

Depending on the application in question, C_t represents the economic risk capital C_t^{EC}, the regulatory capital C_t^{Reg}, or some other kind of capital. For example, RORAC is usually determined on the basis of cash flows, for which C_t represents the economic capital C_t^{EC}. In contrast, for fair value calculations of liabilities using the cost of capital approach, the capital notion C_t^{FV}, which is consistent with the risk aversion of the market, is usually selected.

Performance measurement of insurance companies is complicated by the fact that technical provisions are a central mechanism used for risk mitigation, and must be accounted for in measuring performance.

6.1 Performance Measurement Based on Balance Sheets

Balance sheets are, in general, not much use for performance measurement. Their purpose is to provide shareholders and other interested parties with as *objective* a picture as possible of the financial situation of a company. To attain this objectivity certain (partial) standardizations are adhered to, which poses limits on the interpretation of performance of these quantities. Furthermore, many accounting rules, like, for instance, those of the statutory accounting, are conservative. Conversely, estimates that depend on interpretations (e.g., of future loss developments) play a role in quantities that are used for performance measurement.

M. Kriele, J. Wolf, *Value-Oriented Risk Management of Insurance Companies*, EAA Series, DOI 10.1007/978-1-4471-6305-3_6, © Springer-Verlag London 2014

Performance measurement based on balance sheet data (and on the profit & loss account) can therefore only provide a distorted view of reality. However, for a third party who does not have access to company-internal information, these data are often the only pieces of information that are accessible. For this reason the company will communicate its performance using balance sheet data as proxies. Reciprocally, companies react to the feedback that they get from the market, so that managing a company based on the balance sheet can certainly make economic sense.

We start below under the assumption that internal information about the company is known, so that one is not thrown back upon balance sheet quantities.

6.2 Profit Measurement

Profit measurement for individual business areas presents no mathematical difficulties, since the expected profit of the whole company, as an expected value, is the linear superposition of profits of the individual corporate divisions.[1] Practically, however, there is the problem that some quantities that influence profits do not arise within these individual business areas. In this section, we will illustrate this with the example of costs.

It is clear to all that good cost accounting is needed for value-oriented management. Any business activity generates costs, and conversely one can assign costs to any product. While the costs generated by an activity are in principle measurable, the assignment of costs to products depends on additional factors. In the simplest case one can distinguish *direct costs* (or *variable costs*) and *fixed costs*. The fixed costs do not depend on whether a product is made or not, while direct costs depend on the volume produced.

Definition 6.1 By a *cost unit* we mean a product or activity which must be assigned costs in cost accounting. In *full-cost accounting* all costs are assigned to a cost unit. In *partial-cost accounting* only the direct costs (those costs directly resulting from production) are assigned to a cost unit.

If the profitability of the entire company is being examined, then full-cost accounting is appropriate, since otherwise not all costs are taken into account. On the other hand, when the profitability of individual products are considered, partial-cost accounting can be suitable, as otherwise approximative allotted fixed costs can skew the results.

We consider a highly simplified example. Suppose an insurance company has fixed costs of 180,000 € that should be allotted to the insurance policies in the portfolio according to the sums insured. Now we would like to compare the profitability of two policies. Policy A has an insured sum of 1,000 € and leads to directly assignable costs of 10 €. Policy B is for an amount of 5,000 € and has directly allocatable costs of 15 €. (The handling of the policies is the same, but clients with

[1] In contrast, risk capital is not linear, see Chap. 5.

higher policy amounts contact the company more often, which makes for 5 € in extra costs.) On average the benefit paid is 97 % for small insured amounts like policy A and 98 % for larger ones like policy B. If one considers only the directly allocatable costs, one obtains for the profit of policy A

$$(1-97\%) \times 1000\,€ - 10\,€ = 20\,€$$

and for policy B a profit of

$$(1-98\%) \times 5000\,€ - 15\,€ = 85\,€.$$

Thus policy B seems, with our partial-cost accounting, to be more profitable. Let the full amount insured in the total portfolio be 10,000,000 €. From this there are 18 € in fixed costs allotted to policy A and 90 € in fixed costs allotted to policy B. Using full-cost accounting, we find policy A is still just profitable by 2 €, while policy B shows a loss of 5 €. Using full-cost accounting one could conclude that policy B, in contrast to policy A, loses money. Management could therefore entertain the idea of canceling policy B (if possible). The result would be a worsening of the situation since now 85 € are missing, with which part of the fixed costs, naturally unchanged by canceling the policy, were covered.

Another difficulty in accounting for costs is that there is no clear difference between variable costs and fixed costs. A classical fixed cost is, for example, the rent on the administration building of the insurance company. As long as the variations in the volume of the business are not too large, the rental costs are indeed independent of production. However, if the company grows or shrinks a lot, the administration building will not be appropriate for the changed volume of business. We thus see a change in the rental costs resulting from a change in the company's business volume. On the other hand, the policy handling expenses are often considered as variable costs. But one will not hire or fire staff for each change in the volume of business. Even in the face of a large turn-down in business one must very carefully consider if one wishes to announce operations-related redundancies, since this does have an effect on the morale of the remaining staff, means costs for severance packages, and could produce additional costs when business picks up later, for new employees will have to be hired and trained.

The division into fixed and variable costs depends therefore on the company's strategy and the scenarios considered.

6.3 Absolute Performance Measures

In this section we will describe the Economic Value Added (EVA)[2] and related concepts.

---

[2]EVA is a registered trademark of the firm Stern Stewart & Co. The notion that a risk-appropriate interest rate is decisive as to whether one should invest in a company is however much older and is present, for example, in the work of Eugen Schmalenbach [5, pp. 49–50].

Table 6.1 Abbreviations for the components of net profit

Variable	Meaning	Time
P_t	Premium income	Period start
Com_t	Commissions paid	Period start
L_t	Claims payments	Period close
r_t	Relative yield on capital	During the period
A_t	Investment volume incl. risk capital	Period start
$I_t = r_t A_t$	Investment profit	Period close
E_t	Other profit or loss	Period close
K_t	Costs	Period close
Tax_t	Taxes	Period close
V_t	Technical provisions	Period close

Definition 6.2 The *net profit after tax* N_t is the difference between the income and the expenses after taxes. It is given by

$$N_t = P_t - \text{Com}_t - L_t + I_t + E_t - K_t - \text{Tax}_t - (V_t - V_{t-1}),$$

where we have used the abbreviations given in Table 6.1. See also Fig. 6.1.

The final term, the change in reserves, $(V_t - V_{t-1})$, can be interpreted as an "expected claims expenditure" for the business year. This term is not always counted toward the net profit (in particular, in applications in the area of banking). We subsume the change in reserves under the net profit, since making provisions plays a central role in the insurance business.

Capital costs are, as a rule, not included in N_t. Exceptions are real interest costs that arise from capital instruments such as subordinated loans. In contrast EVA (see Definition 6.4 below) explicitly takes into account cost of capital.

Definition 6.3 The *hurdle rate* h_t is the lowest interest rate that investors would require for providing economic capital.

The hurdle rate therefore depends on the risk to which the risk capital C_t is exposed. Often $h_t = s_t + k_t$ is written where s_t denotes the risk-free interest rate. k_t is then the spread to compensate for the risk.

Remark 6.1 An objective determination of the hurdle rate is seldom possible. If one knows the probability for losing the risk capital C_t then one can take as an approximation for the spread the spread of a corporate bond with the same probability of failure. This approximation is not exact even in liquid markets without arbitrage, because in practice only part of the risk capital is lost, and there is no reason that the remaining amount should be equal to the recoverables for the corporate bond in the event of a failure.

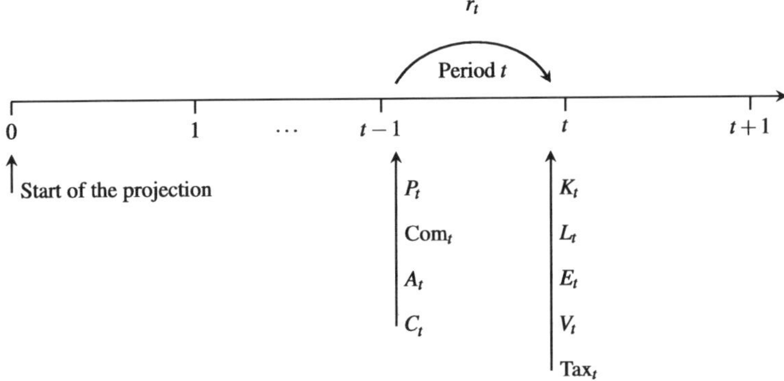

Fig. 6.1 A simple temporal cash flow model. All payments are made either at the start or at the close of a period. Thus the first premium P_1 arrives at time $t = 0$ at the start of period 1

Another possibility for determining the hurdle rate for companies listed on the stock exchange is to use the Capital Asset Pricing Model (CAPM, see Appendix A). This method does, however, have a large uncertainty in its modeling since it assumes a strong coupling between the real risks and share prices. Even more important is that external investors who determine the share price have at their disposal much less information about the risk position of a company than the company itself.

Hence often the pragmatic assumption is selected (in particular for subsidiary companies) that the hurdle rate is simply given (by the owner).

Remark 6.2 Another controversial question in determining the hurdle rate is whether the hurdle rate should be taken uniformly for all the business areas of a company or organization, or chosen individually.

Definition 6.4 If the feedback effect from investing the risk capital (see Sect. 4.5.6) is taken into account in calculating it, then the *Economic Value Added* (*EVA*) is given by

$$\mathrm{EVA}_t = N_t - h_t C_t^{\mathrm{EC}}.$$

One may interpret this formula as saying that in order to maintain the business one must raise the economic risk capital C_t^{EC}. The quantity $h_t C_t^{\mathrm{EC}}$ gives the *interest costs* (or *cost of capital*) for the capital required which are to be subtracted from the net profit. Only when there remains a profit after paying the cost of capital is there a so-called Economic Value Added (cf. Definition 6.5).

If the risk capital is determined without taking into account the feedback effect, then the yield realized on the risk capital must be included explicitly in the calculation. In this approximation one is assuming that the risk capital is invested risk-free. Thus one has

$$\mathrm{EVA}_t = \widetilde{\mathrm{EVA}}_t = \tilde{N}_t - k_t \tilde{C}_t^{\mathrm{EC}},$$

where $\tilde{N}_t$ does not account for the investment return on $\tilde{C}_t^{\mathrm{EC}}$. If the assumption that the risk capital is in a risk-free investment is not correct, then we naturally have $\mathrm{EVA}_t \neq \widetilde{\mathrm{EVA}}_t$.

Definition 6.5 A company *creates value* if it achieves a positive EVA, and *destroys value* if the EVA is negative.

It follows that a company can destroy value, even if it makes a profit, provided that the profits are not sufficient to pay the cost of capital. In this case it would have been more advantageous to invest the capital in a different company which with the same risk realizes a profit that exceeds the cost of capital. (If there is no such company then the hurdle rate determined is probably wrong.)

6.4 Relative Performance Measures

The Economic Value Added gives an absolute yield, and is therefore not suited to the comparison of companies of different sizes. For that purpose it is better to define a risk-adjusted relative yield.

The notation in the literature is not uniform. It is thus easily possible to find alternative definitions for the yield measures introduced in this section which are incompatible with our definitions.

Definition 6.6 The *Return on Capital* (ROC) is given by

$$\mathrm{ROC}_t = \frac{N_t}{EK_t},$$

where EK_t denotes the balance-sheet capital.

The ROC is therefore not a performance measure that is adjusted for risk and only suitable in a very limited way in managing companies. Depending on the accounting method adopted (e.g, statutory accounting, IFRS, US-GAAP) different values result for the ROC.

Definition 6.7 The *Risk Adjusted Return on Capital* (RAROC) is given by

$$\mathrm{RAROC}_t = \frac{N_t - h_t C_t^{\mathrm{EC}}}{EK_t} = \frac{\mathrm{EVA}_t}{EK_t},$$

where EK_t denotes the balance-sheet capital.

Just like the ROC, the RORAC depends on the accounting system used. Since the balance-sheet capital is not a direct economic quantity the RORAC used as a performance measure skews the true risk-adjusted return and is thus only suitable in a limited way for company management. Often, however, the risk-adjusted relative yield defined in Definition 6.8 below is also called the RAROC. In doing so it is implicitly suggested that the available risk capital corresponds exactly to the capital required.

Definition 6.8 The *Return on Risk Adjusted Capital* (RORAC) is the relative quantity corresponding to EVA,

$$\mathrm{RORAC}_t = \frac{N_t}{C_t^{\mathrm{EC}}} = \frac{\mathrm{EVA}_t}{C_t^{\mathrm{EC}}} + h_t,$$

where we again assume that the feedback effect from investing the economic capital C_t^{EC} has been taken into account.

If the feedback effect from investing the economic risk capital $\tilde{C}_t^{\mathrm{EC}}$ is not taken into account it must be assumed that $\tilde{C}_t^{\mathrm{EC}}$ is invested at the risk-free interest rate s_t. We then have

$$\mathrm{RORAC}_t = \widetilde{\mathrm{RORAC}}_t = \frac{\tilde{N}_t + s_t \tilde{C}_t^{\mathrm{EC}}}{\tilde{C}_t^{\mathrm{EC}}} = \frac{\tilde{N}_t}{\tilde{C}_t^{\mathrm{EC}}} + s_t.$$

For an investment of the risk capital that is not risk-free there holds $\mathrm{RORAC}_t \neq \widetilde{\mathrm{RORAC}}_t$.

Another advantage of the RORAC is that this definition does not involve the hurdle rate. However, the hurdle rate does have a significance here as well:

Corollary 6.1 *A company creates value exactly when* $\mathrm{RORAC}_t \geq h_t$.

The structure of RORAC calculation is illustrated in Fig. 6.2.

RORAC results must always be seen in context since niche products, which would not by themselves be viable, can nonetheless generate particularly high returns. For German companies supplementary accident insurance often falls in this category. Their natural market is, however, small and their production cannot be indefinitely expanded. It is necessary also to have mass-market products, with perhaps smaller return, but which can cover the fixed costs of the company by virtue of their volume. The decision, taken on the grounds of the good RORAC of supplementary accident insurance, to cancel all other product lines, would undoubtedly lead directly to ruin for the company in question.

6.5 A Numerical Example

We consider three divisions whose business written at the start of the year t are to be compared with each other. All three divisions have the same nominal insurance results (see Table 6.2). For this we take as *nominal results* (resp. more generally, *nominal values*) values that are neither risk-adjusted nor discounted.

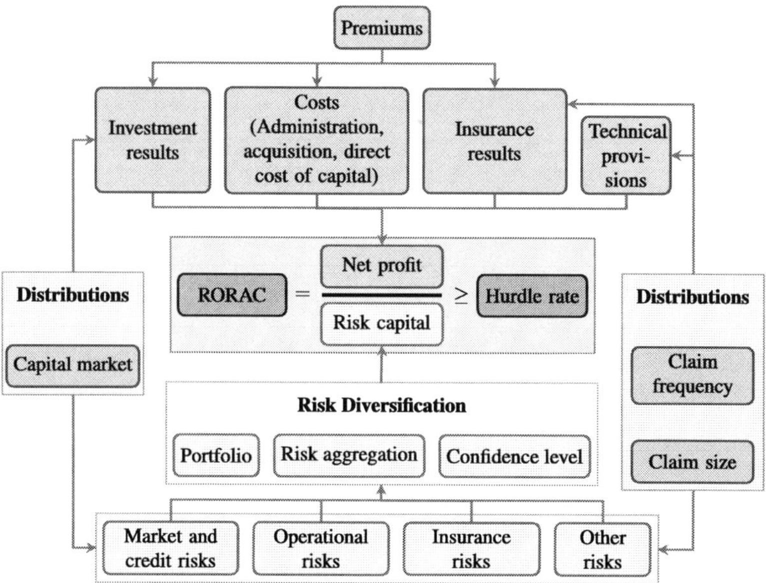

Fig. 6.2 Schematic structure of a RORAC calculation

Table 6.2 Nominal values for divisions A, B, C

Nominal values	Division		
	A	B	C
Premiums	1000	1000	1000
Commissions	50	50	50
Claims (projected)	800	800	800
Other attributed results (projected)	−50	−50	−50
Nominal results	100	100	100

In this description costs and taxes are subsumed under other attributed results. While premiums and commissions apply to the beginning of the year t, claims, claim settlements and other attributed results, occur in the future, so are only projected values. We suppose that the technical provisions and other attributed results for division x display a linear expected run-off pattern $(\alpha^x_{t,\tau})_{\tau \geq t}$ (Table 6.3).

To simplify notation we set $\alpha^x_{t,t-1} = 100\ \%$ for each division. We assume that this run-off pattern also held in the past, so that $\alpha^x_{t-k,\tau-k} = \alpha^x_{t,\tau}$ for all $k \in \mathbb{N}$. The technical provisions have been determined such that (taking account of the other attributed results) neither profits nor losses are expected in the run-off. Assets are invested risk-free and generate an interest of

$$s_{t+k} = 3\ \% \quad \big(k \in \{0, \dots, 4\}\big).$$

Table 6.3 Run-off pattern of divisions A, B, C

Division	Technical provisions at year end				
	t	$t+1$	$t+2$	$t+3$	$t+4$
A	80 %	60 %	40 %	20 %	0 %
B	0 %	–	–	–	–
C	50 %	0 %	–	–	–

Table 6.4 Historical run-off results $Q_{\hat{t}}^x$ for divisions A, B, C

Division	Underwriting year $\hat{t}$				
	$t-10$	$t-9$	$t-8$	$t-7$	$t-6$
A	20 %	150 %	50 %	140 %	40 %
B	85 %	75 %	80 %	85 %	75 %
C	75 %	65 %	80 %	85 %	95 %

For division x and the underwriting year $\hat{t}$ let

- $L_{\hat{t},\tau}^x$ be the total undiscounted claims payment in the year $\tau \geq \hat{t}$,
- $E_{\hat{t},\tau}^x$ be the total undiscounted other attributed results in the year $\tau \geq \hat{t}$.

We denote the undiscounted total claims for the underwriting year $\hat{t}$ by $\mathbf{L}_{\hat{t}}^x = \sum_{\tau \geq \hat{t}} L_{\hat{t},\tau}^x$ and the undiscounted attributed results by $\mathbf{E}_{\hat{t}}^x = \sum_{\tau \geq \hat{t}} E_{\hat{t},\tau}^x$. The single premiums received in the underwriting year are denoted by $P_{\hat{t}}^x$. The (undiscounted) insurance run-off results including costs and taxes, but without technical provisions for the underwriting year $\hat{t}$, can be described by the equation

$$Q_{\hat{t}}^x = \frac{\mathbf{L}_{\hat{t}}^x - \mathbf{E}_{\hat{t}}^x}{P_{\hat{t}}^x}.$$

Divisions A, B, C have experienced very different volatilities in their run-off results during the past few years. The run-off results for the years

$$\hat{t} \in \{t-10, \ldots, t-6\}$$

are given in the Table 6.4. We assume that only the total undiscounted claims $\mathbf{L}_{\hat{t}}^x - \mathbf{E}_{\hat{t}}^x$ are at risk. We also stipulate that for each division both the volume and the typical extent of the underwritten business in the years $\hat{t}$ and in the year t are comparable.

Although the nominal results of divisions A, B, C are identical, the characteristics of the divisions are very different. We will now analyze the consequences of these different properties for the determination of performance. We start from the assumption that risk capital is invested risk-free, and therefore do not take into account the yield on the risk capital when calculating the net profit and other quantities. To indicate this we will mark the corresponding quantities in the rest of the chapter with a tilde $\tilde{}$. In addition, we assume that the actual claims experience of the insurance policies written in the year t is for each division x precisely as expected. With $\alpha^x_{t,t-1} = 1$ and $\Delta\alpha^x_{t,\tau} = \alpha^x_{t,\tau} - \alpha^x_{t,\tau-1}$ this assumption means that for each $\tau \geq t$

$$L^x_{t,\tau} = -\Delta\alpha^x_{t,\tau}\mathbf{L}^x_t, \qquad E^x_{t,\tau} = -\Delta\alpha^x_{t,\tau}\mathbf{E}^x_t.$$

In our projections we will also assume that all profits are withdrawn from the company at each year's end.

The technical provisions at the end of year τ are given by

$$V^x_{t,\tau} = \frac{V^x_{t,\tau+1} + \mathrm{E}(L^x_{t,\tau+1} - E^x_{t,\tau+1})}{1+s_{\tau+1}} = \frac{V^x_{t,\tau+1} - \Delta\alpha^x_{t,\tau+1}\,\mathrm{E}(\mathbf{L}^x_t - \mathbf{E}^x_t)}{1+s_{\tau+1}}. \tag{6.1}$$

The volume of investments of division x at the beginning of the year τ is

$$\tilde{A}^x_{t,\tau} = \delta^t_\tau\left(P^x_t - \mathrm{Com}^x_t\right) + \left(1-\delta^t_\tau\right)V^x_{t,\tau-1},$$

where

$$\delta^t_\tau = \begin{cases} 1 & \text{if } t = \tau \\ 0 & \text{if } t \neq \tau \end{cases}$$

denotes the Kronecker symbol. From (6.1) follows

$$V^x_{t,\tau} - V^x_{t,\tau-1} = \begin{cases} V^x_{t,\tau} & \text{for } \tau = t, \\ s_\tau V_{t,\tau-1} + \Delta\alpha^x_{t,\tau}\,\mathrm{E}(\mathbf{L}^x_t - \mathbf{E}^x_t) & \text{for } \tau > t. \end{cases}$$

The net profit at the end of year $\tau \geq t$ is then given by

$$\begin{aligned} \tilde{N}^x_{t,\tau} &= \delta^t_\tau\left(P^x_t - \mathrm{Com}^x_t\right) - L^x_{t,\tau} + E^x_{t,\tau} + s_\tau\tilde{A}^x_{t,\tau} - \left(V^x_{t,\tau} - V^x_{t,\tau-1}\right) \\ &= \delta^t_\tau\left(P^x_t - \mathrm{Com}^x_t\right) + \Delta\alpha^x_{t,\tau}\left(\mathbf{L}^x_t - \mathbf{E}^x_t\right) + s_\tau\tilde{A}^x_{t,\tau} - \left(V^x_{t,\tau} - V^x_{t,\tau-1}\right), \end{aligned}$$

and we have

$$\begin{aligned} \mathrm{E}\left(\tilde{N}^x_{t,\tau}\right) &= \delta^t_\tau\left(P^x_t - \mathrm{Com}^x_t\right) + \Delta\alpha^x_{t,\tau}\,\mathrm{E}\left(\mathbf{L}^x_t - \mathbf{E}^x_t\right) + s_\tau\delta^t_\tau\left(P^x_t - \mathrm{Com}^x_t\right) \\ &\quad + s_\tau\left(1-\delta^t_\tau\right)V^x_{t,\tau-1} - \delta^t_\tau V^x_{t,\tau} \\ &\quad - \left(1-\delta^t_\tau\right)\left(s_\tau V^x_{t,\tau-1} + \Delta\alpha^x_{t,\tau}\,\mathrm{E}\left(\mathbf{L}^x_t - \mathbf{E}^x_t\right)\right) \\ &= \delta^t_\tau\left((1+s_t)\left(P^x_t - \mathrm{Com}^x_t\right) - V^x_{t,\tau} + \Delta\alpha^x_{t,\tau}\,\mathrm{E}\left(\mathbf{L}^x_t - \mathbf{E}^x_t\right)\right). \end{aligned}$$

A positive net profit is to be expected only in the first year, because the technical provisions have been determined such that they just cover the expected future expenses and the expected claim payments.

From this we get the run-off development shown in Table 6.5.

As a result of the fact that the losses are met much later than the time at which the premiums are paid, there arise interest benefits that are not accounted for in purely nominal considerations. A purely net examination which did consider these effects would prefer the business of division A over the businesses of divisions B and C.

Table 6.5 Run-off development of divisions A, B, C

Quantity	$\tau = t$	$\tau = t+1$	$\tau = t+2$	$\tau = t+3$	$\tau = t+4$	$\tau = t+5$
$\tilde{A}^A_{t,\tau}$	950.0	631.9	480.9	325.3	165.0	0.0
$s_\tau \tilde{A}^A_{t,\tau}$	28.5	19.0	14.4	9.8	5.0	0.0
$L^A_{t,\tau}$	160.0	160.0	160.0	160.0	160.0	0.0
$E^A_{t,\tau}$	−10.0	−10.0	−10.0	−10.0	−10.0	0.0
$V^A_{t,\tau}$	631.9	480.9	325.3	165.0	0.0	0.0
$\tilde{N}^A_{t,\tau}$	176.6	0.0	0.0	0.0	0.0	0.0
$\tilde{A}^B_{t,\tau}$	950.0	0.0	0.0	0.0	0.0	0.0
$s_\tau \tilde{A}^B_{t,\tau}$	28.5	0.0	0.0	0.0	0.0	0.0
$L^B_{t,\tau}$	800.0	0.0	0.0	0.0	0.0	0.0
$E^B_{t,\tau}$	−50.0	0.0	0.0	0.0	0.0	0.0
$V^B_{t,\tau}$	0.0	0.0	0.0	0.0	0.0	0.0
$\tilde{N}^B_{t,\tau}$	128.5	0.0	0.0	0.0	0.0	0.0
$\tilde{A}^C_{t,\tau}$	950.0	412.6	0.0	0.0	0.0	0.0
$s_\tau \tilde{A}^C_{t,\tau}$	28.5	12.4	0.0	0.0	0.0	0.0
$L^C_{t,\tau}$	400.0	400.0	0.0	0.0	0.0	0.0
$E^C_{t,\tau}$	−25.0	−25.0	0.0	0.0	0.0	0.0
$V^C_{t,\tau}$	412.6	0.0	0.0	0.0	0.0	0.0
$\tilde{N}^C_{t,\tau}$	140.9	0.0	0.0	0.0	0.0	0.0

We will now take into account the riskiness of the divisions. As a measure for the economic risk capital we choose the value at risk at the confidence level $\alpha = 99.5\ \%$,

$$\tilde{C}^{\mathrm{EC},x}_{t,\tau} = \mathrm{VaR}_\alpha\left(X^x_{t,\tau}\right),$$

where $X^x_{t,\tau} = -\tilde{N}^x_{t,\tau}$ denotes the loss experienced by division x in the year τ for the underwriting year t. We also assume that the sum of the total loss and other contributions to the results is normally distributed and has a non-negative mean. We can then derive an estimated value for the standard deviation of $X^x_{t,\tau}$ from the (dimensionless) coefficients of variation of the historical values $Q^x_{\hat{t}}$, $\hat{t} \in \{t-10, \ldots, t-6\}$.

$$\begin{aligned}
\sigma\left(X^x_{t,\tau}\right) &= \sigma\left(\delta^t_\tau\left(P^x_t - \mathrm{Com}^x_t\right) + \Delta\alpha^x_{t,\tau}\left(\mathbf{L}^x_t - \mathbf{E}^x_t\right) + s_\tau \tilde{A}^x_{t,\tau} - \left(V^x_{t,\tau} - V^x_{t,\tau-1}\right)\right) \\
&= -\Delta\alpha^x_{t,\tau}\,\sigma\left(\mathbf{L}^x_t - \mathbf{E}^x_t\right) \\
&= -\Delta\alpha^x_{t,\tau}\,P^x_t\,\sigma\left(\frac{\mathbf{L}^x_t - \mathbf{E}^x_t}{P^x_t}\right) \\
&= -\Delta\alpha^x_{t,\tau}\,\frac{\sigma(Q^x_t)}{\mathrm{E}(Q^x_t)}\,\mathrm{E}\left(\mathbf{L}^x_t - \mathbf{E}^x_t\right),
\end{aligned}$$

where we have used that premiums, commissions, and technical provisions are, by our assumptions, deterministic. It follows that

Table 6.6 Mean, standard deviation for the historical Q_t^x of divisions A, B, C

Quantity	Division		
	A	B	C
$\mathrm{E}(Q_t^x)$	80.00 %	80.00 %	80.00 %
$\sigma(Q_t^x)$	60.42 %	5.00 %	15.81 %
$\Phi_{0,1}^{-1}(99.5\ \%)\sigma(Q_t^x)/\mathrm{E}(Q_t^x)$	194.52 %	16.10 %	50.91 %

$$\begin{aligned}\tilde{C}_{t,\tau}^{\mathrm{EC},x} &= \mathrm{VaR}_{99.5\ \%}\big(X_{t,\tau}^x\big) \\ &= -\mathrm{E}\big(\tilde{N}_{t,\tau}^x\big) + \Phi_{0,1}^{-1}(99.5\ \%)\sigma\big(\tilde{N}_{t,\tau}^x\big) \\ &= -\mathrm{E}\big(\tilde{N}_{t,\tau}^x\big) + \frac{\Phi_{0,1}^{-1}(99.5\ \%)\sigma(Q_t^x)}{\mathrm{E}(Q_t^x)}\big(-\Delta\alpha_{t,\tau}^x\,\mathrm{E}\big(\mathbf{L}_t^x - \mathbf{E}_t^x\big)\big).\end{aligned}$$

Numerical evaluation then leads to the values in Table 6.6.

We suppose that the hurdle rate is $h_t = 9\ \%$ and the spread $k_t = 6\ \%$. From our risk capital calculations, we have

$$\widetilde{\mathrm{EVA}}_{t,\tau}^x = \tilde{N}_{t,\tau}^x - k_\tau \tilde{C}_{t,\tau}^{\mathrm{EC},x} \quad \text{and} \quad \widetilde{\mathrm{RORAC}}_{t,\tau}^x = \frac{\tilde{N}_{t,\tau}^x}{\tilde{C}_{t,\tau}^{\mathrm{EC},x}} + s_\tau,$$

where $\tilde{C}_{t,\tau}^{\mathrm{EC},x}$ denotes the economic risk capital in the period τ for the business of division x underwritten in the period t. Summing up, we obtain the results given in Table 6.7.

We have used four different systems for measuring performance: Nominal considerations, net profit examination, $\widetilde{\mathrm{EVA}}$, and $\widetilde{\mathrm{RORAC}}$. In doing so we used the nominal considerations to normalize the divisions, so that comparing the divisions employing our various performance measures is sensible. If net profit is used then division A seems best, using $\widetilde{\mathrm{EVA}}$ division C and using $\widetilde{\mathrm{RORAC}}$ division B. Since the risk is neglected in the net profit it is not surprising that the volatile division A comes out best. The difference between the $\widetilde{\mathrm{EVA}}$-result and the $\widetilde{\mathrm{RORAC}}$-result can be explained by the volatility of division B: For $\tilde{C}_{t,\tau}^{\mathrm{EC},x} \to 0$ we get (for a positive net profit) $\widetilde{\mathrm{RORAC}} \to \infty$. This relationship holds independently of the net profit $\tilde{N}_{t,\tau}^x$, as long as there is a $\delta > 0$, so that in the limit $\tilde{N}_{t,\tau}^x \geq \delta$ holds. The RORAC measure is thus not well suited for very small capital requirements, where the net profit is decisive for overall success. For $\tilde{C}_{t,\tau}^{\mathrm{EC},x} \to 0$ on the other hand we get $\widetilde{\mathrm{EVA}} \to \tilde{N}_{t,\tau}^x$. There is therefore no performance measure that is to be preferred in all applications. One must choose an appropriate measure, or combination of measures, for each specific case.

Remark 6.3 The modeling method used here is very simplified and thus tailored to our example. For real-world applications it is however not appropriate:

Table 6.7 Results for divisions A, B, C

Amount	$\tau=t$	$\tau=t+1$	$\tau=t+2$	$\tau=t+3$	$\tau=t+4$	$\tau=t+5$
$\tilde{N}^{A}_{t,\tau}$	176.6	0.0	0.0	0.0	0.0	0.0
$\tilde{C}^{\text{EC},A}_{t,\tau}$	154.1	330.7	330.7	330.7	330.7	0.0
$k_\tau \tilde{C}^{\text{EC},A}_{t,\tau}$	9.2	19.8	19.8	19.8	19.8	0.0
$\text{EVA}^{A}_{t,\tau}$	167.3	−19.8	−19.8	−19.8	−19.8	0.0
$\text{RORAC}^{A}_{t,\tau}$	117.6 %	3.0 %	3.0 %	3.0 %	3.0 %	–
$\tilde{N}^{B}_{t,\tau}$	128.5	0.0	0.0	0.0	0.0	0.0
$\tilde{C}^{\text{EC},B}_{t,\tau}$	8.3	0.0	0.0	0.0	0.0	0.0
$k_\tau \tilde{C}^{\text{EC},B}_{t,\tau}$	0.5	0.0	0.0	0.0	0.0	0.0
$\text{EVA}^{B}_{t,\tau}$	128.0	0.0	0.0	0.0	0.0	0.0
$\text{RORAC}^{B}_{t,\tau}$	1543.6 %	–	–	–	–	–
$\tilde{N}^{C}_{t,\tau}$	140.9	0.0	0.0	0.0	0.0	0.0
$\tilde{C}^{\text{EC},C}_{t,\tau}$	12.1	153.0	0.0	0.0	0.0	0.0
$k_\tau \tilde{C}^{\text{EC},C}_{t,\tau}$	0.7	9.2	0.0	0.0	0.0	0.0
$\text{EVA}^{C}_{t,\tau}$	140.2	−9.2	0.0	0.0	0.0	0.0
$\text{RORAC}^{C}_{t,\tau}$	1165.9 %	3.0 %	–	–	–	–

- With 5 data points in the neighborhood of the mean one cannot reliably estimate the distribution near the 99.5 % quantile.
- The assumption of a normal distribution is not satisfactory. The central limit theorem cannot be called upon as a rationale since, amongst other things, one cannot check that the distribution to be approximated in fact has a finite standard deviation.
- The assumption that the sum of the total losses and other contributions to the results is the only risk driver, can often not be justified. In practice the timing of the claims payments has an important influence. To include these influences would, however, have made out calculations much more complicated.
- We have assumed that the historical business parameters are almost constant and continue to hold in the present. This is in general not so. Even in our example the historical claims experience of division C suggest a clear trend which we have neglected. In addition, the composition of the portfolio, which we have not looked at, can be important. Few large contracts are riskier, for the same business volume, than many small contracts, since the portfolio is less diversified.

6.6 Basics of Company Valuation

In Sects. 6.3–6.5 we measured the performance on the basis of risk-adjusted profit in a period. Alternatively one can define the performance in relation to changes in the value of the business. This is particularly reasonable when insurance contracts are long-term, as in life insurance, and thus extend over several periods.

Definition 6.9 The *market consistent value* of a company is the price at which the company could be sold to an independent rational investor who knows the company well.

This definition implicitly assumes that different independent investors will arrive at the same price. This is questionable since the price depends on the risk profile of the buyer [4]. Furthermore, the buyer will as a rule be another insurance company so that individual synergies will be essential. Definition 6.9 should therefore be interpreted more abstractly:

- In order to arrive at as objective a result as possible we assume that synergies can be neglected.
- The risk profile of the buyer is accounted for through an explicit normalization. An example for such a normalization is provided by the concept of the cost of capital (see Sect. 6.6.3) for a chosen risk measure and confidence level. Another example of normalization is the risk profile induced by risk neutral valuation (see Sect. 6.6.4).

Definition 6.10 A *deterministic cash flow* for n projection periods is a mapping

$$\mathrm{Cf}\colon \{0,\dots,n\} \to \mathbb{R}, \quad t \mapsto \mathrm{Cf}_t,$$

where each period t is assigned a monetary result Cf_t, evaluated at the end of the period. If $\mathrm{Cf}_t > 0$, then this value is considered income in this period, otherwise it is a disbursement.

In particular, we assume that the cash flow Cf_t is immediately withdrawn from the company, so it does not influence future cash flows Cf_{t+k} $(k > 0)$.

To describe uncertain cash flows we need filtrations (see Sect. 2.4.1).

Definition 6.11 Let $\mathbb{T} = \{1,\dots,n\}$ and $(\mathcal{F}_t)_{t\in\mathbb{T}}$ be a filtration on Ω. A *cash flow* for n projection periods is an adapted stochastic process

$$\mathrm{Cf}\colon \Omega \times \mathbb{T} \to \mathbb{R}, \quad (\omega, t) \mapsto \mathrm{Cf}_t(\omega),$$

where $\mathrm{Cf}_t(\omega)$ describes the income (or disbursement) during period t.

Because it is adapted Cf_t is known at time t. This is especially clear when $(\mathcal{F})_{t\in\mathbb{T}}$ is a product filtration (see Corollary 2.2).

An insurance cash flow expresses the income and outgo, for each time period (see Fig. 6.3). There are two versions of insurance cash flows, depending on whether the changes in technical provisions are included in the cash flow or not.

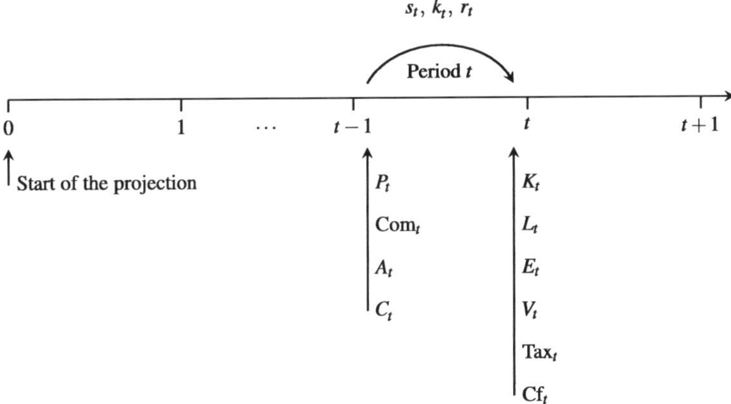

Fig. 6.3 A simple temporal cash flow model

Definition 6.12 In terms of the notation in Definition 6.2 a *pure insurance cash flow* is given by

$$\widehat{\mathrm{Cf}}_t = r_t A_t + P_t - \mathrm{Com}_t - L_t - K_t - (s_t + k_t)C_t + E_t - \mathrm{Tax}_t,$$

where C_t is the risk capital and k_t the spread in period t.

For a pure insurance cash flow the setting up of technical provisions is considered to be an exclusively internal process and so not taken into account. In practice this definition of a cash flow has the following disadvantages:

- The pure cash flow in year t, $\widehat{\mathrm{Cf}}_t$, cannot be interpreted as the amount that is due to the owner and policy holders as profit, since the changes in technical provisions have to first be taken into account.
- In life insurance the delay between premium payments and claim payments can be as much as several decades. In particular this is significant for single-premium policies. Because of this it can often occur in modeling that the payment of premiums is on a policy-by-policy basis during the n-year time horizon of the projection, while the associated claim payment is just represented by a lump sum as the remainder $\widehat{\mathrm{Cf}}_n$. The policies in question are therefore not being consistently modeled, so that results may be distorted.

Alternatively the technical provisions can be seen as resources that are not to be assigned to the insurance company but rather to the insurance collective. In this interpretation the cash flow would take into account the changes in technical provisions as incoming or outgoing payments.

Definition 6.13 In the notation of Definition 6.2 an *insurance cash flow* is given by

$$\mathrm{Cf}_t = r_t A_t + P_t - \mathrm{Com}_t - L_t - (V_t - V_{t-1}) - K_t - (s_t + k_t)C_t + E_t - \mathrm{Tax}_t, \quad (6.2)$$

where C_t is the risk capital and k_t the spread in period t.

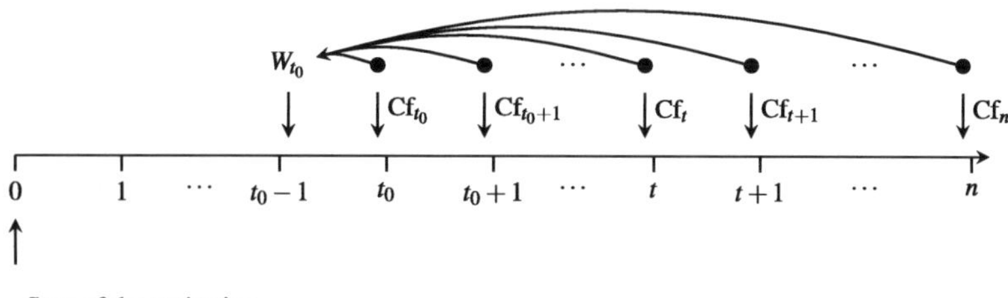

Fig. 6.4 Definition of the value of a company

This definition has the advantage that Cf_t really is available to be used as profit and the delays between premium payments and covering claims is "bridged" by making the corresponding technical provisions. Thereby distortions resulting from differing modeling of cash flow during projection years and of the remainder term are reduced. These advantages are to be compared with the disadvantage that it is no longer a cash flow simply made up of the difference between incoming and outgoing payments. This has to be taken into account in interpreting it. Definition 6.12 and Definition 6.13 differ only by a temporal shift in the results for the periods.

The absolute performance measures introduced in Sect. 6.3 are closely related to our concept of a cash flow:

Proposition 6.1 *Let* Cf_t *be in an insurance cash flow as defined in* (6.2), *and assume* $C_t = C_t^{\mathrm{EC}}$. *Then* $\mathrm{Cf}_t = \mathrm{EVA}_t$.

Proof With the notation of Definition 6.2 we calculate

$$\begin{aligned}\mathrm{Cf}_t &= r_t A_t + P_t - \mathrm{Com}_t - L_t - (V_t - V_{t-1}) - K_t - \left(s_t + k_t^{\mathrm{EC}}\right)C_t^{\mathrm{EC}} + E_t - \mathrm{Tax}_t \\ &= N_t - \left(s_t + k_t^{\mathrm{EC}}\right)C_t^{\mathrm{EC}} \\ &= \mathrm{EVA}_t. \end{aligned}$$

□

One obtains the value from a cash flow by changing to a present value. In doing so we implicitly take it that the cash flow is withdrawn from the company at the end of each period (and does not remain within the company as part of the assets A_{t+1}), as otherwise we would be double counting.

Definition 6.14 Let Cf_t be an insurance cash flow that describes the liquidation of a company in the period n. If r_t^{Discount} is an adapted stochastic process that describes the discount rate, then the *value* at the start of period t_0 is given by

$$W_{t_0}(\omega) = \sum_{t=t_0}^{n} \frac{\mathrm{Cf}_t(\omega)}{\prod_{\tau=t_0}^{t}(1 + r_\tau^{\mathrm{Discount}}(\omega))} \tag{6.3}$$

(see Fig. 6.4).

In making a valuation the cash flow of the company in the year in which it is liquidated includes the run-off result (respectively, residual value in the year n). In the literature the run-off result in (6.3) is often shown separately.

Remark 6.4 The value is not an adapted stochastic process, since future cash flows are required to calculate it.

Remark 6.5 The estimation of the interest r_t^{Discount} for $t \gg 1$ is subject to great uncertainty, but does have much influence on the calculation of the value. To illustrate this, we assume a cash flow representing the investment of 1 Euro for n years with an interest rate of i and that $r_t^{\text{Discount}}(\omega) = r$ is constant for all years t. Then we obtain

$$W_1 = \sum_{t=1}^{n} \frac{\text{Cf}_t}{\prod_{\tau=1}^{t}(1 + r_\tau^{\text{Discount}})} = \left(\frac{1+i}{1+r}\right)^n 1.$$

If $r > i$, then we have for $n \to \infty$ the value $W_1 = 0$; if $r = i$ exactly, then we have $W_1 = 1$; and finally if $r < i$, then the value diverges, $W_1 \to \infty$. If instead each year we subtract the interest i from our investment, then we get

$$\begin{aligned} W_1 &= \sum_{t=1}^{n} \frac{\text{Cf}_t}{\prod_{\tau=1}^{t}(1 + r_\tau^{\text{Discount}})} \\ &= \sum_{t=1}^{n} \frac{i}{(1+r)^t} + \left(\frac{1}{1+r}\right)^n 1 \\ &= i\frac{1-(1+r)^{-n}}{r} + \left(\frac{1}{1+r}\right)^n. \end{aligned}$$

This gives for $i \neq r$ a new result, although it is supposed to be describing an equivalent economic scenario. These differing results can naturally be explained by the difference between the realized interest i and the discount rate r. In our risk-free example there is no reason why i should be different from r. In a more realistic valuation of a portfolio or a company, however, the choice of an appropriate discount rate is less clear.

6.6.1 Various Perspectives for Company Valuation

The cash flow Cf depends crucially on what perspective is used.

- In the *equity method* the value is determined from the point of view of the owner. The valuation arrived at by this method is called the *shareholder value*. The influence of outside capital is therefore directly reflected in the cash flow as inflows, repayments and interest payments.

- In the *entity method* the company is considered from the common point of view of the owner and investors who provide external capital. In a further step then, either the cash flow itself or the value is divided between the owner and the investors. It is suggested that an advantage of this method is that the capital structure is readily seen. However, the valuation of the investors' cash flows, and dividing up the total value require further assumptions.
- In the *stakeholder method* other stakeholders are taken into account in addition to the owner and investors, such as, for example, the policy holders or the company's employees. It is certainly in practice hardly possible to take all stakeholders into consideration. The definition of a notion of performance that fairly takes into account all stakeholders (and their concrete interests) would seem just as difficult. For some management tasks, however, it is advantageous to include specific stakeholders. An example of this is asset liability management in life insurance. For a long-term projection the algorithm that simulates management decisions to maximize shareholder values would reduce the future surplus participation of existing business as much as possible,[3] because the policy holders have only cancellation at their disposal, which generally would lead to a profit for the company in the form of a cancellation fee. The negative effects connected with such a decision are considerable (e.g., loss of reputation with its effect on new business, adverse changes in portfolio, a relative increase in the fixed costs, dissatisfaction on the part of employees because of moral conflicts) but difficult to model. Since real management decisions do in fact consider such negative effects at least qualitatively and thus the old business does participate in surpluses in an appropriate manner, the model would inadequately reflect reality. For strategic asset allocation therefore it is often the total performance of the policy holders and the owner that is optimized.

In the following we choose the equity method, since this method is best suited to the RORAC and EVA methods of Sects. 6.3–6.5.

6.6.2 Deterministic Valuation

For an insurance company it is a natural question, which value it should attach to its portfolio. Therefore it is not surprising that company valuations were done long before stochastic Monte Carlo-based methods could be used in the industry. Since it is difficult without using that sort of stochastic methods to calculate economic capital, the traditional approach to taking account of future risks is different:

Valuation is based on a deterministic (expected) cash flow without economic capital components. However, the costs of capital are taken into account for regulatory

[3]In Germany there is a principle of equal treatment that forbids a reduction in the surplus to policy holders, if it applies only to existing business and not to new business.

capital C_t^{Reg}. The spread chosen is usually the company spread, k_t^{EC}, since regulatory capital must be provided by the company. Therefore $C_t = C_t^{\text{Reg}}$ is not seen as risk capital, but $(s_t + k_t^{\text{EC}})C_t^{\text{Reg}}$ are viewed as additional regulatory costs. As a discount rate $r_t^{\text{Discount}} = s_t + k_t^{\text{Discount}}$ is chosen, where k_t^{Discount} is a spread that is supposed to represent the risk. There are no generally accepted methods to estimate k_t^{Discount}, and in fact very different spreads are used by different companies. This method leads to the *deterministic value*

$$W_t = \sum_{\tilde{t}=t}^{n} \frac{\text{Cf}_{\tilde{t}}}{\prod_{\tau=t}^{\tilde{t}}(1 + (s_\tau + k_\tau^{\text{Discount}}))}$$

with

$$\text{Cf}_{\tilde{t}} = r_{\tilde{t}} A_{\tilde{t}} + P_{\tilde{t}} - \text{Com}_{\tilde{t}} - L_{\tilde{t}} - (V_{\tilde{t}} - V_{\tilde{t}-1}) - K_{\tilde{t}} - (s_{\tilde{t}} + k_{\tilde{t}}^{\text{EC}})C_{\tilde{t}}^{\text{Reg}} + E_{\tilde{t}} - \text{Tax}_{\tilde{t}}.$$

For value-oriented management it is naturally necessary to undertake additional sensitivity analyses in respect of various company strategies (and the corresponding differing cash flows).

Deterministic valuation and the value-oriented management methods employing it are no longer considered "best practice".

6.6.3 *Cost of Capital Based Valuation*

Since future cash flows are uncertain it is attractive to determine the value of a company with stochastic processes.

Let the insurance cash flow Cf_t be defined on a product filtration

$$(\mathscr{F}_t)_{t\in\mathbb{T}} \quad (\mathbb{T} = \{0, \ldots, n\}).$$

Let $\mathbf{P}$ be a probability measure on the σ-algebra $\mathscr{F}_n$ that describes the uncertainty in the real world.[4] The risk-free interest s_t is usually used as the discount rate.

The value (Definition 6.14) is not an adapted stochastic process, but using its conditional expectations (see Definition 2.11) it is still possible to define an adapted stochastic process that provides the value of the company at each point in time.

Definition 6.15 Let $\mathbb{T} = \{0, \ldots, n\}$, $(\Omega, \mathbf{P})$ be a probability space and $(\mathscr{F}_t)_{t\in\mathbb{T}}$ a filtration on Ω. If Cf is an insurance cash flow with respect to this filtration, then the map

$$(t, \omega) \mapsto \text{E}(W_{t+1} \mid \mathscr{F}_t)(\omega)$$

is the *value process* associated to Cf.

---

[4]In Sect. 6.6.4 we will define another "risk neutral" probability measure which does not describe the "real world" directly.

One should not be confused by the index $t+1$. W_{t+1} is the value at the start of the period $t+1$, thus at time t at the end of period t.

Since $\mathrm{E}(W_{t+1} \mid \mathscr{F}_t)$ is obviously an $\mathscr{F}_t$-measurable random variable, the value process is an adapted stochastic process. Proposition 2.7 implies that this definition can be made concrete in a form which is easy to use in our applications, which are based on product filtrations.

6.6.3.1 Monte Carlo Simulation

A practical way of calculating the value process is provided by Monte Carlo simulation.

In a Monte Carlo simulation the distribution F_X of a random variable $X \colon \tilde{\Omega} \to \mathbb{R}^k$ is simulated by the m values $X(\tilde{\omega}_1), \ldots, X(\tilde{\omega}_m)$ with $\tilde{\omega}_1, \ldots, \tilde{\omega}_m \in \tilde{\Omega}$, so that for each $x \in \mathbb{R}^k$ we have[5]

$$F_X(x) \approx \frac{1}{m} \#\bigl\{i \in \{1, \ldots, m\} \mid X(\tilde{\omega}_i) \le x\bigr\}.$$

The values $\tilde{\omega}_i$ in this are pseudorandom numbers provided by a random generator. We therefore are transforming $(\tilde{\Omega}, \mathbf{P})$ approximately into the discrete probability space

$$\bigl(\{1, \ldots, m\}, \mathbf{P}_{\text{uniform}}\bigr), \quad \text{where } \mathbf{P}_{\text{uniform}}(i) = \frac{1}{m} \text{ for all } i \in \{1, \ldots, m\}.$$

The random variable is transformed by this into

$$X^{\approx} \colon \{1, \ldots, m\} \to \mathbb{R}^k, \quad i \mapsto X(\tilde{\omega}_i).$$

Now let X_t $(t \in \{0, \ldots, n\})$ be a stochastic process which is adapted relative to a product filtration and $\Omega = \tilde{\Omega} \times \cdots \times \tilde{\Omega}$. We also assume that $\mathbf{P} = \widetilde{\mathbf{P}} \otimes \cdots \otimes \widetilde{\mathbf{P}}$ is a product measure on Ω. To approximate X_t we draw mn random numbers $\tilde{\omega}_{t,i}$ $((t,i) \in \{1, \ldots, n\} \times \{1, \ldots, m\})$ from $(\tilde{\Omega}, \widetilde{\mathbf{P}})$. This gives the approximation

$$X_t^{\approx} \colon \{1, \ldots, m\} \to \mathbb{R}^k, \quad i \mapsto X_t(\tilde{\omega}_{1,i}, \ldots, \tilde{\omega}_{n,i}),$$

where $X_t(\tilde{\omega}_{1,i}, \ldots, \tilde{\omega}_{n,i})$ does not depend on the drawn values $\tilde{\omega}_{t+1,i}, \ldots, \tilde{\omega}_{n,i}$, since X_t is an adapted stochastic process.

Example 6.1 In practice one will mostly want to determine $\mathrm{E}(W_1 \mid \mathscr{F}_0)$, the value at the start of the projection, and $\mathrm{E}(W_2 \mid \mathscr{F}_1)$, the random variable for the value at the end of the first period. We will describe the Monte Carlo simulation and the difficulties associated with the example $\mathrm{E}(W_2 \mid \mathscr{F}_1)$. To make the structure of our

[5] For $a, b \in \mathbb{R}^k$ the inequality $a \le b$ means that the inequality holds in every component.

formulas clearer we will omit those variables that are not needed for the calculation of an adapted process at time t.

In a Monte Carlo simulation the equation in Proposition 2.7 simplifies to

$$\mathrm{E}(W_2 \mid \mathscr{F}_1)^{\approx}(i) = \frac{\sum_{j_2,\dots j_n=1}^{m} W_2^{\approx}(i, j_2, \dots, j_n)}{m(n-1)}$$

with

$$W_2^{\approx}(i, j_2, \dots, j_n) = \sum_{t=2}^{n} \frac{\mathrm{Cf}_t^{\approx}(i, j_2, \dots, j_t)}{\prod_{\tau=t_2}^{t}(1 + s_\tau^{\approx}(i, j_2, \dots, j_\tau))}.$$

To find the $(n-1)$ interest rates $s_2^{\approx}(i, j_2), \dots, s_n^{\approx}(i, j_2, \dots, j_n)$ is fairly straightforward. The cash flow is given by

$$\begin{aligned}
\mathrm{Cf}_t^{\approx}(i, j_2, \dots, j_t) = {} & r_t^{\approx}(i, j_2, \dots, j_{t-1}) A_t^{\approx}(i, j_2, \dots, j_{t-1}) + P_t^{\approx}(i, j_2, \dots, j_{t-1}) \\
& - \mathrm{Com}_t^{\approx}(i, j_2, \dots, j_{t-1}) - L_t^{\approx}(i, j_2, \dots, j_t) \\
& - \big(V_t^{\approx}(i, j_2, \dots, j_t) - V_{t-1}^{\approx}(i, j_2, \dots, j_{t-1})\big) - K_t^{\approx}(i, j_2, \dots, j_t) \\
& - \big(s_t^{\approx}(i, j_2, \dots, j_t) + k_t^{\mathrm{EC},\approx}(i, j_2, \dots, j_t)\big) C_t^{\approx}(i, j_2, \dots, j_t) \\
& + E_t^{\approx}(i, j_2, \dots, j_t) - \mathrm{Tax}_t^{\approx}(i, j_2, \dots, j_t).
\end{aligned}$$

Modeling most of these terms is as unproblematic as modeling the interest rates. But the technical provisions $V_t^{\approx}(i, j_2, \dots, j_t)$ and the risk capital $C_t^{\approx}(i, j_2, \dots, j_t)$ cannot be directly read from the quantities calculated in the projection steps up to time t. Both the technical provisions and the risk capital are results of stochastic functionals that depend on future uncertain cash flows.

Conceptually the technical provisions are the value of the obligations. In practice, however, the technical provisions are determined approximately:

- In the simplest (and in practice most often seen) case the technical provisions are recursively calculated from a single deterministic "expected" scenario. In that case the time to calculate $V_t^{\approx}(i, j_2, \dots, j_t)$ is of the order $O(n-t)$.
- If the technical provisions are defined based on the cost of capital, their calculation has a comparable complexity to the valuation of the company (based on a pure insurance cash flow). We do not wish to go into that here since the calculation is analogous to valuing the company through a run-off portfolio, up to a missing term for technical provisions, possible use of a different notion of capital, and the treatment of profits.

To compute the capital there are also several methods in common use:

- Sometimes risk capital is approximated using a simple factor-based formula. In that case the computation time is of the order $O(1)$. The method brings with it however a very high uncertainty.
- Particularly in life insurance the approximation derived from deterministic stress tests for a run-off portfolio is popular. The computation time for

$$C_t^{\approx}(i, j_2, \dots, j_t)$$

  is of order $O(n-t)$. In calibrating stress tests there is uncertainty, since the future stochastic evolution is being represented by a deterministic scenario.
- The most consistent method is to determine the risk capital stochastically. In Sect. 2.3.2 we saw that in practice one replaces the calculation by evaluating a suitable random variable on the $\tilde{m}$-fold product of the basic probability space. This leads to a nested stochastic simulation (see Fig. 6.5). At time t an additional $\tilde{m}$ 1-period simulation are done, with which the random variable for the loss $-N_t$ by the company in the period t can be approximated. This gives the discrete random variable

$$-N_t^{\approx}(\tilde{i} \mid i, j_2, \dots, j_t) \quad \tilde{i} \in \{1, \dots, \tilde{m}\},$$

  from which the risk capital $C_t^{\approx}(i, j_2, \dots, j_t)$ can be found numerically. The computation time is of the order $O(\tilde{m})$. Since for this computation additional random numbers $\{\omega(\tilde{i}_t)\}_{\tilde{i}_t=1}^{\tilde{m}}$ must be drawn, at each time t the probability space to be modeled increases by $\tilde{m}$ factors $\tilde{\Omega}$.

To get an idea of the computational requirements of the above, we assume that in the Example 6.1 the liabilities of the company can be described with 100,000 model points. We further suppose that the evolution of each model point takes 20 years and that we neglect any new business. In order to have quantile statements with sufficient confidence we wish to generate at least 100 events beyond the quantile. In order to make quantile assertions about the value of the company at the start of the second period with a confidence level of 1 %, we need $m = 10{,}000$ Monte Carlo scenarios. Without taking into account the cost of capital we therefore have to have $20 \times 10{,}000 = 200{,}000$ scenarios for the portfolio. Let the risk capital be calculated with a confidence level of 99.5 %, which requires $\tilde{m} = 20{,}000$ scenarios. In total then we have 4 million 1-year-projections of the portfolio. Since we have 100,000 model points, that makes 400 billion individual contract 1-year projections.

The estimated computational volume for the example thus exceeds by a long way the capacity of present-day computers.[6] So for value-oriented risk measures approximations must be made, which we do not wish to explore further here.

Remark 6.6 In addition to the practical problem of the computational time necessary this method also has a conceptual difficulty in respect of Definition 6.9: The

[6] Personal computers that are commercially available in 2013.

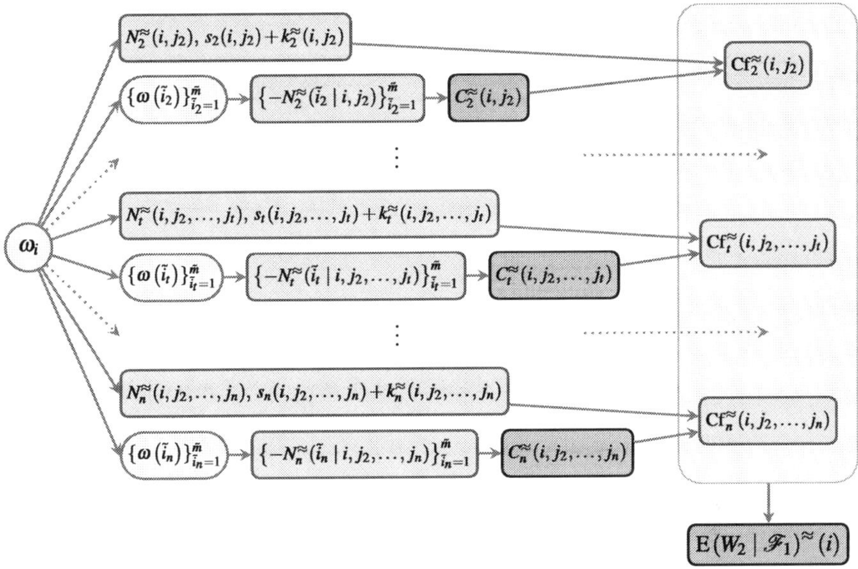

Fig. 6.5 The calculation of $\mathrm{E}(W_2 \mid \mathscr{F}_1)^{\approx}(i)$

value depends on the risk capital C_t and thus on the confidence level chosen. These confidence levels are as a rule consistent with the rating which the company wishes to achieve, however, it cannot be assumed that a buyer of the company would wish to achieve the same rating. Therefore there is a lack of comparability for this measure as well. It should nonetheless be emphasized that — in contrast to the discount rate in deterministic valuations — the confidence level is operationally defined and so can be objectively determined. We will sketch a method of solving this problem in Sect. 6.6.4.

6.6.3.2 Value-Based Risk-Adjusted Profit

In RORAC, the profit realized in a period is seen in relation to the risk capital required for the period. The risk capital also takes into account future changes in the payment liabilities of the company that can be traced back to changes in the period under consideration, but not a future reduction in profit. In contrast, the value-based risk-adjusted profit does take into account such a future loss in profits.

Let us first look at the change in the value process $\mathrm{E}(W_t \mid \mathscr{F}_{t-1})$:

$$\begin{aligned}
\mathrm{E}(W_t \mid \mathscr{F}_t) &= \mathrm{E}\left(\sum_{\tilde{t}=t}^{n} \frac{\mathrm{Cf}_{\tilde{t}}}{\prod_{\tau=t}^{\tilde{t}}(1+s_\tau)} \mid \mathscr{F}_t\right) \\
&= \mathrm{E}\left(\frac{\mathrm{Cf}_t}{1+s_t} \mid \mathscr{F}_t\right) + \mathrm{E}\left(\frac{1}{1+s_t} \sum_{\tilde{t}=t+1}^{n} \frac{\mathrm{Cf}_{\tilde{t}}}{\prod_{\tau=t+1}^{\tilde{t}}(1+s_\tau)} \mid \mathscr{F}_t\right)
\end{aligned}$$

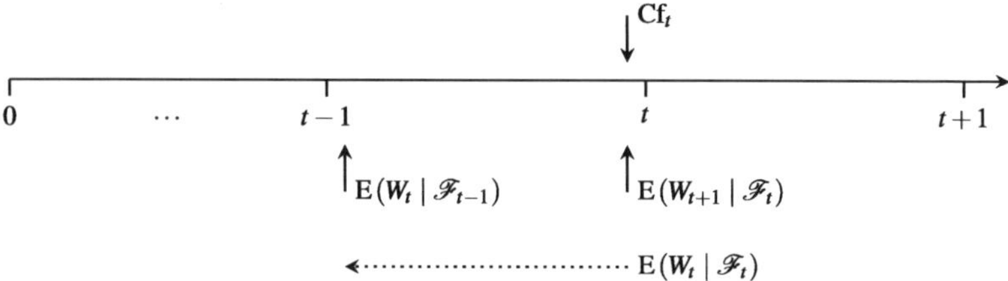

Fig. 6.6 The computation of increase in value. We use that the risk-free interest rate for the period t is already known at the start of the period

$$= \frac{\mathrm{Cf}_t}{1+s_t} + \frac{1}{1+s_t}\mathrm{E}(W_{t+1} \mid \mathscr{F}_t)$$
$$= \frac{1}{1+s_t}\big(N_t - (s_t + k_t)C_t\big) + \frac{1}{1+s_t}\mathrm{E}(W_{t+1} \mid \mathscr{F}_t),$$

where we have used that s_t and Cf_t are $\mathscr{F}_t$-measurable since they are adapted stochastic processes.

Corollary 6.2 *Under the assumption $C_t = C_t^{\mathrm{EC}}$ it follows, with the hurdle rate $h_t = s_t + k_t$, that*

$$\mathrm{EVA}_t = \mathrm{E}(W_{t+1} \mid \mathscr{F}_t) - (1+s_t)\mathrm{E}(W_t \mid \mathscr{F}_t).$$

This corollary shows that the EVA really represents the added value, if the hurdle rate reflects the actual cost of capital. Note that the term $\mathrm{E}(W_t \mid \mathscr{F}_t)$ is not simply the value estimate at time $t-1$ but rather the retrospective value at time $t-1$, given the information at time t. This subtle difference makes the EVA less intuitive than the modification

$$\mathrm{EVA}_t^{\mathrm{Value}} = \mathrm{E}(W_{t+1} \mid \mathscr{F}_t) - (1+s_t)\mathrm{E}(W_t \mid \mathscr{F}_{t-1}),$$

which directly links the value estimates at the beginning and the end of period t (see Fig. 6.6). We will use this definition for the period gain and define the value based loss distribution through

$$X_t^{\mathrm{Value}} = -\mathrm{EVA}_t^{\mathrm{Value}} = -\big(\mathrm{E}(W_{t+1} \mid \mathscr{F}_t) - (1+s_t)\mathrm{E}(W_t \mid \mathscr{F}_{t-1})\big).$$

Consider the dynamic value at risk (see Definition 2.18) for the filtration up to time t as the risk measure. Then we obtain for the value based capital

$$C_t^{\mathrm{Value}} = \mathrm{VaR}_{\alpha,t-1}\big(X_t^{\mathrm{Value}}\big) = (1+s_t)\mathrm{E}(W_t \mid \mathscr{F}_{t-1}) - \mathrm{VaR}_{1-\alpha,t-1}\big(\mathrm{E}(W_{t+1} \mid \mathscr{F}_t)\big).$$

Observe that C_t^{Value} is $\mathscr{F}_{t-1}$-measurable, as it should be.

From this we obtain as a candidate for a risk-adjusted performance measure

$$\mathrm{RORAC}_t^{\mathrm{value}} = \frac{\mathrm{E}(W_{t+1} \mid \mathscr{F}_t) - (1+s_t)\mathrm{E}(W_t \mid \mathscr{F}_{t-1})}{(1+s_t)\mathrm{E}(W_t \mid \mathscr{F}_{t-1}) - \mathrm{VaR}_{1-\alpha}(\mathrm{E}(W_{t+1} \mid \mathscr{F}_t))}.$$

Table 6.8 Claims payment spectrum for the unit-linked life insurances in the portfolio considered. Here $p \in]0, 1[$ is a constant that fixes the profit of the company, and i is a previously agreed interest rate

Event	Payment
Maturity in year n	$\max(pS_n^1, 1)B$
Cancellation in year $t \leq n$	$\max(pS_t^1, (1+i)^{t-n})B$
Death in year $t \leq n$	$\max(pS_{t,}^1, 1)B$

6.6.4 Market Consistent Valuation

A valuation in the sense of Definition 6.9 should be calibrated against the financial market. Since there exists no liquid market for insurance portfolios, a valuation entirely relying on market values is impossible. But one can at least determine a value for those parts of the total portfolio in a market consistent way for which there are markets, and value the other components with a standardized, generally accepted procedure. For this the risks are divided into *hedgeable risks* and *non-hedgeable risks*. Hedgeable risks are those that can be replaced by replication with financial instruments that are traded in liquid markets. Non-hedgeable risks are insurance risks, operational risks and financial risks for which there are no liquid markets, e.g., fixed-interest bonds with very long maturities. The Market Consistent Embedded Value (MCEV) of the CFO Forum [1, 2] is (in Europe[7]) a commonly accepted and widespread method (with focus on life insurance). Sensitivity analyses on the basis of MCEV are in practice employed in value-oriented management.

A proper description of MCEV, or market consistent valuation in general, is beyond the scope of this book. We will thus restrict ourselves to describing some of the fundamental ideas of market consistent valuation with a very simple example.

Example 6.2 We consider a much simplified portfolio that consists in similar unit-linked life insurances, each with single premium B and a duration of n years, that were all taken out at the same time. We assume that the insurance company puts all the cash into an equity fund S_t^1, where we normalize so that $S_0^1 = 1$. The agreed claims payment spectrum is given in Table 6.8.

From the company's point of view there are three risks:

1. Market risk. The company is providing a guarantee for the performance of the equity fund, in that in the case of a payment it pays out in addition to the policy holder's share of the fund $\max(0, 1 - pS_t^1)B$ (resp. $\max(0, (1+i)^{t-n} - pS_t^1)B$). Even, in the case of a maturity payment, this is no pure put option, since the maturity payment is actually paid only if the policy holder is still living and has not already cancelled the policy. The market risk is a (largely) hedgeable risk.
2. Cancellation risk. This risk is that the policy holder cancels when the value of the cancellation payment exceeds the policy holder's share in the equity fund. Strictly speaking this is not a hedgeable risk, since the policy holder can cancel at

[7] Presently (2013) market consistent methods (and Solvency II too) enjoy only limited acceptance in the USA.

any time because of a sudden need for money, which may not seem rational if one does not consider the policy holder's individual situation. The risk is however less pronounced than the hedgeable risk for which a rational approach to the market is assumed of the policy holder. Often, either this hedgeable risk is modeled, or the non-hedgeable risk is approximated by a hedgeable risk in which the rational behavior is functionally corrected.
3. Mortality risk. This is a non-hedgeable risk, since one cannot trade mortality. If a policy holder dies there is an additional risk resulting form the market risk described in 1.

There are two risk drivers here, market risk and mortality risk, since we think of the cancellation risk as a function of the market risk.

In actual modeling we will only consider the hedgeable market risk as stochastic over a probability space $\Omega = \tilde{\Omega} \times \cdots \times \tilde{\Omega}$, which is equipped with a filtration $\mathcal{F}_t$.

The cancellation risk will be modeled by taking the cancellation probability as a function $r(t, S_t^1)$ of time t, and the performance of the equity fund S_t^1. With this we have led the cancellation risk back to the market risk. The risk of personal grounds for cancellation, such as unemployment or other unexpected monetary need, can naturally only be estimated roughly in this way.

Mortality risk is included by a cost of capital method, where we assume that the risk capital for the mortality risk, normalized for the sum at risk, has already been independently calculated. It appears reasonable to choose the regulatory minimum as the confidence level for the calculation of economic capital.[8] Let us assume that the capital for the loss of one Euro because of mortality is given by the constant value $C_{\text{norm}}^{\text{Mort,Reg}}$ and the associated cost of capital by $(s_t + k_t^{\text{Reg}})C_{\text{norm}}^{\text{Mort,Reg}}$.[9] We also suppose that the death probability for any year is a constant q.

From this we obtain for $t \in \{1, \ldots, n\}$ the following cash flow:

$$\begin{aligned}
\text{Cf}_0(\omega) &= B \\
\text{Cf}_t(\omega) &= \prod_{\tau=1}^{t-1} \left(1 - q - r\left(\tau, S_\tau^1(\omega)\right)\right) \\
&\quad \times B(-q \max\left(pS_t^1(\omega), 1\right) \\
&\quad - \left(s_t + k_t^{\text{Reg}}\right)C_{\text{norm}}^{\text{Mort,Reg}} \max\left(0, 1 - pS_t^1(\omega)\right)
\end{aligned}$$

[8]In jurisdictions where such a regulatory minimum is not prescribed, one could use the confidence level prescribed by Solvency II as a proxy.

[9]Since we obviously have a binomial distribution at our disposal one may wonder why we have not explicitly computed the capital. This is because it makes no sense to calculate the risk capital for an individual personal risk: either the policy holder dies and the full amount of the claim must be paid, or he lives and no loss occurs. The value of risk would thus be either the full sum insured or nothing (depending on the confidence level). The concept of capital only makes sense at the portfolio level. We assume here that our policy holder is a member of a larger insurance portfolio, and that the amount $C_{\text{norm}}^{\text{Mort,Reg}}$ is just the portion of the total capital for the mortality risk that is assigned to him. That we have chosen capital to be constant in time is a further simplification.

$$
\begin{aligned}
&- r\big(t, S_t^1(\omega)\big) \max\big(pS_t^1(\omega), (1+i)^{t-n}\big) \\
&- \big(1 - q - r\big(\tau, S_t^1(\omega)\big)\big)\delta_t^n \max\big(pS_t^1(\omega), 1\big)\big),
\end{aligned}
$$

where

$$
\delta_t^n = \begin{cases} 1 & \text{if } n = t \\ 0 & \text{otherwise} \end{cases}
$$

is the Kronecker delta, and we have used the convention that the product over the empty set gives 1. For each t one can obviously consider the cash flow Cf_t as derived from the interest rate process s_t and the stock index process S_t^1. To calculate the value associated with the cash flow (and thus determine the worth of the portfolio) one can call upon the techniques of financial mathematics, and in particular upon methods for valuing derivatives.

We will now briefly sketch how one may value a derivative. We consider the two-dimensional adapted stochastic process

$$
S_t(\omega) = \big(s_t(\omega), S_t^1(\omega)\big)
$$

on our product filtration $(\mathscr{F}_t)$ and put $v_t = \prod_{\tau=1}^{t}(1+s_\tau)^{-1}$. Let H_t be a derivative of this process and assume it corresponds to a single payment $\tilde{H}(S_t)$ at time t, where $\tilde{H}$ is measurable. We need to find the value of H_t at time $\tau = 0$. To do so we need the following notation.

Definition 6.16 Let M_t be an adapted stochastic process relative to the filtration $\mathscr{F} = (\mathscr{F}_t)_{t\in\mathbb{T}}$ on Ω, and $\mathbf{P}$ a probability measure on Ω. M_t is called a $(\mathscr{F}, \mathbf{P})$-*martingale*, if for each t

$$
\begin{aligned}
&\mathrm{E}\big(|M_t|\big) < \infty, \\
&\mathrm{E}(M_t \mid \mathscr{F}_{t-1}) = M_{t-1}.
\end{aligned}
$$

We now assume that W_τ is an adapted stochastic process[10] which at each time $\tau \leq t$ describes the value of the derivative H_t. If $v_\tau W_\tau$ were a $(\mathscr{F}, \mathbf{P})$-martingale, the valuation would be very simple, since by successive applications of the martingale property we would obtain

$$
\begin{aligned}
v_0 W_0 &= \mathrm{E}(v_1 W_1 \mid \mathscr{F}_0) = \mathrm{E}\big(\mathrm{E}(v_2 W_2 \mid \mathscr{F}_1) \mid \mathscr{F}_0\big) \\
&= \mathrm{E}(v_2 W_2 \mid \mathscr{F}_0) = \ldots \\
&= \mathrm{E}(v_t W_t \mid \mathscr{F}_0) = \mathrm{E}\big(v_t \tilde{H}(S_t)\big).
\end{aligned}
$$

[10]This stochastic process is not to be confused with the value for the cost-of-capital-based valuation as defined in Sect. 6.6.3, since that is not an adapted stochastic process. Also the value process defined in that chapter is not identical to the notion used here, since there we explicitly used the cash flow structure and the cost of capital method. The economic interpretation of both value processes is the same.

This would mean that the value at time $\tau = 0$ would simply be the expectation

$$W_0 = \mathrm{E}\big(v_t \tilde{H}(S_t)\big).$$

Naturally, in general, the process W_τ is not a martingale. However, in many cases finding the value of the derivative can be reduced by a trick to calculating an expected value for a suitable probability measure. To explain this we first need to introduce the concept of a self-financing trading strategy.

Definition 6.17 If $\mathscr{F}_t$ is a filtration, then a stochastic process θ_t is called *predictable*, if for every $t \in \{1, \dots, n\}$ the random variable θ_t is measurable with respect to $\mathscr{F}_{t-1}$.

Remark 6.7 Predictable stochastic processes are also called *previsible* stochastic processes.

A predictable process is, speaking intuitively, a process whose value θ_t in the period t is already known. We assume that the investor at time $t-1$, thus at the start of period t, has the portfolio $\theta^0_{t-1} S^0_{t-1}, \dots, \theta^d_{t-1} S^d_{t-1}$, where $\theta_{t-1} \in \mathbb{R}^{d+1}$ is his portion of the corresponding financial instrument. At that time there is the possibility of reshuffling the portfolio, whereby the total worth should naturally remain unchanged. Therefore $S_{t-1} \cdot \theta_t = S_{t-1} \cdot \theta_{t-1}$ holds, where $\cdot$ denotes the standard scalar product in $\mathbb{R}^{d+1}$. This motivates the following definition:

Definition 6.18 Let the $\mathbb{R}^{d+1}$-valued adapted, stochastic process S_t describe the dynamics of $d+1$ securities. A *self-financing trading strategy* is a predictable process θ_t with $S_{t-1} \cdot \theta_t = S_{t-1} \cdot \theta_{t-1}$ for any t.

We would now like to calculate the result of a trading strategy. For this we need the following lemma.

Lemma 6.1 *For a $\mathscr{F}_{t-1}$-measurable function h there holds*

$$\mathrm{E}(h W_t \mid \mathscr{F}_{t-1}) = h\,\mathrm{E}(W_t \mid \mathscr{F}_{t-1})$$

Proof Let Z be a $\mathscr{F}_{t-1}$-measurable function. Then hZ is also $\mathscr{F}_{t-1}$-measurable and we have

$$\mathrm{E}\big(\mathrm{E}(h W_t \mid \mathscr{F}_{t-1})Z\big) = \mathrm{E}(h W_t Z) = \mathrm{E}\big(\mathrm{E}(W_t \mid \mathscr{F}_{t-1})hZ\big)$$

The assertion now follows from the uniqueness of conditional expectations. □

The value of the portfolio at time t is obviously $\theta_t \cdot S_t$. We assume further that the 0-th component S^0_t is a risk-free security with return s_t for the period t and the process $\bar{S}_t = (v_t S_t^k)_{k\in\{1,\dots,d\}}$ is a martingale in each component. Then

$$\sum_{k=1}^{d} \theta_1^k \bar{S}_0^k = \sum_{k=1}^{d} \theta_1^k \operatorname{E}(\bar{S}_1^k \mid \mathscr{F}_0) = \operatorname{E}\left(\sum_{k=1}^{d} \theta_1^k \bar{S}_1^k \mid \mathscr{F}_0\right)$$
$$= \operatorname{E}\left(\sum_{k=1}^{d} \theta_2^k \bar{S}_1^k \mid \mathscr{F}_0\right) = \operatorname{E}\left(\sum_{k=1}^{d} \theta_2^k \operatorname{E}(\bar{S}_2^k \mid \mathscr{F}_1) \mid \mathscr{F}_0\right)$$
$$= \operatorname{E}\left(\operatorname{E}\left(\sum_{k=1}^{d} \theta_2^k \bar{S}_2^k \mid \mathscr{F}_1\right) \mid \mathscr{F}_0\right) = \operatorname{E}\left(\sum_{k=1}^{d} \theta_2^k \bar{S}_2^k \mid \mathscr{F}_0\right)$$
$$= \cdots = \operatorname{E}\left(\sum_{k=1}^{d} \theta_t^k \bar{S}_t^k \mid \mathscr{F}_0\right).$$

Let H_t be a derivative, that is a financial instrument that can be considered to be a function of S_t. If there are no possibilities of arbitrage in the market and H_t can be replicated by a trading strategy θ_τ, so that $\theta_t(\omega) \cdot S_t(\omega) = H_t$ holds for almost all $\omega \in \Omega$, then for any other replicating trading strategy $\tilde{\theta}_\tau$ we have

$$\tilde{\theta}_\tau(\omega) S_\tau(\omega) = \theta_\tau(\omega) S_\tau(\Omega)$$

almost everywhere. A proof can be found in [3, Lemma 2.2.3]. Thus, if

- $\bar{S}$ is a martingale for the filtration $\mathscr{F}_t$ and **P** is a probability measure,
- there exists a self-financing trading strategy θ_τ which replicates H_t,
- there are no possibilities for arbitrage,

then the value $\pi(H_t)$ of H_t at time 0 can be obtained simply from taking expectation values. This is because on the one hand this value corresponds exactly to the costs of a replicating strategy,

$$\pi(H_t) = \sum_{k=1}^{d} \theta_1^k S_0^k,$$

and on the other, with $v_0 = 1$, we have

$$\sum_{k=1}^{d} \theta_1^k S_0^k = \sum_{k=1}^{d} \theta_1^k \bar{S}_0^k = \operatorname{E}\left(\sum_{k=1}^{d} \theta_t^k \bar{S}_t^k\right) = \operatorname{E}\left(\sum_{k=1}^{d} v_t \theta_t^k S_t^k\right) = \operatorname{E}(v_t H_t).$$

So far we have *assumed* that $\bar{S}$ is a martingale for the filtration $\mathscr{F}_t$ and the probability measure **P**. It can now be shown that the market does not allow any arbitrage exactly when there exists a measure **Q** with the same null sets as **P** for which $\bar{S}_\tau$ is componentwise a martingale. A proof for the case that $\mathscr{F}_\tau$ is finitely generated is to be found in [3, Theorem 3.2.2]. The theorem is also given for the general case in [3], however the authors refer the reader to the original papers for a proof. Therefore we can replace our assumption that $\bar{S}_\tau$ is a martingale by the assumption that there are no arbitrage possibilities. In that case we construct the expectation values with $\mathbf{Q}$ in place of $\mathbf{P}$.

We have seen how the value of a derivative can be found, if there are no arbitrage possibilities in the market and the derivative can be replicated by a trading strategy. If the derivative cannot be replicated then it is possible that several equivalent martingales exist that each give a different value. This is (in particular, for valuing insurance liabilities) no purely academic problem, and even in the simple case of a mixed Poisson distribution the value can remain largely undetermined [4].

We now *assume* that for each t the cash flow Cf_t may be replicated and that there are no possibilities for arbitrage. There exists an equivalent measure $\mathbf{Q}$. The value of the portfolio is then given, using

$$v_t^q = \left(\prod_{\tau=1}^{t} \left(1 - q - r\left(\tau, S_\tau^1(\omega)\right)\right) \right)^{-1},$$

by

$$\begin{aligned}
\sum_{t=0}^{n} \mathrm{E}_{\mathbf{Q}}(\mathrm{Cf}_t) = B - pBS_0^1(\omega) \\
&+ B \sum_{t=0}^{n} \mathrm{E}_{\mathbf{Q}} \left(\frac{v_t}{v_{t-1}^q} (-q \max\left(pS_t(\omega), 1\right) \right. \\
&- (s_t + k_t^{\mathrm{Reg}}) C_{\mathrm{norm}}^{\mathrm{Mort,Reg}} \max\left(0, 1 - pS_t^1(\omega)\right) \\
&- r\left(t, S_t^1(\omega)\right) \max\left(pS_t^1(\omega), (1+i)^{t-n}\right) \\
&\left. - \left(1 - q - r\left(\tau, S_t^1(\omega)\right)\right) \delta_t^n \max\left(pS_t^1, 1\right)) \right),
\end{aligned}$$

where $\mathrm{E}_{\mathbf{Q}}$ is the expected value with respect to $\mathbf{Q}$.

To compute this value one naturally requires an equivalent martingale measure $\mathbf{Q}$. In principle, one could obtain $\mathbf{Q}$ from the Radon-Nikodym derivative, $\frac{d\mathbf{Q}}{d\mathbf{P}}(\omega)$, which is in our case bounded and well defined. The Radon-Nikodym derivative can be interpreted as a weight that expresses the risk aversion of the market, and we have

$$\mathrm{E}_{\mathbf{Q}} \left(\sum_{t=0}^{n} v_t \mathrm{Cf}_t \right) = \sum_{t=0}^{n} \int_\Omega v_t(\omega) \mathrm{Cf}_t(\omega)\, d\mathbf{Q} = \sum_{t=0}^{n} \int_\Omega v_t(\omega) \mathrm{Cf}_t(\omega) \frac{d\mathbf{Q}}{d\mathbf{P}}(\omega)\, d\mathbf{P}.$$

In practice it is simpler to find the equivalent martingale measure directly. First we choose a parameterized version of the stochastic process S_τ (relative to the probability measure $\mathbf{Q}$) so that it only remains to find suitable values for the parameters of this process. Since, for each replicable derivative H_t of S_τ, the price is given by $\mathrm{E}_{\mathbf{Q}}(H_t)$, one gets with the present price $\pi(H_t)$ of this derivative in the market an equation of the form

$$\pi(H_t) = \mathrm{E}_{\mathbf{Q}}\big(H_t(S)\big).$$

On the left-hand side there is a known number, while on the right-hand side (after taking the expected values) there is a function of the parameters $p_1, \ldots, p_r$ of the pair $(S, \mathbf{Q})$. The expectation value on the right is computed numerically by a Monte Carlo simulation. The probability measure $\mathbf{Q}$ is uniformly distributed in this representation, analogously to the numerical representation in Sect. 6.6.3.1. From a purely mathematical point of view it would, as a rule, suffice to use r prices in calculating the parameters $p_1, \ldots, p_r$, but since choosing the model is an uncertain procedure this would lead to not very stable results. Thus it is advisable to consider more prices and to choose the parameters $p_1, \ldots, p_r$ so that all the prices are well approximated by the model. The real probability measure $\mathbf{P}$ remains undetermined in this method, but is also not needed for just a valuation.

Remark 6.8 One often calls the probability measure $\mathbf{Q}$ *risk neutral*. This is because in a world in which the probability space $(\Omega, \mathbf{Q})$ were real, so that $\mathbf{P} = \mathbf{Q}$ held, the investors would not be averse to risk, since the risk aversion of the market would be $\frac{d\mathbf{Q}}{d\mathbf{P}}(\omega) = 1$ for any state ω. Therefore one also speaks of "risk neutral valuation" and, often in relation to $(\Omega, \mathbf{Q})$, of a risk neutral world. At the same time, one speaks of $(\Omega, \mathbf{P})$ as the "real world"" and calls $\mathbf{P}$ the "real world probability measure".

This descriptive way of talking has, however, the disadvantage that it leads one to believe in a parallel "risk neutral" world, which does not exist. In particular, one reads in the literature that a risk neutral investor uses a different probability measure than a typical investor in the market. In such cases the metaphor is being extrapolated beyond its range of validity, since any real investor, whatever his personal risk preferences, operates in the same real market with the same real probability measure $\mathbf{P}$.

6.7 Key Performance Indicators and Constraints

It is the object of company management to maximize the performance of the company. But what is "performance"? We saw in Sects. 6.3 and 6.4 that there are various possibilities in defining risk-adjusted performance. The example in Sect. 6.5 showed that different definitions can lead to different guidelines for performance optimization. It is thus a part of the company strategy to define a suitable performance measure.

Definition 6.19 The *key performance indicator* (KPI) is a performance measure that can be chosen as the primary control element and thus reflects the company's strategy.

If the key performance indicator is determined using the concept of cost of capital then associated with the definition are the performance measure, the risk measure

Table 6.9 Illustrative example for the definition of a risk profile

Safety level α	70 %	90 %	95 %	99.5 %
Risk appetite (in % of the available capital)	5 %	25 %	50 %	100 %

and the confidence level. For example, a company could be characterized as using the RORAC concept with the risk measure "expected shortfall" and the confidence level 99 %.

Along with the key performance indicator which is to be maximized come constraints that also must be fulfilled. These constraints can be of a regulatory type or serve to ensure the continuation of the business. Examples of direct constraints are:

- Covering the fixed costs (One cannot just concentrate on profitable niches, since money covering the fixed costs of the company must also be generated.)
- Maintaining the reputation of the company
- Regulatory capital requirements (e.g., solvency and stress tests)
- Maintaining those ratings thought essential
- A sufficient long-term yield

A further class of constraints arises because the appetite for risk depends on the safety level under consideration. A company may be ready to suffer a loss up to 10M Euro with a probability of 1 %. The same company would certainly not be prepared to suffer a loss of 1M Euro with a probability of 50 %. This leads to the introduction of a risk profile in which a maximal tolerable value for the risk measure is given for each safety level. An example of a risk profile is given in Table 6.9. The key performance indicator could be a function of the expected yield N_t and of all values of the risk measure for the different safety levels, e.g.,

$$\mathrm{KPI}_t = \frac{N_t}{a_1 \mathrm{ES}_{70\,\%}(t) + a_2 \mathrm{ES}_{90\,\%}(t) + a_3 \mathrm{ES}_{95\,\%}(t) + a_4 \mathrm{ES}_{99.5\,\%}(t)},$$

where a_1, a_2, a_3, a_4 are appropriate constants. The risk profile itself would then be a constraint.

6.8 Differing Requirements in Life Insurance and Non-life Insurance

Life and non-life insurance, as they are practiced today, have fundamentally different properties.

In non-life insurance policies are ordinarily issued for a year, while for life insurance long-term policies with a time to maturity of 20 years or more are the norm. For annuities the term can be even longer. Therefore the danger of trends that do not agree with the calculations underlying the original policy is more pronounced in life insurance than in non-life insurance. By contrast, in non-life insurance there are typically larger variations in the size of the portfolio.

Another difference is that sizes of claims in non-life insurance (and in health insurance) are not fixed as they are in life insurance. Since in non-life insurance there is no possibility of a retrospective correction, reinsurance is here particularly interesting as tool to mitigate risk.

In non-life insurance one would typically first concentrate on the 1-year time horizon with regard to risk based management. The emphasis is on the analysis of annual claims processing and the optimal reinsurance structure. Longer-term effects such as the run-off of reserves or the periodic hardening of the reinsurance market can be considered in a next step.

In life insurance the one-year perspective is only reasonable with restrictions because policies are written for a term of many years and the policy holders are afforded options and guarantees reaching far into the future. Furthermore, there are shifts in costs and income, since commissions are usually paid at the beginning of the policy term, but are amortized over the full term of the policy. A radical single-year point of view could, taken to the extreme, lead to not writing any new business, although new business is essential to the long-term survival of the company.

References

1. CFO Forum, Market consistent embedded value — basis for conclusions. www.cfoforum.nl, October 2009 (p. 261)
2. CFO Forum, Market consistent embedded value — principles. www.cfoforum.nl, October 2009 (p. 261)
3. R.J. Elliott, P.E. Kopp, *Mathematics of Financial Markets* (Springer, New York, 1999) (p. 265)
4. M. Kriele, J. Wolf, On market value margins and cost of capital. Blätter DGVFM **28**(2), 195–219 (2007) (pp. 250, 266)
5. E. Schmalenbach, *Finanzierungen*, 3rd edn. (Gloeckner, Leipzig, 1922) (p. 239)

Chapter 7
Value-Oriented Company Management

7.1 The Concept of Value-Oriented Company Management

The basic question in value-oriented management is how opportunities and risks can be managed most efficiently. The general objective of value-oriented management can be viewed from several standpoints with in each case a slightly different focus.

Enterprise Risk Management (ERM): In ERM the focus is on running the company as a whole and on the risks connected with its activities — a holistic risk and process view.
Value Based Management (VBM): The focus is on managing the profits in relation to the corresponding risks from the standpoint of the whole company — a holistic risk and profit view.
Risk and Capital Management: Here the focus is on optimal allocation of capital in relation to the risks assumed — a holistic risk and capital view.

Since in reality none of these points of view can be seen as isolated from the other two, often aspects of one standpoint are included in another. Here we view value-oriented company management as a unification that includes all three points of view in the same measure.

The interests of investors, supervision and rating companies are the drivers for the introduction of value-oriented management.

Investors: Using economic risk capital the quantitative expectations of investors regarding risk and return can be integrated into risk management. It also addresses their qualitative expectations, since the decision making in the company is based on robust evaluation of the risks and capital requirements. Furthermore risk capital is such a general concept that on the basis of it the risk and profit in different industries can be compared. This is, in particular, of importance for insurance, since in that way mark-down as a result of its lack of transparency for many investors could be reduced.
Supervisors: In Europe a direct incentive is produced by Solvency II (and especially the use test, see Sect. 8.2.4) for the use of value-oriented control based on economic

M. Kriele, J. Wolf, *Value-Oriented Risk Management of Insurance Companies*,
EAA Series, DOI 10.1007/978-1-4471-6305-3_7, © Springer-Verlag London 2014

risk capital. The IAIS Solvency Project acts largely parallel to Solvency II and produces similar direct incentives on the global scale.

Rating companies: Rating companies calculate the capital that is the basis for their ratings with their own capital models. They are beginning to take into account internal risk capital calculations by companies, if the companies have a very good enterprise risk management procedure.

The introduction of value-oriented company management has four principal components:

Strategic component: What risk profile for the company is to be achieved?

Measurement component: How can profits, risks and opportunities be taken account of as fully as possible?

Organizational component: How must responsibility be distributed in the company so that value-oriented management can be carried through to all levels of the company?

Process component: What processes must be implemented so that value-oriented control is carried out in practice? The challenge is to bring together different processes to a unified system that manages the risks in the way they affect the entire company, instead of processes in compartmentalized silos.

7.1.1 The Strategic Component

Determining the appetite for risk sets the strategic framework for value-oriented management.

Definition 7.1 The *risk appetite* specifies what level of risk a company wishes to achieve.

The risk appetite can be expressed in the simplest case as a risk measure (with a fixed confidence level) or simply as a rating. A more detailed specification of the risk appetite can be given in the form of a risk profile (see Sect. 6.7).

The risk tolerance is to be distinguished from the risk appetite.

Definition 7.2 The *risk tolerance* expresses the maximum tolerable amount of risk. If this level is exceeded then the level of risk to the company must immediately be reduced.

Just like the risk appetite, the risk tolerance can be given as a confidence level, a rating or a risk profile.

Example 7.1 The company XYZ Inc. has a risk appetite given in terms of the Expected Shortfall $\mathrm{ES}_\alpha(X)$ with $\alpha = 99.5\ \%$. Because of fluctuations and the nature

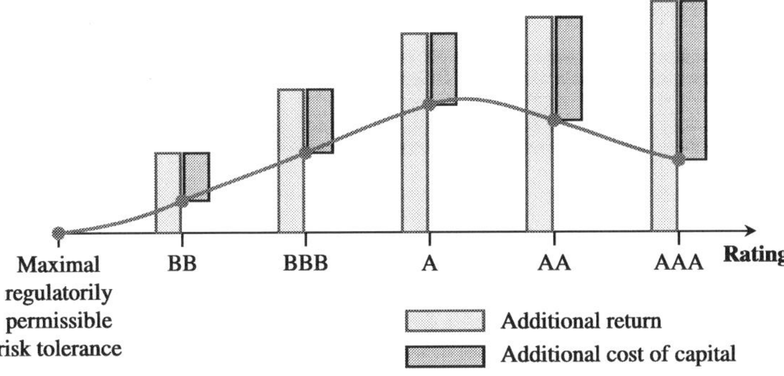

Fig. 7.1 Optimizing the risk tolerance

of risk it is impossible to manage the company such that at each point in time the available capital K obeys the equation $K = \mathrm{ES}_{99.5\,\%}(X)$.

The company's management must therefore live with certain variations around the desired amount of risk. In order to get on top of all the variability, the company defines a risk tolerance of $\mathrm{ES}_{99\,\%}(X)$. The risk tolerance has a lower confidence level and so corresponds to a higher risk. If the risk rises to the level so that $K = \mathrm{ES}_{99\,\%}(X)$, then immediately countermeasures are taken to contain the risk.

The difference between risk appetite and risk tolerance can be understood as a sort of risk buffer. The larger this risk buffer is, the less often immediate countermeasures must be undertaken, which is often very costly. On the other hand, a large risk buffer means that the available capital has not been employed optimally. The optimal risk buffer therefore represents a compromise.

Remark 7.1 In the literature the notions of risk appetite and risk tolerance are not used uniformly. Sometimes the notions are even viewed as synonyms.

From a strategic point of view, the next question is how to sensibly define the risk appetite and risk tolerance. Schematically, one can proceed in the following way:

1. Determining the risk tolerance: The risk tolerance fixes the rating that the company can expect. The higher the rating, the easier it is to acquire (non-retail) new business or to get low-cost credit, but the higher is the need for capital. Therefore neither an extremely low nor an extremely high risk tolerance is acceptable. It follows that there must exist an optimal risk tolerance. In order to find this risk tolerance the company can develop scenarios for each rating which describe the benefits and the capital costs for the rating. An economic capital model can take these scenarios as input and compute the businesses performance and the associated costs of capital. From this one gets a relationship, as shown in Fig. 7.1. The optimal risk tolerance would correspond to a rating of A in this case.
2. The risk buffer is determined in a second step, in which it is approximately found (on the grounds of the fluctuation of risks) how often the risk tolerance is reached

for a given risk buffer and what operational costs are incurred by taking countermeasures immediately. Determining these costs is usually done by working through representative scenarios.

The method we have demonstrated is subject to considerable uncertainty, so that only an "approximate" optimization of the risk appetite and risk tolerance can be expected. On the other hand, great precision is not required for this application. A main benefit of this method is that the company is analyzing its market niche in a largely quantitative framework.

In everyday business it is not realistic to run an economic capital model for each decision to be made, to find out if it is compatible with the risk tolerance. To integrate processes there need to be simple *key risk indicators* (KRI). Then limits are defined based on these, and respecting these limits ensures consistence with the risk tolerance. Typical limits are, for example, a maximum investment into each stock, or maximum concentration limits in underwriting. That staying within these bounds (or nearly so) leads to compliance with the risk tolerance must be checked periodically with an economic capital model. For the risk appetite similar quantities can be defined, but in this case they are not to be interpreted as strict limits but as recommended values.

7.1.2 The Measurement Component

A modern method consists in taking account of opportunities and risk quantitatively with the concepts of economic capital, and to base management on the key indicators which are based on these concepts. In Chap. 6 we saw that it is necessary to measure both the expected return and its associated risks for individual business areas to arrive at a risk-adjusted performance measurement. For value-oriented management we will mainly choose economic risk capital for our idea of a risk. Management of economic risk capital sets the following objectives:

Protecting the company: This is in the first place protection from insolvency. Capital is also needed to fulfill the regulatory requirements (solvency guidelines) and to ensure the target rating deemed necessary is achieved.

Efficient allocation of resources: Risk capital is a risk measure, using which the risks of several activities can be compared. Return on risk adjusted capital,

$$\text{RORAC} = \frac{\text{Gain}}{\text{Economic risk capital}},$$

provides a simple risk-adjusted measure of performance, with which several activities can be compared from an economic point of view (see Chap. 6).

Price setting: To stay in business over the long term, insurance companies must price their products according to the risks they assume. This is possible using the concept of economic risk capital and the associated costs of capital.

In trying to achieve these objectives a balance must be found: A larger available risk capital leads on the one hand to better protection of the company, a smaller likelihood of ruin, and a better rating. On the other hand, as a rule it means higher costs for capital and thus a smaller risk-adjusted return.

There are risks that cannot be accounted for in risk capital, e.g.:

Strategic risks: Strategic risks are those risks that generally have a longer time horizon, because of which they are not taken account of in the 1-year view of economic capital. Since strategic decisions, as a rule, are based on individual expert opinions, strategic risk can hardly be modeled stochastically at all.

Trend risks: Trend risks have similar properties to strategic risks, but reach even further into the future. With this the real problem is that historical data can provide no predictions about future trends. For example the life expectancy in Germany has risen in the last hundred years mainly because an ever-widening sector of the population has had access to good medical care. This trend has however come to its natural end. Other factors, such as, for example, advances in medical genetic engineering, could lead to an additional future increase in life expectancy. These advances cannot be estimated on the basis of past data, because they arise from other factors. Consequently, a stochastic modeling is hardly possible without involvement of expert opinions.
Using capital is not an appropriate way of protecting against trend risks, because trend risks typically affect whole insurance collectives and have an impact only in the long run. It would be simply too expensive to set aside capital for all risks whose effects are without time limits and impact a large section of the collective.

Liquidity risks: Liquidity risk arises if an insurance company does have at its disposal financial resources, but cannot always produce the needed cash flow quickly. This danger can be avoided by good management of liquidity. Holding cash is, as a rule, too costly.

Remark 7.2 Operational risks have a special role. In principle, they can be accounted for in risk capital, but identifying them is fraught with great uncertainty. For many types of operational risk (e.g., mistakes in processing a contract) methods from non-life insurance can be used provided that the damages can be quantitatively specified and there are enough data on hand. If there is not sufficient data then one can use data from industrial consortia or from third parties. This path has been followed with success by some major banks. There are nonetheless operational risks that are statistically described only with difficulty. For example, a company whose head office happens to be in the flight path of an airport has the operational risk that an aircraft could crash into the headquarters building. The probability of this happening is so small that there are no statistical data about it. On the other hand, the loss resulting from such an event would be so large that there is the possibility that this risk is actually relevant. A perhaps a bit less exotic example is the danger of fraud by a senior manager. Such risks can be estimated using scenario-based stochastic methods.

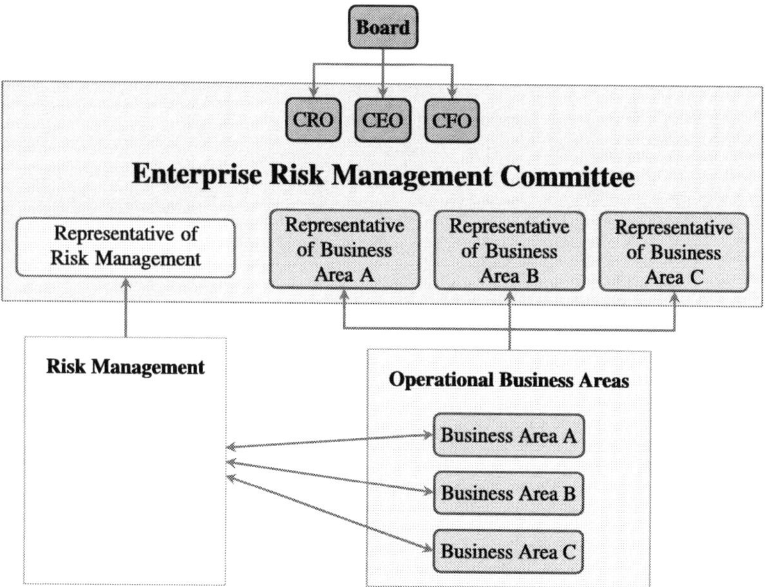

Fig. 7.2 The Enterprise Risk Management Committee

7.1.3 The Organizational Component

Since value-oriented management extends through the whole company, the presence of a clear governance structure that connects all the important parties involved is a necessary condition for its successful introduction. There are different ways to meet this requirement in practice. Figure 7.2 sketches one possible structure that does it. The nerve center is an Enterprise Risk Management Committee. It sets

- objectives and incentives for the management,
- risk limits, for example maximum exposures for asset classes,
- the allocation of capital taking into account results from internal economic capital models,
- risk management processes where appropriate.

Because of the wide sphere of activity of the Enterprise Risk Management Committee it is often practical to form specialized subcommittees, e.g., an ALM Committee, that looks at the investment strategy.

Since Enterprise Risk Management is one of the responsibilities of the board, the CEO, CFO and, as appropriate, other members of the board are also members of the Enterprise Risk Management Committee. In Fig. 7.2 Risk Management has the tasks of both assessing and reporting on all risk, as well as calculating key risk indikators (KRI) such as, for example, the economic risk capital or the risk-adjusted return. In the Enterprise Risk Management Committee the risk managers have the

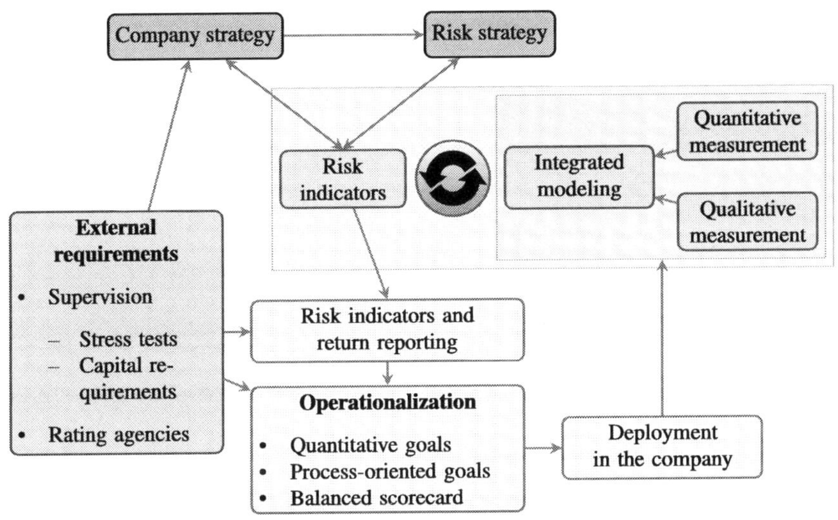

Fig. 7.3 The cycle of value-oriented management

function of technical experts. It is often desired that risk management be organizationally separated form the operational management of business and independent. Nonetheless risk management and the operating business areas must cooperate in the assessment of risks since the operational areas have the necessary raw data and specialist know-how. The operating business areas are also represented in the Enterprise Risk Management Committee since its decisions have a direct impact on them.

7.1.4 The Process Component

In Fig. 7.3 the cycle of the primary processes in value-oriented management is described. First the risk strategy is derived from the company's general strategy. The company strategy is, amongst other things, influenced by external factors such as regulatory guidelines on solvency or criteria from rating agencies. Company strategy and risk strategy are represented by the risk indicators (and particularly the risk measures) used in the company. Naturally in doing so it must be ensured that the technical departments are in fact able to adequately model the risk measures and risk indicators through suitable quantitative and qualitative methods. This cycle is described in Fig. 7.3:

1. Measurement of the present state using the risk indicators.
2. Preparation of reports that describe the risk and return. The basis for these reports is provided by the current risk indicators. The reports should if possible also include an analysis of the changes which have occurred since the last report. Risk

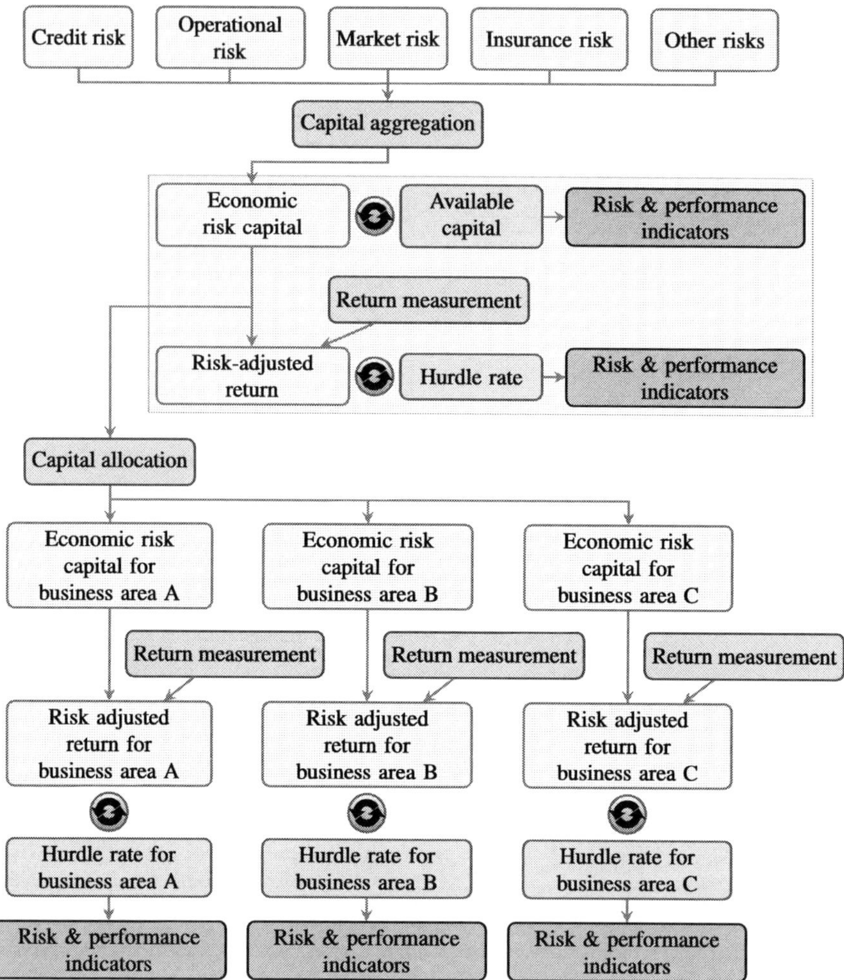

Fig. 7.4 Basic design of value-oriented management from the risk capital point of view

and return reporting is also dependent on external requirements, for examples, regulatory requirements in the framework of Solvency II.

3. The insights represented in the reports are made operational, in that they are turned into quantitative or qualitative objectives for management.
4. When management implements measures to achieve these objectives, risk and performance change, and the measurement process has to begin again.

Of course, this cycle is a continuous process. One cannot wait to begin a phase until the previous phase is finished. Therefore it is necessary, in the analysis in step 2, to be aware what management objectives are the basis for measured activities.

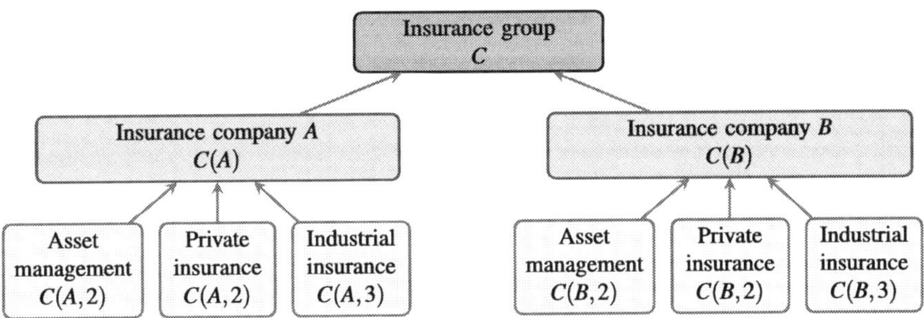

Fig. 7.5 Simple example with a three-level hierarchy of business areas. We have $C = C(A) + C(B)$ and $C(x) = \sum_{i=1}^{3} C(x, i)$ for $x \in \{A, B\}$. The total capital is thus constant for all three hierarchy levels

We would like to analyze the measurement process in Fig. 7.3 a bit more closely. In doing so we assume the risk indicators are based on the economic risk capital and on quantities derived from this risk capital (Fig. 7.4). The individual risks are first separately measured. It should be remarked that these risks will typically affect several business areas simultaneously. For example, operational risk is relevant in every business area. The individual risks are then aggregated and the total economic risk capital is calculated. This aggregation can be done with the tools described in Chap. 3.

As an initial check the risk capital should be compared with the available capital. The capitalization and the risk capital are well matched if they are nearly equal and the risk capital does not exceed the risk appetite. In this case there is little excess capital, which would lead to unnecessary costs for capital.

The risk-adjusted return is the quotient of the return and the economic risk capital (see Sect. 6.4). These key indicators can be compared, for a second check, with the hurdle rate, i.e., the return for the company required by the owner.

In order to consider risk and performance indicators both at the level of the entire company and at the level of individual business areas, one must break down the economic risk capital as fairly as possible to these business areas. This is achieved through capital allocation, which is considered in more detail in Chap. 5. The result is an economic risk capital for each business area. With this it is possible to compute a risk-adjusted return for each business area and compare it with the return required. In the simplest case there is a single hierarchy level of business areas in the company. However, in particular for groups, there can be several levels in the hierarchy. The sum of the risk capitals for business areas per level does not depend on the level (see Fig. 7.5).

Suitable key indicators for risk and performance are defined in Sect. 6.7. Important criteria in choosing them are how easy they are for management to interpret, and whether they can be calculated with sufficiently small error limits that the risk and performance indicators really reflect the management criteria that are supposed to be modeled.

7.1.5 Objective Setting and Supervision: Balanced Scorecard

The implementation of a company procedure is only then successful if those affected know what is expected of them and are motivated to facilitate the new process. To this end it is common practice to formulate clear objectives for employees (and managers). These objectives should also be formulated in such a way that it can be verified how closely the objectives have been approached, so that employees can later be given feedback on how successful they were.

In value-oriented management the company is seen as a whole. This holistic view naturally has an impact on defining objectives, in that, for example, the visions and strategies for the company are put into operation at the employee level. The balanced scorecard (BSC) is a concept that incorporates both visions and strategies into the measurement of performance [1, 2]. To get a "balanced picture" of the whole, four complementary perspectives are considered.

1. *Financial perspective*. A key indicator could be the risk-adjusted return per unit of economic capital or another economic indicator. In accordance with the hierarchy level in the company, where a balanced scorecard is being used it can be preferable to use simpler indicators. For the application processing department, for example, the average processing cost per application could be a reasonable indicator. It is, however, important that the actual objectives of the company be reflected. So such an indicator would not itself be enough, since the quality of application checking also plays an essential role in regard to profitability. Often indicators based on the statutory balance sheet are also included. However, in our context such indicators are less relevant than economically based indicators.
2. *Customer perspective*. A possible indicator for sales representatives is the cancellation rate. The quality of the marketing materials could, for example, be deduced from the number of complaints resulting from misunderstandings by clients in interpreting policies.
3. *Internal business processes perspective*. Here, as a rule, the quality and the efficiency of a process are separately judged. In accounting, for example, the quality of the process could be measured by the number of or by the seriousness of the corrections required by the external auditor. The process efficiency could be measured from the amount of work required to produce the balance sheet.
4. *Learning and growth perspective*. Here it is all about achieving the long-term objectives of the company. Special focus is upon learning and applying what has been learned. Even retaining good employees is part of this perspective. For a large actuarial service the number of employees who are members of the national actuarial society (and, in particular, the time which entrants to the profession require to become actuaries) could be a simple indicator.

For each of these perspectives objectives and quantitative indicators are formulated and put together in a scorecard (cf. Fig. 7.6). Overall the following process results:

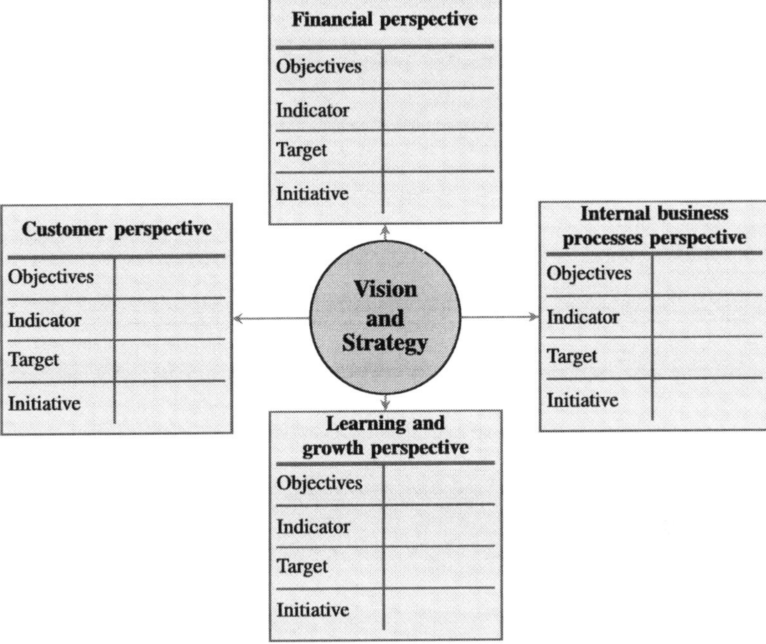

Fig. 7.6 Balanced scorecard. Diagram based on [3]

- Translation of the visions into operational objectives for the company
- Communication of the vision and relating the operational objectives to individual performance targets
- Planning and control
- Feedback and adjusting of strategies

In this general form the conceptual frame is applicable to any industry. For the insurance business or other specific addressees one would tailor the categories correspondingly. So, for example, the introduction of a perspective of its own called "risk management" is an obvious idea, since herein lies the core competency of an insurance company.

There is a certain amount of arbitrariness in how one turns the objective setting into measurable indicators. Here lies the danger that the real objectives are replaced by indicators which are only chosen because they can easily be measured. Consider, for example, the internal business processes perspective regarding accounting. If, as suggested above in item number 3, the quality of the process is measured by the number of objections from the external auditor, the accounting department could ask the auditor to first explain possible objections to them privately, so that errors can be corrected before the measurement is taken. In addition, this measure can lead to confusing results if the external auditor is changed. The new auditor may take a different point of view over some of the details, and consequently would then raise objections that might be about the interpretation of the old auditor and not reflect on the quality of the accounting.

Risks are also hardly considered in the traditional balanced scorecard, but are of special importance in the insurance business.

The success of a balanced scorecard approach stands or falls on the quality of the operational descriptions of strategies. For this, however, there is hardly any help. If the strategy is not a known one (and has not been tested in models to see its effects) it is also not possible to formulate corresponding visions and operational objectives.

7.2 An Example Company

In this section we describe the building up of a value-oriented management structure for a fictional simplified company, XYZ Inc. The company XYZ Inc is a property insurance business that has the business areas

- Fire,
- Liability,
- Theft.

The company is exclusively in retail underwriting. In addition, the company is EU-based and does business in the Euro zone within which it restricts its investment, so as to avoid exposure to any currency exchange risk.

Remark 7.3 Our example company XYZ Inc is not to be seen as a template, but just as an illustration of the considerations and relationships to be taken into account in value-oriented management. A real company would set up many processes quite differently. For example, restricting assets to the Euro zone is not an effective investment strategy. Another example of a simplification that would not be made by real companies is the decision to not differentiate between high-frequency losses and major losses in the economic capital model (Sect. 7.2.3.1). Furthermore, XYZ Inc weighs the effort required for, and the uses of, components of value-oriented management against each other. Because of this not every risk is handled in a way fully appropriate to that risk.

7.2.1 Definition of Business Areas Subject to Risk Oriented Management

XYZ Inc wishes both to manage the divisions on a risk basis, and to evaluate the achievements of important company functions on a risk basis. Therefore two viewpoints on business areas are defined, the division view and the function view.

7.2.1.1 Divisions

The division view divides the company into four business areas which do not overlap, where three of them are the divisions and a further business area describes the rest of the company. The division view provides information for the future strategic direction of the company, and helps to answer the question "In which divisions of the company should we invest more heavily?"

1. *Fire*: This division includes those parts of the business that can be directly attributed to the Fire division. The asset management of those assets that cover the liabilities of this division also belong here.
2. *Liability*: This division includes those parts of the business that can be directly attributed to the Liability division. The asset management of those assets that cover the liabilities of this division also belong here.
3. *Theft*: This division includes those parts of the business that can be directly attributed to the Theft division. The asset management of those assets that cover the liabilities of this division also belong here.
4. *Other*: This division includes those parts of the business that cannot be directly attributed to the divisions Fire, Liability or Theft. The accounting and human resource departments belong here.

7.2.1.2 Functions

The function view divides the company up into non-overlapping business areas that are described purely by their functions. A principal motivation for the functional point of view is that it allows compensation based on performance. The function aspect has two layers. Economic capital and risk-adjusted return are only computed for the upper layer.

1. *Asset Management*: Asset Management is considered as a profit center. They must provide a pre-defined liquidity and, in addition, credit the divisions for their deposits with short-term risk-free interest.
2. *Insurance*: Insurance (excluding asset management) is considered another profit center. The Fire, Liability and Theft divisions represent different markets with different competitive pressures and different insurance behaviors. These differences are not explicitly taken into account in remunerations.

    (a) *Sales*: Each policy that is taken on by Underwriting is compared with an expected (risk-adjusted) profit calculated for that product. There is a target for the sum of all such profits. The risk is that this target will not be reached.
    (b) *Underwriting*: Underwriting is judged on the basis of its effectiveness and its adherence to the guidelines for Underwriting.
    (c) *Claims Processing*: Claims Processing purely serves a control function. Judging its performance takes the same form as for Underwriting.
    (d) *Product Pricing*: The results from Product Pricing influence the performance measurements for Sales, Marketing, and Underwriting.

(e) *Marketing*: The performance-dependent compensation for this function follows balanced scorecards, without regard to the risk-adjusted profits of the company.

(f) *Administration & Other Insurance Business*: Performance-dependent compensation for this function follows balanced scorecards, without regard to the risk-adjusted profits of the company.

3. *Central functions*: Human Resources, Accounting, Controlling, Product and Portfolio Controlling etc. The performance-dependent compensation for this function follows balanced scorecards. The economic capital for central functions is based entirely on operational risks.

7.2.2 *Mitigation of Risks for which Economic Capital is Only Partly Suitable*

7.2.2.1 Trend Risk

It is not possible to directly quantify trend risk. Being based in Europe, the legal department of XYZ Inc follows the American case-law regarding liability claims in order to give estimates when, and to what extent, a similar development is to be expected in Europe. In doing so they also consider industrial liabilities, since expansion of the business to industrial clients is being considered by the board. Product Pricing uses these estimates both in developing products and in designing guidelines for underwriting. Product Pricing reports to the lawyers, to what extent and how their appraisals were implemented.

7.2.2.2 Liquidity Risk

In order to mitigate liquidity risk the Product and Portfolio Controlling department maintains a simple, multi-period, stochastic ALM model. The expected cash flows from the portfolio of insurance contracts and the expected volatilities of these cash flows serve as inputs for the liability side of the ALM model. These data are generated with an assumed log-normal distribution by a multi-period stochastic liability model. The assumptions about new business in this model are set by the product development department. Using a Hull-White model, $\mathrm{d}r(t) = (\theta(t) - \alpha)r(t)\,\mathrm{d}t + \sigma\,\mathrm{d}W(t)$, the dynamics of short-term interest rates is projected. For the Hull-White model time steps of a month are used. Spreads and default probabilities are taken to be constant. Defaults are modeled through a binomial distribution. To this end the credit portfolio is strongly compressed, which tends to produce an overestimate of the default risk. Other asset classes are modeled using a log-normal distribution and are assumed to be independent of each other. All input parameters regarding investments, as well as the investment strategy, are provided by the department Asset Management.

Asset Management has the liquidity requirement that, as measured by this model, in the next 5 years the accumulated probability that there is a shortage of liquidity is at most 5 %. The model is also used by Asset Management to underpin the strategic asset allocation.

In addition to the ALM modeling, scenarios of liquidity shortage are played through and so the liquidity needed is estimated. The scenarios are so defined that each scenario has about a 1 % probability of happening. The probability of occurrence is based on qualitative estimates from Product and Portfolio Controlling.

Using the results from both methods, liquidity limits are derived as constraints for Asset Management.

7.2.3 The Economic Capital Model of XYZ Inc

Before the introduction of an economic capital model many doubts were uttered that capital allocation algorithms would reflect realistic diversification only insufficiently and therefore would lead to unfairness. Since XYZ Inc wants to use the risk-adjusted profit as a compensation measure, it was a priority to find a method that was acceptable to everyone. Easiest to communicate turned out to be the system in which each business area received exactly the proportion of the economic capital attributable to it. From this XYZ Inc infers that it should use the expected shortfall as a risk measure and the capital allocation as described in Proposition 5.6 (see also the discussion following Proposition 5.6).

7.2.3.1 A Simplified Economic Capital Model

As a first step, XYZ Inc introduces a (very much simplified) economic capital model V1.0. If it shows itself useful and is adopted in the organization, it is intended to be incrementally refined (see Sect. 7.2.4). Doing things this way avoids basing the company's management on a black box that is only understood to a limited extent within the company.

In the division view, the business areas Fire, Liability, Theft, Other have been chosen as an exhaustive partition of the company, and Asset Management is seen as part of each of these 4 divisions. For modeling reasons, the economic capital model views Asset Management as a business area on its own. When applying results from the economic capital model to the division view, the capital for Asset Management will be mapped to the other four business areas.

It is assumed in the model that all premiums are due at the beginning of a year, and that the insurance coverage continues for one year. No cash flows within a year are modeled. The business area Asset Management is modeled by a normally distributed random variable r that describes the return from the investments for one year as an average rate of return over all asset classes. Investment costs are modeled as a fixed percentage of the assets at the start of the year.

Table 7.1 Data specific to the divisions of XYZ Inc

Quantity	Variable	Business area A		
		F	L	T
Premiums	$P(A)$	600	300	100
Loss ratio	$c(A)$	75 %	75 %	75 %
Cost ratio	$k(A)$	5 %	5 %	5 %
Quota share reinsurance: Assigned	$Q_{\text{zed}}(A)$	25 %	20 %	20 %
Quota share reinsurance: Provision	$Q_{\text{prov}}(A)$	6 %	6 %	6 %
Coefficient of variation	$\text{vc}(A) = \sigma(S(A))/\text{E}(S(A))$	50 %	60 %	70 %

The aggregate loss distributions of the business areas Fire, Liability and Theft are described by log-normally distributed random variables $S(F)$, $S(L)$, $S(T)$. The expected losses and costs are respectively given as multiples of the premiums. Reserves are not modeled.

A number $n = 10{,}000$ scenarios are generated (see Remark 7.4). The company has assets of $K_0 = 1400$ at the beginning of the year and the fixed costs are $k_{\text{fix}} = 20$. The risk-free interest rate is taken to be $r_0 = 3\ \%$. The risk measure is $\text{ES}_{99\ \%}(X_{\text{net}})$, where the net profit is denoted by $-X_{\text{net}}$.

Remark 7.4 In general many more than 10,000 scenarios are required to reliably estimate the risk capital or RORAC. This is relevant in our example: the quantitative results are significantly influenced by the choice of scenarios.[1]

The (relative) investment return r is assumed to be normally distributed with mean $\text{E}(r) = 5\ \%$ and standard deviation $\sigma(r) = 2\ \%$. The investment costs are 0.5 % of the investment volume.

The data for the divisions are summarized in Table 7.1.

Assume $A \in \{F, L, T\}$. Then, gross,

$$\text{E}\big(S(A)\big) = c(A)P(A).$$

If $m(A), s^2(A)$ are the parameters of the log-normal distribution

$$S(A) = \exp(N\big(m(A), s^2(A)\big),$$

where $N(m, s^2)$ is a normally distributed random variable with mean m and standard deviation s. Then the expected loss and its standard deviation are given by

$$E\big(S(A)\big) = \exp\left(m(A) + \frac{s(A)^2}{2}\right),$$

[1] The qualitative effects that we discuss in this section are already apparent for 10,000 Monte Carlo scenarios. We chose this small number so that all the computation can be completely carried out with tolerable speed on a notebook computer from the year 2006.

$$\sigma\big(S(A)\big) = \sqrt{\exp\big(s(A)^2 - 1\big)} \exp\left(m(A) + \frac{s(A)^2}{2}\right).$$

In this way one gets for $A \in \{F, L, T\}$ as a parameterization of the loss distribution $S(A)$

$$s(A) = \sqrt{\ln\big(1 + \mathrm{vc}(A)^2\big)}, \qquad m(A) = \ln\big(c(A)P(A)\big) - \frac{1}{2}\ln\big(1 + \mathrm{vc}(A)^2\big),$$

where $\mathrm{vc}(A)$ denotes the coefficient of variation.

The dependency structure of the random variables $(r, S(F), S(L), S(T))$ is described by the Gaussian copula with Kendall's τ

$$\tau_{\text{Kendall}} = \begin{pmatrix} 1 & 0 & 0 & 0 \\ 0 & 1 & 0.3 & 0.2 \\ 0 & 0.3 & 1 & 0.6 \\ 1 & 0.2 & 0.6 & 1 \end{pmatrix}.$$

In the following, for consistency, we will often put an index $_{\text{gross}}$ on quantities that are different in their gross and net amounts, if we mean the amount before any reinsurance has been applied. The net premium is $P_{\text{net}}(A) = (1 - Q_{\text{zed}}(A))P_{\text{gross}}(A)$, and the net loss distribution obviously is

$$S_{\text{net}}(A) = \big(1 - Q_{\text{zed}}(A)\big)S_{\text{gross}}(A).$$

The profit to be assigned to the business area Asset Management (K) must be corrected with the risk-free interest rate. There results for $g \in \{\text{gross}, \text{net}\}$

$$-X_g(K) = (r - r_0)\left(K_0 + \sum_{A\in\{F,L,T\}} P_g(A)\right) - k(K)\left(K_0 + \sum_{A\in\{F,L,T\}} P_g(A)\right).$$

Let

$$\delta^g_{\text{net}} = \begin{cases} 1 & \text{for } g = \text{net}, \\ 0 & \text{otherwise} \end{cases}$$

be the usual Kronecker symbol. Then the profit of business area $A \in \{F, L, T\}$ is

$$-X_g(A) = (1 + r_0)P_g(A) + \big(\delta^g_{\text{net}} Q_{\text{prov}} Q_{\text{zed}} - k(A)\big)P_{\text{gross}}(A) - S_g(A).$$

In particular, we assume that policy management remains with the direct insurer, so that cost ratios do not depend on the reinsurance ratio. The total profit is the sum of the profits from each business area less the fixed costs, plus the risk-free interest on the assets

$$-X_g = -\sum_{A\in\{K,F,L,T\}} X_g(A) + r_0 K_0 - k_{\text{fix}}.$$

Table 7.2 The expected profits for each business area and for the whole company

	$g = \text{gross}$	$g = \text{net}$
$-\mathrm{E}(X_g(K))$	35.9	32.5
$-\mathrm{E}(X_g(F))$	139.9	106.4
$-\mathrm{E}(X_g(L))$	67.0	54.2
$-\mathrm{E}(X_g(T))$	22.0	17.8
$-\mathrm{E}(X_g)$	286.8	232.9

Table 7.3 The economic capital for each business area and for the whole company

	$g = \text{gross}$	$g = \text{net}$
$\mathrm{ES}_{99\,\%}(X_g(K))$	90.8	82.1
$\mathrm{ES}_{99\,\%}(X_g(F))$	827.3	619.0
$\mathrm{ES}_{99\,\%}(X_g(L))$	549.7	439.2
$\mathrm{ES}_{99\,\%}(X_g(T))$	238.5	190.6
$\mathrm{ES}_{99\,\%}(X_g)$	1124.4	856.5

Table 7.2 contains expected values for the profits of the business areas as numerically computed from the data given above.[2] The difference of 22 between the sum for the business areas and the total result is made up of the fixed costs, $k_{\text{fix}} = 20$, and the risk-free return on the initial assets, $r_0 K_0 = 42$. These quantities cannot be assigned to the specific business areas of Asset Management, Fire, Liability or Theft since they involve all areas of the business.

Table 7.3 shows the economic capital for each business area. The difference

$$\mathrm{ES}_{99\,\%}(X_g) - \sum_{A \in \{K,F,H,D\}} \mathrm{ES}_{99\,\%}\big(X_g(A)\big) = \begin{cases} -582.0 & g: \text{gross}, \\ -474.3 & g: \text{net} \end{cases}$$

from the non-allocated profit of 22 can be explained by the diversification between business areas.

The RORAC (see Table 7.4) is simply the ratio of the expected profit and the economic capital.

The economic capital model for XYZ Inc is to be found in Appendix E.

Figure 7.7 shows a section of the distribution functions. The dashed horizontal line indicates the VaR for the chosen confidence level of 99 %. The vertical dashed lines indicate the economic capital associated to the distribution functions.

[2] These expected values naturally could be computed exactly, which would lead to different results because of numerical errors. For example, the exact value for the gross result for Fire is $-\mathrm{E}(X_{\text{gross}}(F)) = 138$.

Table 7.4 The risk-adjusted profit for each business area and for the whole company

	$g = \text{gross}$	$g = \text{net}$
$\text{RORAC}_{99\,\%}(X_g(K))$	39.5 %	39.5 %
$\text{RORAC}_{99\,\%}(X_g(F))$	16.9 %	17.2 %
$\text{RORAC}_{99\,\%}(X_g(L))$	12.2 %	12.3 %
$\text{RORAC}_{99\,\%}(X_g(T))$	9.2 %	9.3 %
$\text{RORAC}_{99\,\%}(X_g)$	25.5 %	27.2 %

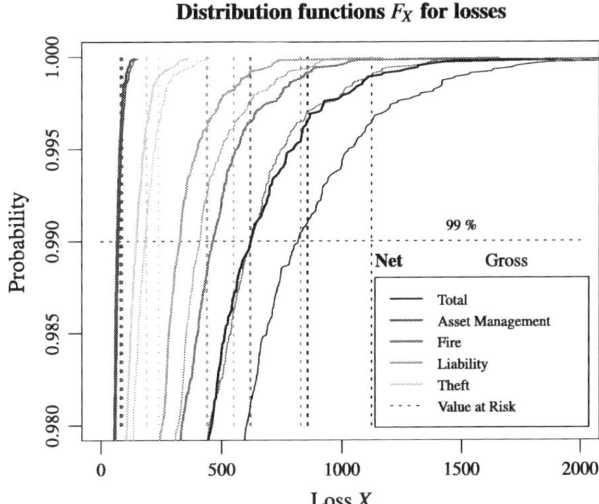

Fig. 7.7 The distribution functions for the business areas Asset Management, Fire, Liability and Theft. The *vertical dashed lines* indicate the economic capital. From left to right the *curves* represent Asset Management (gross), Asset Management (net, almost indistinguishable from gross), Theft (net), Theft (gross), Liability (net), Liability (gross), Fire (net), Fire (gross), Total (net), Total (gross)

Remark 7.5 The calculations shown in the following are intended just as an exposition of methods. To infer general conclusions from such computations requires both more refined models and careful calibration. Since the XYZ Inc we have here uses a much simplified model, we have intentionally chosen the calibration of the insurance divisions to be unrealistic. For example, we chose nearly the same characteristics for the Fire and Liability divisions, which is definitely not the case in the real world.

7.2.3.2 Risk-Adjusted Price Setting

Fire and Liability have the same loss ratio and the same cost ratio. Fire is, in spite of this, more profitable than Liability, which can be traced back to the following factors:

- The volatility for Fire is somewhat smaller than that for Liability.
- The net Liability volume is smaller than the net Fire volume, so that the effect from diversification with Asset Management comes out smaller although the two Kendall's τ are equal.

Table 7.5 Risk-adjusted return for a relative price increase of liability insurance

Increase for liability	$\mathrm{RORAC}_{99\,\%}(X_{\mathrm{gross}}(F))$	$\mathrm{RORAC}_{99\,\%}(X_{\mathrm{gross}}(L))$
−5.00 %	19.13 %	9.13 %
0.00 %	16.90 %	12.20 %
2.50 %	15.82 %	13.79 %
3.75 %	15.29 %	14.61 %
4.37 %	15.02 %	15.02 %
4.69 %	14.89 %	15.23 %
4.53 %	14.96 %	15.13 %
4.45 %	14.99 %	15.08 %
4.41 %	15.01 %	15.05 %
4.39 %	15.02 %	15.04 %
4.38 %	15.02 %	15.03 %
4.38 %	15.02 %	15.03 %
4.38 %	15.02 %	15.03 %
4.38 %	15.02 %	15.03 %
4.38 %	15.02 %	15.02 %

The reinsurance volume can easily be altered. Therefore XYZ Inc uses the gross RORAC to determine the risk-adjusted price. The price of liability insurance is incrementally raised (and the price of fire insurance correspondingly reduced) until both divisions show about the same gross RORAC.

For this XYZ Inc uses a simple iterative bisection algorithm. Let $p_{\min}(0) = 0$, $p_{\max}(0) = 5\ \%$ and $p(0) = 0$ be our initial price increase. If one has reached the price increase $p(i)$ in step i, if we have

$$\mathrm{RORAC}_{99\,\%}\big(X_{\mathrm{gross}}(F)\big) > \mathrm{RORAC}_{99\,\%}\big(X_{\mathrm{gross}}(L)\big),$$

then we set

$$p_{\min}(i+1) = p(i), \qquad p(i+1) = \frac{p(i) + p_{\max}(i)}{2}, \qquad p_{\max}(i+1) = p_{\max}(i)$$

and otherwise set

$$p_{\min}(i+1) = p_{\min}(i), \qquad p(i+1) = \frac{p(i) + p_{\min}(i)}{2}, \qquad p_{\max}(i+1) = p(i).$$

The process is terminated when $p_{\max}(i) - p_{\min}(i)$ is sufficiently small. This algorithm leads to the result shown in Table 7.5.

The new economic capital has the value $\mathrm{ES}_{99\,\%}(X_{\mathrm{net}}) = 855.9$, and the new risk-adjusted profit is $\mathrm{RORAC}_{99\,\%}(X_{\mathrm{net}}) = 27.3\ \%$. Since the characteristics of fire and liability insurance (as we assumed to be) are very similar and XYZ Inc has let the total premium amount remain unchanged, it is not surprising that the values for economic capital and RORAC have hardly been altered by adjusting the prices.

Table 7.6 Risk-adjusted profit in dependence on the capitalization

K_0	$\mathrm{ES}_{99\,\%}(X_{\mathrm{net}})$	$\mathrm{RORAC}_{99\,\%}(X_{\mathrm{net}})$	$\mathrm{ROC}_{99\,\%}(X_{\mathrm{net}})$
1400.0	855.9	27.3 %	16.7 %
1300.0	860.3	26.6 %	17.6 %
1200.0	864.7	26.0 %	18.7 %
1100.0	869.2	25.3 %	20.0 %
1000.0	873.6	24.7 %	21.6 %
900.0	878.1	24.0 %	23.4 %
800.0	882.6	23.4 %	25.8 %
700.0	887.2	22.8 %	28.9 %

Remark 7.6 Risk-adjusted pricing is not identical with pricing that optimizes profit, since the first completely ignores customer preferences.

Exercise 7.1 Expand the program printed in Appendix E to handle the risk-adjusted pricing described here.

7.2.3.3 Optimizing Capitalization

Since the initial assets volume is $K_0 = 1400.0 > 855.9 = \mathrm{ES}_{99\,\%}(X_{\mathrm{net}})$, it can be reduced. To find the optimal asset volume, XYZ Inc computes the capital $\mathrm{ES}_{99\,\%}(X_{\mathrm{net}})$, and the RORAC and ROC for various K_0 (Table 7.6).

The RORAC is highest for an initial asset volume of 1,400. This does not mean in our case that it would be advantageous for the company to hold a lot of capital. This is because the extra return from excess capital goes into the numerator of the RORAC but the economic capital in the denominator increases only from the increased risk resulting from additional invested capital. The excess capital itself is not accounted for in the quotient. Therefore the ROC is the decisive factor in the choice of an optimal initial asset volume. In our example the ROC falls with increasing initial asset volume.

XYZ Inc chooses for its capital $K_0 = 900$, since this value lies fairly near the optimum, but is still gives a safety buffer against fluctuations.

$$\mathrm{RORAC}_{99\,\%}(X_{\mathrm{net}}) = 24.0\,\%.$$

The surplus capital can be returned to the owner.

Exercise 7.2 Expand the program printed in Appendix E to handle the risk-adjusted optimization of the initial asset volume described here.

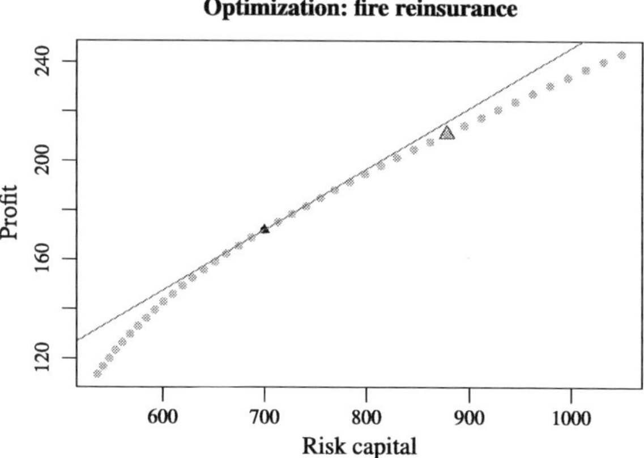

Fig. 7.8 Efficient frontier for fire reinsurance

7.2.3.4 Optimizing the Product Mix with Reinsurance

XYZ Inc sees two ways to influence the product mix. The company can either put in place incentives for sales to sell certain products harder than others, or it can reinsure portions of its portfolio. XYZ Inc decides on the second method because it is easier to control. At present they cede 20 % of both the liability and theft business, but 25 % of the fire business. XYZ Inc thus first works out what the optimal fire reinsurance ratio would be.

To optimize the fire reinsurance XYZ Inc could, as was done in optimizing the capitalization, calculate the $\mathrm{RORAC}_{99\,\%}(X_{\mathrm{net}})$ for different combinations of premiums. However, XYZ Inc prefers instead to determine the *efficient frontier* (see Appendix A). For this the reinsurance ratio for fire is incrementally increased, and each time the expected net profit $\mathrm{E}(X_{\mathrm{net}})$ and also the economic risk capital $\mathrm{ES}_{99\,\%}(X_{\mathrm{net}})$ are calculated and displayed graphically (Fig. 7.8). The efficient frontier is the curve indicated by the points computed. Each point on this curve represents an optimal product mix in the sense that there is no other product mix that promises such a high return for the same risk. The optimal RORAC is given by the slope of the straight line that begins at the coordinate system's origin (Risk capital $= 0$, Profit $= 0$) and is tangent to the "lightly dotted" curve. It is 24.6 %, and 55 % of the fire premiums are to be ceded. The unfilled triangle represents the original fire reinsurance ratio and the filled triangle the new one. The optimal capital is $\mathrm{ES}_{99\,\%}(X_{\mathrm{net}}) = 699.8$.

Remark 7.7 In optimizing the fire reinsurance ratio the differences in investment risks for each division were neglected. For example, if Liability has higher reserves than Fire, then the Liability division is exposed to a greater investment risk than Fire. The error that results from this can easily be significant. Therefore XYZ Inc should allocate investment risk to the Fire and Liability divisions when optimizing the product mix.

It now suggests itself that one should determine an efficient frontier for the simultaneous optimization of reinsurance for Fire, Liability and Theft. There are now

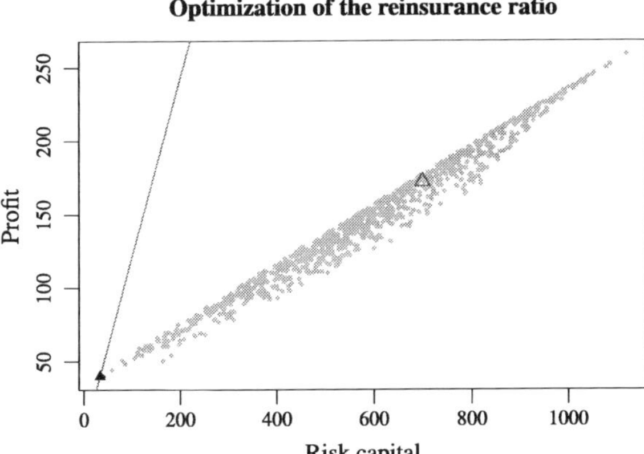

Fig. 7.9 Efficient frontier for the reinsurance structure

three variables that can be varied independently, namely the portions of the Fire, Liability and Theft businesses to be ceded. The resulting points form thus a two-dimensional cloud of points instead of describing a one-dimensional curve. The efficient frontier is now the upper boundary of the point cloud. It is plausible, analogously to the treatment of fire reinsurance, to choose an evenly spaced lattice of reinsurance combinations, e.g.,

$$P_{\text{ceded}}(F) = \frac{i}{N} P(F), \qquad P_{\text{ceded}}(L) = \frac{j}{N} P(L), \qquad P_{\text{ceded}}(T) = \frac{k}{N} P(T)$$

with $i, j, k \in \{0, \dots, N\}$. However then the lattice structure carries over to the figure which can divert attention from the essential properties of the reinsurance. To ensure better communication, XYZ Inc chooses $P_{\text{ceded}}(F) = \omega_F P(F)$, $P_{\text{ceded}}(L) = \omega_L P(L)$, $P_{\text{ceded}}(T) = \omega_T P(T)$, where ω_F, ω_L, and ω_T are uniformly distributed random variables in $[0, 1]$. Figure 7.9 shows the result for $N = 1000$ points.

The unfilled triangle shows the profit-risk position of the original reinsurance program, and the filled triangle corresponds to the new reinsurance program. The slope of the straight line again gives the optimal RORAC and is 121.4 %. This amounts to ceding 95.2 % of the fire premiums, 97.4 % of the liability premiums and 94.3 % of the theft premiums. Because of the limited resolution of 1,000 points these ratios in our calculation cannot be differentiated from ceding all premiums. And, in fact, this does lead to a maximal RORAC. With that XYZ Inc degenerates to a purely sales and administration organization. The only risk would come from the investments. The capital could however be reduced to zero, so that XYZ Inc would receive as an optimal risk-adjusted return $\text{RORAC}_{99\,\%}(X_{\text{net}}) = \infty$. The absolute return would however be very small, since the reinsurance provision, after subtracting the administrative costs, would be only $6\,\% - 5\,\% = 1\,\%$ of the premiums. If sales could be indefinitely scaled up, this would be no problem. In practice that is not possible, and furthermore it is to be remarked that we have not modeled the sales and administrative costs at all, so that the real RORAC is probably very much smaller. Overall one can conclude that a mechanical mathematical optimization of the RORAC delivers, in this case, no economically useful result.

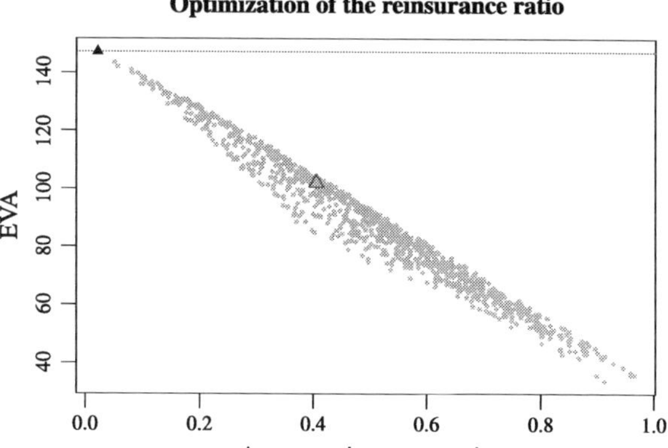

Fig. 7.10 EVA in dependence on the reinsurance structure. A hurdle rate of 10 % is assumed

One can easily see that optimizing the EVA would lead to a similar problem. The optimal EVA would be achieved if absolutely nothing were ceded (see Fig. 7.10). Optimization leads simply to the conclusion that buying reinsurance is not sensible from the EVA point of view.

Neither RORAC optimization nor EVA optimization provide any indication of the optimal portfolio mix. In order to optimize the product mix one must bring in other constraints. An obvious constraint would be specifying the value of the economic capital. However, such a condition would require elaborate iterative computations. To simplify the calculations XYZ Inc simply specifies the ratio of the total net premiums and the total gross premiums. The company chooses a ratio f, that corresponds to the optimal ratio for fire if each of liability and theft are ceded to the tune of 20 %, i.e., $f = 40.54$ %. With this the result of fire reinsurance optimization can be compared with the result of simultaneous optimization of fire, liability and theft reinsurance. Again in total $N = 1000$ points are calculated and N independent uniformly distributed random variables $(\omega_F, \omega_L, \omega_T) \in [0, 1]^3$ generated. Using $f = 40.54$ % we get for $A \in \{F, L, T\}$

$$P_{\text{ceded}}(A) = f\omega_A \frac{\sum_{B\in\{F,L,T\}} P(B)}{\sum_{B\in\{F,L,T\}} \omega_B P(B)} P(A)$$

(Fig. 7.11). XYZ Inc obtains as its best result

$$\frac{P_{\text{ceded}}(F)}{P(F)} = 40.3\ \%, \qquad \frac{P_{\text{ceded}}(L)}{P(L)} = 26.4\ \%, \qquad \frac{P_{\text{ceded}}(T)}{P(T)} = 86.6\ \%.$$

The new risk-adjusted return is $\text{RORAC}_{99\ \%}(X_{\text{net}}) = 25.5$ %, and the risk capital needed is $\text{ES}_{99\ \%}(X_{\text{net}}) = 670.9$.

For application in Sect. 7.2.5 we put together in Table 7.7 the economic capital and the RORAC for the confidence levels $\alpha = 50$ %, $\alpha = 75$ %, $\alpha = 90$ % and $\alpha = 99$ %.

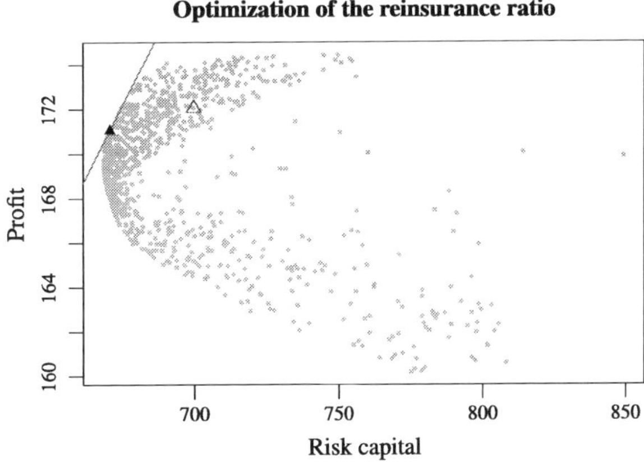

Fig. 7.11 Optimization of reinsurance under the constraint that 40.54 % of the total premiums are ceded

Table 7.7 Economic capital and risk-adjusted profit in dependence on the confidence level

α	50 %	75 %	90 %	99 %
$\mathrm{ES}_\alpha(X_{\mathrm{net}})$	−18.5	107.9	263.5	670.9
$\mathrm{RORAC}_\alpha(X_{\mathrm{net}})$	–	158.5 %	64.9 %	25.5 %

Optimization is an iterative process. For example, after a change in the reinsurance program the capitalization K_0 is no longer optimal, so the RORAC optimization is not yet finished.

Exercise 7.3 Extend the program printed in Appendix E to carry out the optimization of the product mix described here. Optimize the initial asset volume for the new reinsurance structure.

7.2.4 Criticism of the Capital Model for XYZ Inc

The economic capital model described in Sect. 7.2.3 is so simplified that its direct application would lead to badly mistaken decisions. For example, in the version of the economic capital model described above a few essential risks, like the reserve risk or specific risks from certain asset classes (credit risk, interest rate risk), are not considered. As for any model, the results should be interpreted taking into account the limitations of the modeling in order to be employed sensibly. Modeling limitations are often particularly obvious in a simple model, which is why there are also for the economic model of XYZ Inc reasonable uses — for example as a first plausibility test of management decisions that are justified by purely qualitative considerations.

Like any non-trivial model the economic capital model is also embedded in a continuous evolution process. As a first step, XYZ Inc makes a list of those functionalities that the following version of the economic capital model should address.

In doing so risks are also included that cannot be modeled. This increases the transparency and helps to take account of the limitations of the system when applying the results. In addition, the relationship to other models is documented. XYZ Inc maintains a multi-period ALM model (see Sect. 7.2.2.2), that does overlap in modeling and application with the economic capital model. For some modules it may be sensible to use the same code, for other modules the different applications can dictate different approximation or implementations. To interpret the results of both models appropriately overlaps, limitations and, as necessary, alternative implementations must be clearly documented.

The components of V2.0 of the economic capital model are given in Table 7.8.

XYZ Inc plans the following improvements for V2.0 of the economic capital model:

1. $F_{\mathrm{U}}, L_{\mathrm{U}}, T_{\mathrm{U}}$ — *Underwriting:* Normal losses and major losses are separated.

   XYZ Inc has had the experience that major losses have a more pronounced "fat tail" than normal losses. Since the economic capital is particularly sensitive to rare large losses, the major losses must be explicitly handled.

   XYZ Inc also decides to implement major losses in the ALM model because of their effects on liquidity.
2. $F_{\mathrm{U}}, L_{\mathrm{U}}, T_{\mathrm{U}}$ — *Underwriting:* Underwriting risk is modeled through a compound Poisson distribution.

   Compound Poisson distributions are currently used by Product Pricing. The ALM model presently uses a lognormal distribution (like V1.0 of the economic capital model).

   XYZ Inc will investigate if the ALM model should also be changed.
3. $F_{\mathrm{R}}, L_{\mathrm{R}}, T_{\mathrm{R}}$ — *Reserve:* Both the run-off results and the influence of volatility in the risk-free interest rate on the reserve risk are modeled.

   In version V2.0 only the run-off result at the end of the projection year will be stochastically modeled. The risk induced by volatility of the run-off results in later years will not be accounted for in the economic capital model.

   The ALM model does already model the interest rate risk for reserves. Currently the run-off risk is not modeled by ALM. This discrepancy will be pointed out in the documentation concerning interpretation.
4. $K_{\mathrm{Int}}, K_{\mathrm{Sp}}, K_{\mathrm{Eq}}, K_{\mathrm{Cr}}, K_{\mathrm{RE}}$ — *Interest rate, spread, equity, credit and real estate risks:* The modeling is done through the ALM model. In doing so, monthly time steps are modeled to ensure consistency of the parametrizations of the economic capital model and the ALM model.
5. $K_{\mathrm{Cat}}, L_{\mathrm{Cat}}, F_{\mathrm{Cat}}$ — *Scenario-based modeling of catastrophic risks:* m explicit catastrophe scenarios are run through. Denote by Cat_i the loss resulting from the i-th catastrophe and the (estimated) probability of the i-th catastrophe by p_i. The model computes $n > \min(p_i \mid i \in \{1, \ldots, m\})^{-1}$ Monte Carlo scenarios with associated losses X_j^0 $(j \in \{1, \ldots, n\})$. Now $\lfloor p_1 n \rfloor$ indices

$$I_1 = \{i_{1,1}, \ldots, i_{1,\lfloor np_1 \rfloor}\}$$

Table 7.8 Components of the economic capital model V2.0

Symbol	Model part	Module	Model
K_{Int}	Capital market	Interest rate risk	econ. cap. model ALM model
K_{Sp}	Capital market	Spread risk	econ. cap. model ALM model
K_{Cr}	Capital market	Credit risk	econ. cap. model ALM model
K_{Eq}	Capital market	Market risk	econ. cap. model ALM model
K_{RE}	Capital market	Real estate risk	econ. cap. model ALM model
K_{Cat}	Capital market	Risk of catastrophic developments which are not sufficiently described by the analytic capital market model	econ. cap. model
K_{Liq}	Capital market	Liquidity risk	Mitigation ALM model
A_{U} $A \in \{F, L, T\}$	Liability, fire, theft	Underwriting risk	econ. cap. model ALM model
A_{R} $A \in \{F, L, T\}$	Liability, fire, theft	Reserve risk	econ. cap. model ALM model
A_{Cr} $A \in \{F, L, T\}$	Liability, fire, theft	Reinsurance credit risk	econ. cap. model ALM model
A_{Cat} $A \in \{F, L, T\}$	Liability, fire, theft	Risk of catastrophic developments which are not sufficiently described by the analytic insurance model	econ. cap. model
O	Operational risks	Operational risks	econ. cap. model
OR	Other risks	Risk not covered by other risk submodels	Mitigation econ. cap. model
$Trend$	Trend risks	Long-term trend risks	Mitigation ALM model
Fin	Financial results	Economic balance sheet and income statement, risk measures and graphs	econ. cap. model ALM model
Man	Management rules	Rules for reaching management decisions	econ. cap. model ALM model

are chosen randomly from $\{1, \ldots, n\}$, where $\lfloor \quad \rfloor$ indicates taking the integer part. XYZ Inc now puts

$$X_j^1 = \begin{cases} X_j^0 + \text{Cat}_1 & \text{if } i \in I_1, \\ X_j^0 & \text{otherwise} \end{cases}$$

for all $j \in \{1, \ldots, n\}$. This procedure is successively applied for $k \in \{2, \ldots, m\}$ with

$$X_j^k = \begin{cases} X_j^{k-1} + \mathrm{Cat}_k & \text{if } i \in I_k, \\ X_j^{k-1} & \text{otherwise.} \end{cases}$$

In this way XYZ Inc divides the catastrophe scenarios according to their probabilities of occurrence amongst the n Monte Carlo scenarios. In doing so it does happen occasionally that one Monte Carlo scenario can be assigned more than one catastrophe. This reflects the possibility that in a single business year several real catastrophe could happen. The losses from these catastrophes are, in addition, assigned to the corresponding losses for business areas.

For the ALM model no catastrophes are modeled since theses risks are less relevant for applications of ALM. This difference will be pointed out in the documentation concerning interpretation.

6. F_{U}, L_{U}, T_{U} — *Implementation of additional contract forms for reinsurance:* The ALM model should include an analogous implementation.
7. O — *Operational risks:* To determine the associated capital a generalized Pareto distribution (GPD) is assumed. It is calibrated using scenarios.

   Operational risks are presently not modeled in the ALM model. This difference should be pointed out in the documentation on interpretation, since operational risks lead to a somewhat increased need for liquidity.
8. F_{Cr}, L_{Cr}, T_{Cr} — *Credit risk for reinsurance:* The reinsurance credit risk is modeled analogously to credit risk in the business area Asset Management.

   The credit risk for reinsurance is not modeled in the ALM model at present, since in the ALM context it is considered of secondary importance and modeling it would lead to longer computation times. This difference will be remarked in the documentation on interpretation.
9. OR — *Non-material and not explicitly modeled risks:* These risks are given roughly as lump sum with a normal distribution, where the mean and standard deviation are qualitatively estimated by Product and Portfolio Controlling.

   These risks are presently not modeled in the ALM model. This difference will be pointed out in the documentation on interpretation.
10. K, F, L, T — *Introduction of maximal losses:* The current mathematical risk distributions allow arbitrarily large losses, whose size is not relevant if there are no funds available to cover such losses. Therefore a maximal loss is defined for each risk, with the result that the corresponding distribution (after adjustment for catastrophic scenarios) is capped at this amount. The total assets in the last balance sheet are used as the amount of the maximal loss.

    Maximal losses are not presently modeled in the ALM model. This difference will be pointed out in the documentation on interpretation.
11. *Trend* — *Trend risks:* Trend risks are not modeled in the economic capital model. XYZ Inc sees an alignment of the case-law concerning liability with the American practice as a trend risk.

    Trend risks are treated in the ALM model as sensitivities. This difference will be pointed out in the documentation on interpretation.

Exercise 7.4 Discuss the conditions under which the implementation of maximal losses could be appropriate. How would you judge the choice of maximal losses at XYZ Inc?

Since the capital allocation method chosen by XYZ Inc assumes an overall distribution is given for the company's risks, the individual risk distributions are aggregated using Gaussian copulas.

Let

$$\begin{aligned} K &= \{K_{\mathrm{Int}}, K_{\mathrm{Sp}}, K_{\mathrm{Cr}}, K_{\mathrm{Eq}}, K_{\mathrm{RE}}\}, \\ F &= \{F_{\mathrm{U}}, F_{\mathrm{R}}, F_{\mathrm{Cr}}\}, \\ L &= \{L_{\mathrm{U}}, L_{\mathrm{R}}, L_{\mathrm{Cr}}\}, \\ T &= \{T_{\mathrm{U}}, T_{\mathrm{R}}, T_{\mathrm{Cr}}\}. \end{aligned}$$

XYZ Inc constructs a $(5+3+3+3+1+1)$-dimensional random vector

$$\begin{aligned} X_{K,F,L,T,O,OR}(&X_{K_{\mathrm{Int}}}, X_{K_{\mathrm{Sp}}}, X_{K_{\mathrm{Cr}}}, X_{K_{\mathrm{Eq}}}, X_{K_{\mathrm{RE}}}, X_{F_{\mathrm{U}}}, X_{F_{\mathrm{R}}}, X_{F_{\mathrm{Cr}}}, \\ &X_{L_{\mathrm{U}}}, X_{L_{\mathrm{R}}}, X_{L_{\mathrm{Cr}}}, X_{T_{\mathrm{U}}}, X_{T_{\mathrm{R}}}, X_{T_{\mathrm{Cr}}}, X_O, X_{OR})^\top \end{aligned}$$

using a 16-dimensional Gaussian copula C_ρ^{Gauss}, where Kendall's τ has the form

$$\tau_{\mathrm{Kendall}} = \begin{pmatrix} 1 & \tau_2^1 & \tau_3^1 & \tau_4^1 & \tau_5^1 & A & A & A & B & B & B & C & C & C & D & E \\ & 1 & \tau_3^2 & \tau_4^2 & \tau_5^2 & A & A & A & B & B & B & C & C & C & D & E \\ & & 1 & \tau_4^3 & \tau_5^3 & A & A & A & B & B & B & C & C & C & D & E \\ & & & 1 & \tau_5^4 & A & A & A & B & B & B & C & C & C & D & E \\ & & & & 1 & A & A & A & B & B & B & C & C & C & D & E \\ & & & & & 1 & \tau_7^6 & \tau_8^6 & F & F & F & G & G & G & H & I \\ & & & & & & 1 & \tau_8^7 & F & F & F & G & G & G & H & I \\ & & & & & & & 1 & F & F & F & G & G & G & H & I \\ & & & & & & & & 1 & \tau_{10}^9 & \tau_{11}^9 & J & J & J & K & L \\ & & & & & & & & & 1 & \tau_{11}^{10} & J & J & J & K & L \\ & & & & & & & & & & 1 & J & J & J & K & L \\ & & & & & & & & & & & 1 & \tau_{13}^{12} & \tau_{14}^{12} & M & N \\ & & & & & & & & & & & & 1 & \tau_{14}^{13} & M & N \\ & & & & & & & & & & & & & 1 & M & N \\ & & & & & & & & & & & & & & 1 & O \\ & & & & & & & & & & & & & & & 1 \end{pmatrix}.$$

For simplicity it is assumed that the dependency structure between the groups K, F, L, T, O, OR does not depend on the inner structure of these groups.

Exercise 7.5 Discuss how far the assumption that the dependencies between the groups K, F, L, T, O, OR does not depend on their inner structure, also holds for the credit risk.

The components

$$K_{\text{Int}}, K_{\text{Sp}}, K_{\text{Cr}}, K_{\text{Eq}}, K_{\text{RE}}, F_{\text{U}}, F_{\text{R}}, F_{\text{Cr}}, L_{\text{U}}, L_{\text{R}}, L_{\text{Cr}}, T_{\text{U}}, T_{\text{R}}, T_{\text{Cr}}, O, \text{OR}$$

of the 16-dimensional random vector $X_{K,F,L,T,O,OR}$ will later be modified on taking into account the catastrophe scenarios.

The random variable X for the total loss is the sum of the 16 components of the vector obtained:

$$X = \sum_{i \in K \cup F \cup L \cup T} X_i + X_O + X_{OR}.$$

Remark 7.8 To take into account the effects of liquidity risk on the need for capital, XYZ Inc could have introduced a 17th dimension $X_{K_{\text{Liq}}}$: then $X_{K_{\text{Liq}}}$ would be for each scenario the liquidity needed over and above the prescribed liquidity, multiplied by the market rate for fixed-interest bonds with a rating that is typical for companies with liquidity problems, e.g., B^-.

Exercise 7.6 XYZ Inc appraises the dependency of the compound distributions (frequency and extent of losses). Another possibility would have been to base the dependency structure only on the frequency of losses or only on the loss amounts. Discuss the advantages and disadvantages of the alternatives for XYZ Inc.

To allocate economic capital to business areas, in calculating the expected shortfall account is taken only of the results for scenarios that allow themselves to be assigned to these business areas (see Proposition 5.6 and the subsequent discussion). This allocation method works as long as it is possible to assign each of the 16 components of X uniquely to one business area. This is however not the case for our division view (Sect. 7.2.1.1) as investment risk is assigned to the three insurance divisions Fire, Liability, Theft. On these grounds XYZ Inc divides up each investment risk by percentage with respect to the liabilities for Fire, Liability, Theft and Other, and in each case adds these percentage portions to the division risk.

7.2.5 Indicators

XYZ Inc uses as the key indicator the RORAC (Definition 6.8) taking into account the risk profiles (see Sect. 6.7) given in Table 7.9.

Table 7.9 Risk profile of XYZ Inc

Safety level α	50 %	75 %	90 %	99 %
Risk appetite (in % of the available capital)	5 %	20 %	50 %	100 %
Maximal economic capital	45	180	450	900
Actual economic capital (see Sect. 7.2.3.4)	−18	108	264	671

7.2.5.1 Divisions

The division view also uses the key indicator RORAC with the same risk profile as a constraint.

The principal application of the division view is in managing sales and marketing capacities. Sales and Marketing collaborate on plans for sales and marketing campaigns with detailed cost estimates and sales targets. The key indicators are computed anew for different combinations of these campaigns under the assumption that the sales plans are implemented. The associated additional costs are not accounted for in the model but the RORAC is explicitly corrected for them. Decisions about prioritizing marketing campaigns are made using these results.

Example 7.2 It is intended to introduce D&O (Directors and Officers) insurance as a new market in the Liability division. Product Pricing has already a rough sketch of a product, and with data provided by the reinsurer has parameterized it. The necessary product development costs are estimated to be about 10 and the costs for customization of systems about 20. It is assumed that the life cycle of this product will be 5 years, though in doing so possible synergies with successor products have been neglected. The one-time sales and marketing costs are estimated to be 10. The plan provides that each year about 500 D&O policies can be sold, for which the expected loss ratio is 4 % (of the insured sum). 10 % of the premium is assumed to cover running costs. The average premium income per policy is set at 0.15, and the average sum insured at 2. It is assumed that the new product has no effect on the sales of other insurance products. The additional return is taxed at a rate of 30 %.

The economic capital model provides for an additional economic capital of 50 for D&O insurance. There results then a new RORAC of 26.2 %. However, management fears that increased adverse selection effects may lead to a worsening of the risk profile. In fact, it was assumed by Product Pricing that the business managers of smaller medium-sized companies make mistaken decisions more often than the managers of larger companies. In spite of detailed underwriting guidelines, it was assumed for the model that this effect could lead to a loss distribution with small losses more frequently represented. An evaluation of the entire distribution shows the distribution for the expected shortfall (Table 7.10).

This is contrary to the requirement of the strategic risk profile, so that in spite of its high profitability the new product is not introduced in this form.

Table 7.10 Expected shortfall for the confidence levels of the risk profile

Safety level α	50 %	75 %	90 %	99 %
Maximal economic capital	45	180	450	900
ES_α	50	150	280	697

Exercise 7.7 Verify, on the basis of the other details, the result from Sect. 7.2.3.4 that the RORAC is 26.2 % when the D& O product is taken into account.[3]

7.2.5.2 Functions

The function view serves at XYZ Inc mostly to set performance-oriented remuneration.

For Asset Management the RORAC key indicator is used. Asset Management is subject to implicit constraints that are consequences of the risk profile, and must ensure the liquidity calculated with the ALM system.

The key indicator RORAC is used to measure the performance of the insurance business as a whole. In order to permit a better comparison, the RORAC for Asset Management is corrected with hypothetical sales costs as the Asset Management department does not need to do any marketing in order to acquire deposits.

In the Enterprise Risk Management Committee an "imputed sales cost factor" is fixed as a portion of the premium. The imputed sales cost factor is informed by the average costs in the last 5 years,

$$\text{Imputed sales cost factor}(t) \approx \frac{1}{5}\sum_{i=1}^{5}\frac{\text{Sales costs}(t-i)}{P_{t-i}}.$$

(We have only a rough equality since the Enterprise Risk Management Committee modifies the imputed sales cost factor on the basis of qualitative considerations.) Imputed sales cost factors for past years are also called upon in evaluating the present sales performance in the year t. With

$$f = \frac{1}{3}\sum_{i=1}^{3}\frac{\text{Imputed sales cost factor}(t-i)\,P_{t-i}}{\text{Total costs[Insurance]}(t-i)}$$

the quantities

$$\text{RORAC}_{\text{eff}}[\text{Asset Management}](t) = (1-f)\text{RORAC}[\text{Asset Management}](t)$$

and

$$\text{RORAC}_{\text{eff}}[\text{Insurance}](t) = \text{RORAC}[\text{Insurance}](t)$$

[3] Since this is a modeling exercise for which not all the assumptions are given, 26.2 % is not the only plausible solution.

are chosen as performance indicators for Asset Management and Insurance in the year t. They are compared with the hurdle rate. Using such comparisons the board allocates bonuses depending on performance to Asset Management and Insurance.

Exercise 7.8 Discuss to what extent the modification of the RORAC for Asset Management is appropriate.

For the insurance business the performance-dependent bonus is broken down to the second layer of the function view which consists of 6 business sub-areas. For each employee i in the sub-area j there is an imputed bonus $B_S(i, j)$, that the employee deserves if the profit of the insurance business comes out as expected, and both the sub-area j and the employee i produced the desired performance. Each sub-area j is assigned a performance weight $W(j) \in [0, 2]$, where $W(j) = 1$ is chosen, if the sub-area produced exactly the expected performance.

1. *Sales*: For each policy v which is accepted by Underwriting, a risk-adjusted expected profit $g(v)$ is calculated with the help of the formulas coming from Product Pricing, and contrasted with a risk-adjusted target profit G. The performance weight is given by

$$W[\text{Sales}] = \max\left(0, \min\left(\frac{\sum_v g(v)}{G}, 2\right)\right).$$

2. *Underwriting*: Underwriting has the character of a control function, and it is itself checked on a sampling basis by the Product and Portfolio Controlling department. On the basis of these samples Product and Portfolio Controlling estimates the fraction γ of policies that are incorrectly processed. Product and Portfolio Controlling estimates a load factor

$$F = \frac{1}{12} \sum_{\mu=1}^{12} \frac{\#[\text{policies expected in month } \mu]}{\max_{m \in \{1,\dots,12\}} \{\#[\text{policies expected in month } m]\}},$$

that effectively describes the expected yearly load of each employee. There is for each type k of policy a given time, $z(k)$, in which such a policy should be processed. In addition, the expected number $n(k)$ of policies that should come in during the year is prescribed. The number of employee hours available in the year for policy processing is denoted by Z.[4] Then we have

$$W[\text{Underwriting}] = \max\left(0, (1 - 10\gamma) \frac{\sum_{\text{Policy types } k} z(k) n(k)}{FZ}\right).$$

The factor 10 was chosen to give a higher weight to respecting the underwriting guidelines than to speed of working.

[4] Z is smaller than the total work time of employees, since breaks, time for employee meeting etc. must be subtracted from the total work time

3. *Claims Processing*: The weight

$$W[\text{Claims Processing}]$$

is arrived at in a similar way to

$$W[\text{Underwriting}].$$

4. *Product Pricing*: To come up with a performance weight, the extent to which the pricing in the last 3 years correctly anticipated the actual evolution is examined. This determines up to a quarter of the expected performance weight. The remaining performance weight is fixed by considering the extent to which the qualitative objectives of recent years were achieved (balanced score card, for short BSC). We denote the set of policies sold in the year s by $\mathscr{V}(s)$, and by $P(v,t)$, $S(v,t)$, $R(v,t)$, $K(v,t)$ respectively the cumulated premiums, cumulated losses, the reserve and the cumulated costs, that are to be assigned to the policy v at time t. The assignment for the reserve is done roughly proportional to the insured amount. As a target measure we then define

$$Z(v,t) = P(v,t) - S(v,t) - R(v,t) - K(v,t).$$

An index $_{\text{calc}}$ distinguishes the quantities calculated according to pricing of th products. The weight factor is then

$$\text{W[Product Pricing]} = -\frac{1}{4}\min\left(\left|1 - \frac{\sum_{s=1}^{3}\sum_{v\in\mathscr{V}(t-s)} Z(v,t)}{\sum_{s=1}^{3}\sum_{v\in\mathscr{V}(t-s)} Z_{\text{calc}}(v,t)}\right|, 1\right) + \big[\text{Result(BSC)}\big].$$

A weight $W[\text{Product Pricing}] > 1$ could only be arrived at if the result from the balanced score card exceeds expectations.

5. *Marketing*: Marketing participates in the successes of sales. This is taken account of since the Sales results enter at the 25 % level into the performance weight for marketing:

$$W[\text{Marketing}] = \frac{1}{4}W[\text{Sales}] + \frac{3}{4}\big[\text{Result(BSC)}\big]$$

6. *Administration & Other Insurance Business:* All remaining functions are appraised purely qualitatively:

$$W[\text{Administration \& Other Insurance Business}] = \big[\text{Result(BSC)}\big].$$

Let $M(j)$ be the set of employees in sub-area j. The allocation of the bonus for Insurance to sub-areas j is given by

$$B(j) = \frac{\text{Bonus(Insurance)}\,W(j)\sum_{i\in M(j)} B_S(i,j)}{\sum_{k=1}^{6} W(k)\sum_{i\in M(k)} B_S(i,k)}.$$

The sub-areas decide for themselves how the bonus is divided amongst their employees. The bonus of the head of each sub-area is, however, directly set by the board.

Exercise 7.9 For which functions is the remuneration based on the economic capital? Discuss to what extent this should also be done for other functions.

Exercise 7.10 In order to help computing the performance weight for Product Pricing, flesh out in detail how one can assign to each policy v in the portfolio a reserve $R(v)$. Take into consideration the run-off triangles for liability and fire, and distinguish between IBNR (incurred but not reported), IBNER (incurred [and reported] but not enough reserved) and RBNS (reported but not settled).

7.2.6 The Organizational Components for Value-Oriented Management at XYZ Inc

The company is organized by function (see Sect. 7.2.1.2). Each function within Insurance looks after the three divisions Liability, Fire and Theft. Product Pricing has two subgroups, Actuarial and Legal / Business Analysis. The subgroup Actuarial computes in addition all reserves and the risk-adjusted return per insurance product as a function of its parameters.

Risk management is not assigned to any division and is directly supervised by the CRO (Chief Risk Officer), who is not a member of the board. The CRO reports directly to the CEO (Chief Executive Officer) (with copies to the CFO (Chief Financial Officer)). Risk management encompasses the classic department Controlling and the special technical department Product and Portfolio Controlling. Product and Portfolio Controlling is responsible for all models that extend over the whole firm. Product and Portfolio Controlling employs actuaries and mathematicians with the speciality of quantitative asset management.

The Enterprise Risk Management Committee (ERM Committee) of XYZ Inc is illustrated in Fig. 7.12. The function of the ERM Committee members from Controlling and Product and Portfolio Controlling is restricted to helping the other members to interpret the results. The ERM Committee meets quarterly for 2 or 3 hours each time, and usual has the following agenda:

1. *Risk report.* The principal points of the standard risk report are presented and their possible consequences discussed. The report is made available to committee members a week before the meeting. If members of the committee wish, other points of the standard report are examined, or possible risks that are not covered in the report are brought forward.

   For each risk that is discussed an action statement is pronounced, or it is decided that no action is required.

   For risks that are not discussed there is no need for action.

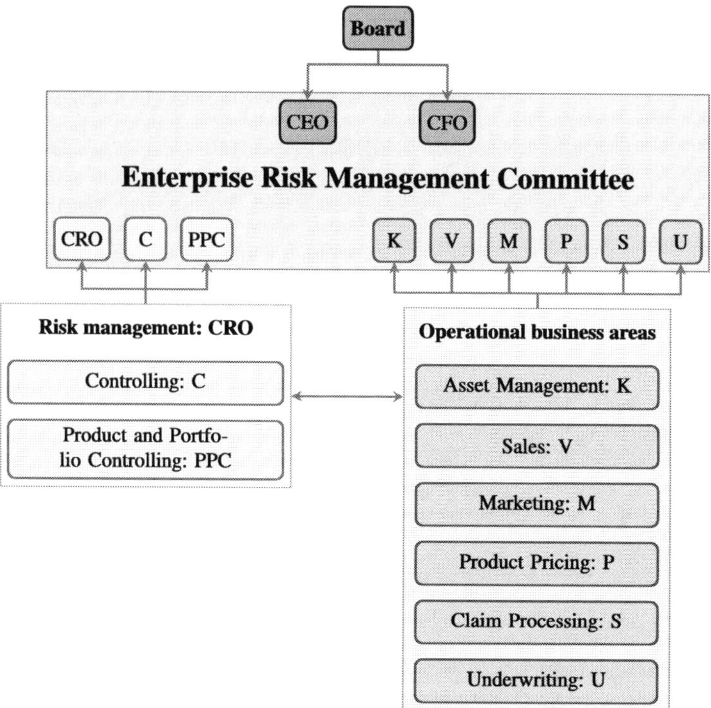

Fig. 7.12 Enterprise Risk Management Committee of XYZ Inc

2. *Report on risk-adjusted performance.* This is just informational. The members of the functions give their views on their risk-adjusted performances and explain possible differences from preceding quarters. If necessary, it is also discussed to what extent the measurement of performance reflects the realities of the functions.

   Example: In the weak summer quarter Underwriting and Claims Processing have each put two experienced employees at the disposal of Product Princing, to help adjusting external data for the appraisal of the potential expansion to D&O insurance. The expected workload will be taken into account during determination of performances, but there is no incentive to encourage this sort of collaboration. Controlling has the task of investigating how such incentives could be built into performance determination. The committee proposes two possible methods, which should be further investigated, that an internal account is kept of employee hours made available to other functions, or a qualitative evaluation using the balanced score card.
3. *The 5 most important risks and opportunities.* The 5 most important risks and opportunities are set by the CEO, and communicated to the participants 3 days before the meeting. Ideas are discussed concerning how these risks and opportu-

nities should be handled. No decision is reached, since these agenda items serve mainly to support the board in formulating risk strategies.

4. *Updating processes.* Suggestions for better integration of risk management and business processes are presented and discussed. For this agenda item contributions are only infrequent, since the ERM Committee members prefer to stick to their core business.
5. *Marketing and Distribution.* Marketing and distribution strategies are explored and discussed. In preparation, Product and Portfolio Controlling has analyzed the strategies and initiatives to be discussed (see Example 7.2).
6. *Bonus.* This agenda item is only treated in the fourth quarter. The performance weights calculated for the whole year are presented and possible qualitative adjustments discussed. A decision is not arrived at, since it is the board that has the final responsibility in setting performance weights.
7. *Miscellaneous.*

Exercise 7.11 How could the agenda item "Updating processes" be revitalized?

7.2.7 The Process Components of Value-Oriented Management at XYZ Inc

The meetings of the ERM Committee are embedded in a process that prepares for the meetings and reports on their results. This is shown in Table 7.11, where all times are in work days (5 per week) relative to the ERM Committee Meeting, which takes place in the last month of the quarter.

7.2.7.1 An Example Scenario: Expansion of the fire Business

Applying the components of value-oriented management is best illustrated with a concrete scenario.

The company considers expanding the fire insurance business to medium-sized firms.[5] A preliminary decision is to be reached during the ERM Committee (ERMC) meeting in 3 months (at time t). Job assignments are handed out as gathered in Table 7.12.

The status reports to the members of the ERM Committee have the function of allowing an early intervention in the project if there should be early indications of problems that would be difficult to overcome.

---

[5]The actual numerical values in this example scenario are completely arbitrarily chosen and have *not* been aligned with real expectations for claims. In particular, it cannot be concluded from this example that the damages expected in the industrial fire insurance business are in fact 20 % higher than in private fire insurance.

Table 7.11 Schedule for preparation for the ERM Committee meeting

Time	Action	Who
up to −22	Developing marketing and distribution initiatives	Marketing
−20 to −10	Risk-adjusted performance evaluation of suggested marketing and distribution initiatives	Product and Portfolio Controlling
−20 to −10	Calculation of the economic capital, risk-adjusted performance etc.	Product and Portfolio Controlling
−15 bis −10	Collection of qualitative risk and performance indicators	Controlling
−10 to −8	Preparation of a report on risks and performance	Controlling, Product and Portfolio Controlling
−5	Proposal for the top 5 opportunities and risks to CEO	CRO
−5	Risk management and performance reports distributed to ERM Committee members	CRO
−3	Definition of the top 5 opportunities and risks	CEO
0	Meeting of the ERM Committee	ERM Committee
$+x$	Assignments resulting from the ERM Committee meeting	

Table 7.12 Planning for industrial fire insurance

Time:	$t-90$	$t-60$	$t-30$	$t-25$	$t-10$
Product Pricing:	Start product development	Brief status report to ERMC members	Brief final report to ERMC members		
Marketing:	Support of Product Pricing in product development				
		Start development of marketing strategy	Brief status report to ERMC members	Finalization of marketing strategy	
Product and Portfolio Controlling:				Calculation of economic capital and risk-adjusted return with and without the new business area	

Shortly after $t-90$ it turns out that there is hardly any data available on the frequency of claims made by mid-sized firms and how large they are. There are only some cumulated data, that are about the whole fire insurance business and are split into private and industrial parts. According to these sparse data the combined ratio for private business has a countrywide average of 75 %. The combined ratio

Table 7.13 Parameters for the relative claim distribution S_0

Parameter	Best case	Likely case	Worst case
Mean	50 %	50 %	75 %
Standard deviation	25 %	38 %	50 %

Table 7.14 Parameters for claims frequency distribution

Parameter	Best case	Likely case	Worst case
Mean	5 %	10 %	15 %

for industrial business countrywide is 20 % higher[6] and is thus 90 %.[7] On the other hand, the combined ratio, according to the company, for private clients is 10 % less than the countrywide average and is 67.5 %. The Product Pricing department has not found any countrywide data about volatility or frequency distributions, or claim distributions.

Product Pricing models the industrial fire insurance business with a mixed Poisson distribution and lognormal claim distribution.

For the distribution of the sizes of claims the following considerations are taken into account:

- In the company's experience claims from private clients for fire damage run on average about 75 % of the total insured sum.
- Product Pricing assumes that for industrial business an intensive loss adjustment process will take place which should reduce the loss quota to about 50 % per individual case.
- For private clients the loss distribution has a standard deviation of 25 %. Since private clientele seems more homogeneous, a higher volatility is expected in the industrial business. Doubling the standard deviation is, according to a qualitative estimate by Product Pricing, a conservative worst case. As a "best estimate" is taken the average of this worst case and the standard deviation for private clients.
- The parameters derived by Product Pricing for the loss distribution S_0 (relative to the sum insured) are itemized in Table 7.13. For private clients a frequency of claims of 5 % is measured. A rough calculation on the basis of the combined ratios of the countrywide industrial business and the company's own private business gives an estimated value of 10 %. Table 7.14 shows the parameters derived by Product Pricing for the frequency distribution.

Exercise 7.12 What implicit assumptions has Product Pricing made if it compares the combined ratios for the company's private business and the countrywide industrial business? Discuss how far this can be justified. How should Product Pricing have proceeded in this case?

[6] See footnote 5.

[7] 20 % of 75 % makes 15 %.

Table 7.15 The parameters σ, μ for the relative claim size distribution S_0

Parameter	Best case	Likely case	Worst case
μ	−80 %	−92 %	−47 %
σ	47 %	67 %	61 %

At time $t-60$ Product Pricing gives a brief presentation of their results, during which the uncertainties of the assumptions at their root are examined. One of the results is a provisional price calculation, which does not yet include the costs of capital. This serves Marketing as an initial basis for a plan for product launch. Because of the bad situation as far as data are concerned, it is decided to introduce the product at first in only one region and then to expand distribution later if this proves successful. Product Pricing gets the job of developing product prices while taking into account risk capital and the probabilities of the *Best Case*, *Likely Case* (e.g., Best Estimate), *Worst Case*.

The density function for a lognormal distribution S is given by

$$f_{\mu,\sigma}(x) = \frac{\exp - \frac{(\ln x - \mu)^2}{2\sigma^2}}{\sqrt{2\pi} x \sigma}.$$

For lognormal loss distribution $S \sim f_{\mu,\sigma}$ we have

$$\mathrm{E}(S) = \mathrm{e}^{\mu+\sigma^2/2}$$

$$\mathrm{var}(S) = \left(\mathrm{e}^{\sigma^2} - 1\right)\mathrm{e}^{2\mu+\sigma^2} = \left(\mathrm{e}^{\sigma^2} - 1\right)\mathrm{E}(S)^2$$

and therefore

$$\sigma = \sqrt{\ln\left(1 + \frac{\mathrm{var}(S)}{\mathrm{E}(S)^2}\right)}$$

$$\mu = \ln \mathrm{E}(S) - \sigma^2/2 = \ln \mathrm{E}(S) - \frac{1}{2}\ln\left(1 + \frac{\mathrm{var}(S)}{\mathrm{E}(S)^2}\right),$$

so that Product Pricing obtains as the parameters for the relative claim sizes per insured case described by the lognormal distribution S_0 the values given in Table 7.15.

Exercise 7.13 Why would it be incorrect to interpret the quantities in the table as average relative values per individual policy?

We wish to derive the absolute claim size distribution S from the relative claim size distribution S_0. In order to do so we need to scale S_0 with the portfolio weighted average insured sum, c. This gives $S_0 \mapsto S = cS_0$, $\mathrm{E}(S_0) \mapsto c\,\mathrm{E}(S_0)$, $\mathrm{var}(S_0) \mapsto c^2\,\mathrm{var}(S_0)$. Since

$$\mu + \frac{1}{2}\sigma^2 = \ln E(S_0)$$

Table 7.16 The parameters σ, μ for the claim size distribution S_i

Parameter	Best case	Likely case	Worst case
μ	0.29	0.18	0.62
σ	47 %	67 %	61 %

we obtain for μ and σ the corresponding transformations

$$\mu \mapsto \ln\big(c\,\mathrm{E}(S_0)\big) - \frac{1}{2}\ln\left(1 + \frac{c^2\,\mathrm{var}(S_0)}{c^2\,\mathrm{E}(S_0)^2}\right) = \ln c + \ln \mathrm{E}(S_0) - \frac{1}{2}\ln\left(1 + \frac{\mathrm{var}(S_0)}{\mathrm{E}(S_0)^2}\right)$$
$$= \mu + \ln c$$
$$\sigma \mapsto \sigma.$$

The average insured sum c to be expected in the pilot region is worked out with the Marketing, and comes out as $c = 3$. Product Pricing is using the claim size model represented in Table 7.16 for the absolute size of losses $S_i = c_i\,S_0$.

Exercise 7.14 Discuss the implicit simplifications that are made in this model.

The parameter combinations for the frequency distribution N_j and the loss distribution S_i $(i, j \in \{1, 2, 3\}$ are summarized in Table 7.17.

The resulting total distribution is therefore

$$X(\omega) = \sum_{k=1}^{N_{j(\omega)}(\omega)} S_{i(k,\omega)}(\omega).$$

This equation is to be understood in the following way:

1. The parameter of the Poisson distribution $j(\omega)$ is chosen from one of the three possibilities "Best Case", "Likely Case", and "Worst Case", where their probabilities are taken to be, respectively, {10 %, 80 %, and 10 %}.

Table 7.17 Summary of parameter combinations for the frequency distribution N_j and the claim size distribution S_i

				Claim size: $S_i = cS_{0,i}$		
				Best Case	Likely Case	Worst Case
			μ_i	0.29	0.18	0.62
			σ_i	47 %	67 %	61 %
		$\mathrm{E}(N_j)$	p_{ij}	20 %	70 %	10 %
Frequency: N_j	Best case	5 %	10 %	2 %	7 %	1 %
	Likely case	10 %	80 %	16 %	56 %	8 %
	Worst case	15 %	10 %	2 %	7 %	1 %

2. The number of losses $N_{j(\omega)}(\omega)$ is drawn according the Poisson distribution $j(\omega)$.
3. For each $k \in \{1, \dots, N_{j(\omega)}(\omega)\}$ a corresponding loss distribution $i(k, \omega)$ is drawn from the three possibilities "Best Case", "Likely Case", and "Worst Case", with probabilities $\{20\,\%, 70\,\%, 10\,\%\}$.
4. For each $k \in \{1, \dots, N_{j(\omega)}(\omega)\}$ a loss $S_{i(k,\omega)}(\omega)$ is drawn from the loss distribution $i(k, \omega)$.
5. Finally the resulting losses are added.

This distribution models both the actuarial uncertainty and the uncertainty over parameters for the new business branch. There are additional uncertainties (such as, for example, uncertainty over the success of the sales team) that are not taken into consideration here. Using this distribution Product Pricing determines the corresponding economic capital $\mathrm{ES}_{99.5\,\%}(X)$. The hurdle rate of the company is $h = 17\,\%$ and the risk-free rate is $s = 4\,\%$. So in total at least $P = \mathrm{E}(X) + (h - s)\mathrm{ES}_{99.5\,\%}(X)$ must be taken in as a premium to achieve a risk-adjusted break-even. On top of this there is an additional loading p that should reflect the administrative and distribution costs, costs for risks that are not covered explicitly, and the profit margins. If IS is the insured sum for a policy and $\mathrm{IS}_{\mathrm{Total}}$ the accumulated insured sum of the portfolio being considered, then we obtain the equation

$$P(\mathrm{IS}) = \big(\mathrm{E}(X) + (h - s)\mathrm{ES}_{99.5\,\%}(X)\big)\frac{(1 + p)\mathrm{IS}}{\mathrm{IS}_{\mathrm{Total}}}$$

as a simplified basis for pricing.

Exercise 7.15 Discuss the extent to which this simplified basis for pricing is appropriate.

Exercise 7.16 Work out suggestions for improvements of the individual components of value-oriented management in this example. In doing so, both simplifying certain aspects and a more careful consideration of risks in other aspects can constitute improvements. According to the situation that the company is in one will arrive at different recommendations.

References

1. R.S. Kaplan, D.P. Norton, The balanced scorecard measures that drive performance. Harv. Bus. Rev., 71–79 (1992) (p. 280)
2. R.S. Kaplan, D.P. Norton, *Balanced Scorecard: Translating Strategy Into Action* (Harvard Business School Press, Harvard, 1996) (p. 280)
3. W. Foundation, Balanced Scorecard. German edn. (December 2006) (p. 281)

Chapter 8
Solvency and Regulatory Questions

In this chapter we treat the regulations in Germany which affect risk management and risk capital (solvency). While current regulation considers these two aspects separately (see Sects. 8.1 and 8.2.3), they will be considered together under Solvency II (see Sect. 8.2.4).

On the surface, this chapter addresses German risk regulation which is currently in flux.[1] Other European countries are also in the process of anticipating Solvency II and may also have a transitional regulatory system. Extensive parts of this chapter apply to these countries as well. However, the purpose of this chapter is more far-reaching than simply describing some European supervisory systems:

- this chapter shows how an entrenched, outmoded regulatory system can be modernized
- this chapter shows to what extent modern risk supervision requires a change in management. It may be especially surprising for readers in jurisdictions where regulators will not be subject to Solvency II and are critical of recent publications by the IAIS (International Association of Insurance Supervisors), to what extent the organizational and operational structure of management is affected by this new type of regulation. Even more surprising may be the fact that these new laws were *not* highly controversial within the industry. On the contrary, many companies welcomed these laws as a further incentive for improved risk management. As these laws also apply to the local competition, they made it easier to invest in better risk management up front and to reap benefits in the mid to long term.

---

[1] We write this in 2013, when Solvency II has not yet been implemented. At the same time, some laws anticipating Solvency II are already in effect.

M. Kriele, J. Wolf, *Value-Oriented Risk Management of Insurance Companies*, EAA Series, DOI 10.1007/978-1-4471-6305-3_8, © Springer-Verlag London 2014

8.1 Law Regulating Control and Transparency in Business (KonTraG)

8.1.1 Goals of KonTraG

Against the background of manifold changes in the financial environment, increasingly intense competition, and enhanced complexity efficient company management becomes ever more important. The law regulating control and transparency in business (KonTraG), which came into force 1 May 1998, had as its goal to achieve quality in management by improved risk management, transparent corporate disclosures, and ensuring the quality of information available to investors and other interested parties. To this end KonTraG strengthens existing and introduces new internal and external control mechanisms. KonTraG is a so-called omnibus act that mainly changes a few items in the German Corporation Act (AktG) and the German Commercial Code (HGB). The preamble of the law states that KonTraG is supposed to have consequences for other parts of the law, in particular that for limited corporations (GmbH). This means that, taking into account their size and complexity, the regulations are intended to be carried over to other legal forms. KonTraG addresses several target areas:

- Introduction of an early warning system for risk jeopardizing the existence of a company
- Increased company transparency to the capital market
- Better information for the non-executive board of directors from the executive board of directors, as well as closer supervision of the executive board of directors by the non-executive board of directors
- Improving the quality of annual external audits
- Improving the collaboration between the non-executive board of directors and the auditors
- Improved internal management control structures and more effective control through the annual shareholders' meeting
- Approval of modern financial and remunerative instruments
- Eradication of voting rights discrepancies and avoidance of conflicts of interest in proxy votes

In the following section we shall explain the regulations in KonTraG in relation to those involved in control systems of German stock corporations (non-executive board of directors, executive board of directors, annual shareholders' meeting, auditors, the capital market).

8.1.2 Regulations

According to § 91 paragraph 2 AktG, the *executive board of directors* has to take suitable steps so that any developments that endanger the continuation of the company or essentially affect its assets, yield or financial status are noticed in good time. The preamble for this law points out that with this requirement of an early warning system, it is also clear that the executive board has the responsibility to provide an appropriate risk management and appropriate internal checks. In this way the KonTraG clarifies that the managerial function required of the executive board of directors according to § 76 paragraph 1 AktG and the due diligence according to § 93 paragraph 1 sentence 1 AktG also extend to risk management. § 93 paragraph 2 AktG makes the executive board of directors liable for damages if they shirk their duties. In this case the burden of proof is upon them. KonTraG provides no concrete advice for the design of an early warning system to give the company the opportunity to implement the best methods in a way tailored to the company. Early recognition of risks does not mean being able to exclude existential risks to the company completely from the outset, but recognizing them early enough to be able to take steps to help ensure the continued existence of the company.

In addition, KonTraG expands the requirements that the *executive board of directors* must report to the *non-executive board of directors*. According to § 90 paragraph 1 number 1 AktG the executive board of directors has also to report to the non-executive board concerning the intended business policies and other fundamental questions of business planning (in particular, financial, investment, and personnel planning). This means conversely that the non-executive board of directors cannot restrict itself to a purely retrospective control function, but must also be involved in the short, middle and long-term future planning under its control. The Law on transparency and publicity (TransPuG) from 2002 extends the reporting requirements of the executive board of directors to include any discrepancies in the actual developments from the reported targets together with justifications for them. In this way the reporting duties of the executive board of directors include both the past and the future.

To enhance the professionalism of the non-executive board of directors, KonTraG restricts the number of seats on the non-executive board to 10 plus 5 company positions (§ 100 paragraph 2 AktG) and requires at least two meetings of the non-executive board in each calendar half-year (§ 110 paragraph 3 AktG). The statutory supervisory role of the non-executive board of directors extends, according to § 111 AktG, to ensuring the legitimacy, the compliance, the efficiency and the expedience of the company's management. This includes setting up a functioning early warning system.

Furthermore, KonTraG targets improvement of the cooperation between the *non-executive board* and the *auditors*. § 111 paragraph 2 sentence 3 AktG specifies that the non-executive board, and not as earlier the executive board, hires the auditor. The non-executive board can set its own priorities in doing this and ensure the necessary support for the auditor in carrying out her/his duty.

KonTraG abolishes any right to multiple votes in the annual shareholders' meeting by removing § 12 paragraph 2 sentence 2 AktG. This is supposed to avoid distortions between the equity structure and the rights to votes. § 128 paragraph 2 AktG regulates anew the responsibilities of financial institutions in exercising depositor proxy votes. The institutions must provide holders of share deposit accounts with comprehensive information and put forward suggestions for the exercise of their votes, and in doing so must be guided by the interests of the shareholders. If a financial institution has more than 5 % of the total equity and exercises its own right to vote, then it can exercise proxy votes only at the express wish of holders of share deposit accounts (§ 135 paragraph 1 sentence 3 AktG). This is intended to avoid conflicts of interest. To enhance the control structure of the annual shareholders' meeting, § 147 paragraph 3 AktG allows an individual complaint to be brought against the executive board or non-executive board, if there is a suspicion of dereliction of duty, if the complainant holds 5 % of the total equity or has an investment of at least 500,000 Euros.

By § 289 paragraph 1 HGB the *annual report* (resp. by § 315 paragraph 1 HGB, the *group annual report*) must address expected developments and the risks that come with future developments. In doing so, all risks with a relevant probability of occurring must be presented, not just those endangering the existence of the company. Discussion of measures that are being taken, or for certain developments planned measures for particular risks, should round out the risk section of the annual report, so that it provides the reader with a comprehensive picture of the risk position.

KonTraG makes further requirements upon the *annual audit* and upon the auditor's report. By § 317 paragraph 2 HGB the auditor must investigate whether the annual report of the company is consistent with the yearly balance sheet, and if the state of the company and the risks of its future evolution are correctly presented. According to § 317 paragraph 4 HGB it must be checked that the early warning system prescribed in § 91 paragraph 2 AktG is fully functional, or if there are measures that must be undertaken to improve it. §§ 319, 319a HGB target ensuring the independence of the auditor. If an auditor has signed off on the audit on more than 6 occasions, that auditor can only be re-appointed after a waiting time of at least 2 years. One cannot appoint an auditor who has earned, in the last 5 years, more than 30 % of the auditor's income from the company to be audited. In order to ensure that the board is fully informed, § 170 paragraph 3 sentence 2 AktG prescribes the delivery of the files and the audit report to each member of the board.

Finally, for the purpose of *harmonizing with international regulations* KonTraG opens up the additional possibility of acquiring one's own shares (§ 71 paragraph 8 AktG). After authorization by the annual shareholders' meeting, the company's own shares can be bought, within a period of 18 months and up to a limit of 10 % of the total equity, to encourage market trading or to improve the position of the shares as an investment. § 192 paragraph 2 number 3 AktG allows a conditional increase in capitalization to permit offering share options to management, to motivate them to orient themselves toward the long-term goals of increasing the value of the company.

8.1.3 Implementation

Even though KonTraG does not make concrete recommendations, in practice, not least because of the requirements of the audit standards IDW PS 340, basic methods and requirements for the implementation of an early warning system have established themselves.

The requirements of KonTraG to be aware of existential risks early on, and thus to be able to undertake countermeasures in good time, affects all phases of the risk management process. First it has to be established that all potential risks have been recorded and evaluated. To do this standardized risk recording forms (risk maps, risk registers) are used These forms are used to recorded estimates of the possible losses for the individual risks and their probabilities of occurrence, as well as values for risk indicators, risk limits and risk mitigation measures and corresponding responsibilities. For this purpose, risk indicators can be financial numbers and "soft" factors such as, e.g., a market trend or the image of the company. Risk limits define upper bounds for certain risk positions (e.g., credit volume per creditor) exceeding which leads to introduction of established countermeasures.

In order to arrive, on the basis of the risk recording forms, at a holistic risk analysis for the company, the risks must be grouped and systemized taking into account their mutual dependencies. In the financial sector the following risk areas are often distinguished:

- Market risk
- Credit risk
- Liquidity risk
- Insurance risk
- Operational risk
- Legal risk

Evaluation of the risks aims for a classification in the form of a two-dimensional risk matrix (see Fig. 1.2) where the axes indicate the size of loss and the probability of occurrence. While for quantifiable risks it is possible to assign to each risk a numerical probability and a corresponding loss amount, for qualitative risks (e.g., political risk, views of the evolution of the market) only a qualitative risk matrix can be constructed that, say, distinguishes the losses as small, medium or heavy, and the probabilities as improbable, possible, likely, and very likely.

The potential extent of losses is the crucial quantity in judging whether a risk endangers the viability of a company, since risk management must also take precautions against scenarios with very low probabilities which are threats to the company's existence.

The risk matrix is a component of the risk handbook which is treated as a guideline for the company and will be examined by the financial auditor. In addition to the definition of fundamental risk policies the risk handbook should identify the business areas and activities where existential threats could appear, as well as the competences and rules for risk reporting. The IDW auditing standards require that existential risks be reported in a verifiable form to the board.

8.2 Solvency

8.2.1 The Task of Solvency Supervision

Insurance companies should always be in a position to fulfil their contractual obligations. They must always be able to pay, i.e., be *solvent*. Since the insurance business is exposed to numerous uncertainties, such as an unknown development of losses, an insurance company has to have its own funds to be able to meet adverse circumstances.

Against a background of deregulation the notion of a level of own funds has gained in importance for risk management in insurance and for the insurance supervisor. The goals of requirements for a suitable level of own funds are

- protection of policy holders,
- an adequate capitalization for the insurance company (financial strength), and
- stability of the financial market (financial stability).

Seen altogether the idea is that the company's own funds will act as a buffer against adverse developments. While the first goal, policy holder protection, should ensure the basis necessary for a client's trust that having advanced his premium payments he can rely on future benefits from the insurance company, the other two goals recognize the important role of insurance companies in the financial market.

The formulation of the solvency requirements raises the questions how large this buffer against adverse developments must be, how long the time horizon to be considered is, and what investments are appropriate in covering the buffer? The answers to these questions depend on the objectives being pursued, and must be developed consistently with the modeling method which is chosen.

Further to these quantitative requirements, for the solvency of the company there are qualitative requirements on the management of relevant existential risks. While the solvency guidelines currently in force[2] (Solvency I, Sect. 8.2.3) were given purely quantitatively, Solvency II (Sect. 8.2.4) includes both quantitative and qualitative demands.

Thus, the notion of solvency is no longer restricted to quantitative capital requirements, but extends to the whole supervisory system.

8.2.2 Definitions

The basic idea that own funds represent a buffer against adverse developments, is already reflected in the definition of the *solvency margin SM* as the difference between assets A and liabilities L due to Pentikainen (1952) [14]:

$$SM = A - L$$

[2]We write this in 2013.

That portion of the solvency margin that is covered by qualified investments, that is those recognized by the regulators, is called the *available solvency margin*.

The smallest amount of available solvency margin compliant with the statutory regulations is called the *minimum solvency margin*.

In making a concrete definition of solvency there are two extreme positions conceivable in regard to the time horizon. While the "going-concern" assumption asks that the continued existence of the company is ensured, so that the company can meet its obligations at any time in the future, for the "run-off" assumption it is enough that the obligations can be met upon an immediate liquidation of the company. (See also Sect. 4.3.1.2.)

Protecting policy holders does not necessarily imply the continued existence of the company. The claims of the policy holders are also guaranteed if, in the case of a crisis, the portfolio can be transferred to another insurer. Therefore the supervisory requirements, under Solvency I and II, upon own funds and the valuation of the insurance liabilities demand that continuation of the company only for a short time, usually one year, so that the transfer of the portfolio can be carried out.

The continued existence of the company is the primary goal of the owner and the management, who will manage the company according to the going-concern principle. Looking at it from the policy holder's expectation, that their often long-term contracts will be fulfilled by the contracting partner, and with a view to the goal of stability of the financial market, supervisory authorities have an interest in the long-term financial stability of insurance companies. This explains why it is that in contrast to the focus of the concrete requirements on own funds over a horizon of one year, the supervisory regulations define solvency using the going-concern principle.

IAIS (2002) [12] defines an insurance company to be solvent if it is able "to fulfil its obligations under all contracts under all reasonably = foreseeable circumstances". In a later version in 2007 [13] the IAIS refines this definition in that it demands the fulfillment of all contractual obligations "at any time".

8.2.3 Solvency I

In this section we will present the statutory requirements on equity capital for insurance companies currently in force, and examine the background for their emergence.

The EU Guidelines for Solvency I (2002) [10, 11] start from the going concern assumption and define the solvency margin as a buffer against variations in business results.

- Solvency I non-life directive: "The requirement that insurance undertakings establish, over and above the technical provisions to meet their underwriting liabilities, a solvency margin to act as a buffer against business fluctuations is an important element in the system of prudential supervision for the protection of insured persons and policyholders."

- Solvency I life directive: "It is necessary that, over and above technical provisions, including mathematical provisions, of sufficient amount to meet their underwriting liabilities, assurance undertakings should possess a supplementary reserve, known as the solvency margin, represented by free assets and, with the agreement of the competent authority, by other implicit assets, which shall act as a buffer against adverse business fluctuations."

Solvency of an insurance company is thus measured with the amount by which the assets exceed the liabilities, and which are available to compensate for adverse developments. The requirements of the Solvency I Life directive point to additional requirements on the quality of the investments covering the solvency margin.

8.2.3.1 Historical Development

Solvency I can at its core be traced back to the first Non-Life EU Guidelines of 24. 07.1973 and the Life Guidelines of 05.03.1979, which in turn were much influenced by the work of the Dutch professor Campagne [2].

Campagne followed essentially a VaR method. Since he held that a probability of insolvency of 0.001 over 3 years was acceptable, he chose the level of approximately 0.9997 for the 1-year VaR. Campagne remarks that as a result of wide-ranging simplifications and unfounded assumptions about distributions the results of his models cannot serve as measures of solvency but rather just as early warning indicators of the need for a deeper analysis tailored to the company itself.

In non-life insurance Campagne noted an average cost rate of 42 % of the gross premium after reinsurance. He fitted a beta distribution to the observed loss quotas and determined the 0.9997 quantile to be 83 %. On this basis he suggested that $42\,\% + 83\,\% - 100\,\% = 25\,\%$ of the gross premiums be seen as the solvency margin needed. For the premium volume passed on to reinsurance he postulates a solvency margin of 2.5 % as a rough approximation. In addition to all this, he suggested a basic amount of 250,000 European currency units.[3]

Campagne considers the principal risk in life insurance to be investment risk and in this light suggests a fixed percentage of the provisions be defined as the required solvency margin. For this he models the ratio of the investment losses within a year relative to the provisions by using a Pearson Type IV distribution.

$$f(x) = c\left(1 + \frac{x^2}{a^2}\right)^m \exp\left(\nu \arctan\left(\frac{x}{a}\right)\right)$$

with the coefficients $a = 5.442$, $c = 31.73$, $m = -4.850$ and $\nu = 2.226$. Referring to the 95 % VaR, Campagne suggests 4 % of the technical provisions as the required solvency margin.

[3]This is a predecessor of the Euro.

The work of Campagne and other groups on behalf of the OECD and the CEA[4] led to the guidelines in the EU Guidelines for Life [10] and Non-Life [11].

The Life Guidelines require as a minimum solvency margin essentially 4 % of the technical provisions and 0.3 % of the capital at risk. Reinsurance can reduce the assessment basis for the technical provisions by up to 15 % and the assessment basis for the capital at risk by up to 50 %.

The Non-Life Guidelines require as a minimum solvency margin the maximum of the premium basis and the claims basis. The premium basis is 18 % of the gross premiums up to 10 million currency units, and 16 % of the gross premiums that exceed 10 million. The claims basis is 26 % of the actual losses up to 7 million and 23 % of the total of losses in excess of 7 million. Reinsurance is taken into account up to 50 %.

In addition, the 1st EU Guidelines regard a third of the minimum solvency margin as guarantee funds, and fix an absolute minimum amount that depends on the line of business.

After the solvency rules in the 2nd and 3rd Guidelines (1992) Life and Non-Life underwent no changes, the insurance committee of the European Union mandated a working group under the chairmanship of Helmut Müller to examine the solvency regulations. The Müller Report [3] identified the risks for insurance companies and offered the opinion that essentially the solvency system had proved itself useful in practice. From among the suggestions of the Müller Report the EU Commission took up the one to increase the absolute minimum amount in the Solvency I guidelines in order to compensate for inflation [6–8]. Already in 1999 it decided [5] to begin working on a new fundamental approach, since it held that it was possible, against the backdrop of changes in the insurance business, that the solvency rules would in the future no longer function satisfactorily. This decision can be seen as a first step toward the Solvency II project.

8.2.3.2 Solvency Margins According to Solvency I

The Solvency I guidelines [6–8] undertake only slight modifications of the solvency rules. The absolute minimum amounts required of the individual divisions are raised. In non-life, the requirements on minimum solvency margins are partly strengthened. In life, regulations are specified for unit-linked types of insurance, and quality requirements expanded to all assets that are to cover technical provisions and the solvency margin.

The basis for the German implementation of the Solvency I guidelines is the ordinance on capital resources, KapAusstV (Kapitalausstattungsverordnung) [4] that specifies the computation of the required solvency capital (minimum solvency margins), § 53c VAG that, in particular, contains a catalog of admissible types of capital with which to cover the solvency margin, and the explanatory BaFin Memo

[4]Comité Européen des Assurances, a federation of European insurers. In 2012 the CEA changed its name to Insurance Europe.

4/2005(VA) [1]. In the sequel here the rules for calculating the solvency margin will be sketched in their dependence on the various lines of business.

The *minimum solvency margin SM for non-life insurance companies* (those in liability/accident insurance, health insurance, reinsurance) is based on the maximum of the premium and claims indexes. The premium index *BI* is the higher of the booked or of the earned gross premiums for the year. The claims index *SI* is defined as the average yearly expenditure for insurance claims, where the average is taken over the last three years, with the exceptions that the last 7 years are used for companies that are essentially in the credit, storm, hail and frost insurance businesses. Reinsurance is taken into account to a maximum of 50 %. For liability insurance, except for motor liability insurance, the premium index and the claims index are each increased by 50 %. If *RV* denotes the ratio of the net expenditures for insurance claims to the gross expenditures for insurance claims in the last three years, the solvency margin becomes

$$\begin{aligned}\mathrm{SM} = \max[&18\ \% \times \min(BI, 50,000,000) + 16\ \% \times \max(0, BI - 50,000,000);\\ &26\ \% \times \min(SI, 35,000,000) + 23\ \% \times \max(0, SI - 35,000,000)\big]\\ &\times \max(RV, 0.5).\end{aligned}$$

For substitute health insurance the percentages in the formula above should be halved. The guarantee funds should be set as one third of the solvency margin. The absolute minimum amount is 2,000,000 Euros and is to be raised to 3,000,000 Euros if the business is being written in the branches of credit, liability or bond insurance. For mutual insurance companies some premium-dependent relief is prescribed (see §§ 2,3 KapAusstV, § 156a paragraph 1 VAG[5]).

The minimum solvency margin SM for life insurance companies is calculated to be the sum of 4 % of the technical provisions, and of the unearned premiums reduced by the cost portion, and 0.3 % of the capital at risk. Negative capital at risk for any individual contract is to be reset to zero in doing this. In the case of unit-linked insurance for which the company carries no investment risk, the rate is reduced from 4 % to 1 % for policies with maturities of over 5 years, and for policies with maturities under 5 years 25 % of the administrative costs that can be assigned to the policy is used instead. For death insurance, the rate of 0.3 % is reduced to 0.1 % for policies with maturities of at most 3 years, and for maturities between 3 years and 5 years to 0.15 %. The guarantee funds are to be one third of the solvency margin, with an absolute minimum of 3,000,000 Euros. For mutual insurers, in particular for pension and burial funds, premium-dependent relief is foreseen (§§ 7,8,8a KapAusstV, § 156a paragraph 1 VAG).

Solvency I can be characterized as a rule-based system, which is simple to understand and implement, but shows only limited sensitivity to risks, in contrast to the individual risk profiles of insurance companies. For example, higher safety margins in technical provisions paradoxically lead to a higher solvency requirement.

[5]The VAG is the German Insurance Supervisory Act.

Going below the minimum solvency margin is seen as an early warning signal. The supervisory authorities then will ask for a solvency plan to restore the necessary capital resources of the company. If there is a shortfall in the guarantee funds, supervisory authorities will demand a financial plan for improving the financial situation in the short term. If this does not seem possible, the supervisory authority will dissolve the company and try to transfer the portfolio to another firm.

8.2.3.3 Allowable Equity Capital

From the definition of the minimum solvency margin there arises the question as to what capital positions can be used to cover it. Since the solvency margin is imagined to be a buffer against adverse developments, it is pertinent how easily a capital position can be used to meet losses. Under Solvency I the question is finally answered in the form of the catalog of own funds in § 53c paragraph 3 VAG , and further explained in BaFin Memo 4/2005(VA).

In grading their potentials for meeting losses the types of own funds listed in § 53c paragraph 3 sentence 1 can be divided into three groups. Positions numbered 1 to 3 can always be used, those numbered 3a to 4 can be employed only under certain circumstances described in the law. These two groups together make up so-called own funds of type A. Own funds of type B are enumerated under 5a) to d) and can only be counted in upon application to and approval by the supervisory authority.

Positions numbered 1 to 3 comprise equity in a narrow sense, capital which is unrestrictedly available:

- for shareholder corporations the paid-up share capital less the amount of own stock, for mutual insurances the effective initial fund
- capital reserves and retained earnings
- profit carried forward after subtraction of the dividends to be paid

Profit-sharing certificates and subordinated liabilities (No. 3a resp. 3b) are obligations upon the company, but have, to a certain extent, an equity character.

To ensure sufficient equity character § 53c paragraph 3 VAG insists on, for both the above positions, as prerequisites that the capital must be made available for a minimum of 5 years, that a claim for refund cannot become due in less than 2 years, and that in case of insolvency or liquidation it can only be paid after satisfying all non-subordinate creditors. Profit-sharing certificates must in addition participate to its full sum in a loss, and if there are losses the insurance company delays paying interest. For subordinate obligations compensation for redemption claims with the company's receivables is not allowed. For both profit-sharing certificates and subordinate obligations the amount eligible to cover the solvency margin is limited to 25 % of the equity according to numbers 1 to 3, and to 50 % of the minimum solvency margin.

By § 56a VAG future policy holder dividends which have not already been promised can be called on, with the approval of the supervisory authorities, to meet losses.

This means that the collective rights of the policy holders to dividends from the free premium refunds (RfB[6]) do not have to be satisfied if there is a crisis meeting the conditions of § 56a VAG. Because of this the liabilities of the company are reduced by the amount of the free premium refunds, so that the free premium refunds can be considered to be own funds of the company that has been made available by the policy holders. This is the basis for § 53c paragraph 3 number 4 VAG, which recognizes the free premium refunds as part of the own funds.

For the capital positions mentioned in § 53c paragraph 3 number 5a) to d) there is some uncertainty that they really will be available in the case of losses. Therefore they are only recognized on application to and approval of the supervisory authorities:

- half the unpaid share capital, insofar as at least 25 % of the share capital is paid up,
- in the case of mutual insurers which have a clause to require policy holders to make additional payments during a crisis for the insurer, half of the difference between the maximum additional payment within a year and any such payments that have already been made.
- hidden net reserves, as long as they are not specially excluded.

Unpaid share capital or additional payments can cover only up to half of the minimum solvency margin, and also make up only up to half of the own funds.

The constraints on recognition of capital positions go back to risk-political consideration of the legislator. In the case of a mutual insurer, whether policy holders will actually make additional payments in crisis situations strongly depends on their awareness and their social conscience. In general, the policy holders of a mutual insurer which emphasizes competitiveness over community will be less willing to make such payments than the policy holders of a more traditional mutual insurer. Therefore recognition of additional payments is subject to approval of the authorities, who must consider in each case the probability that payments will actually be made. While doubt as to the availability of additional payments during crises seems reasonable, a look at history shows that additional payments have saved insurance companies. Examples: The oldest German mutual (Gothaer) was saved in the 19th century after a catastrophic fire that took place in Hamburg. More recently, British insurance collectives for ships stated that their (corporate) policy holders have made such payments during particularly bad years.

Hidden reserves comprise unrealized profits that are in danger of being eaten up quickly if the market develops adversely. Therefore their recognition as own funds is subject to the constraint that they are not the consequence of exceptional market movements, i.e, their value seems lasting. The Circular 4/2005(VA) mentions amongst others the following criteria for checking if it is a case of lasting value. To determine hidden reserves in listed securities one uses the minimum of the value at the balance sheet date and the average of the values reported at the current and

[6]The RFB is basically a reserve for future policy holder dividends peculiar to the German insurance market.

the three preceding balance sheet dates. Recognition of hidden reserves in unlisted securities is in general not possible. In the balance sheet, fixed income investments are either classified as 'held to maturity' or as 'held for trading'. For fixed income investments that are held to maturity, hidden reserves are not considered. For each fixed income investment which is held for trading, the minimum of its market value and the market value of an analogous investment with a duration that is 3 years shorter is determined. The difference between this minimum and the book value can then be counted as the hidden reserves for this investment.

Hidden reserves in real estate must be verified with reports by independent experts. These verifications may not be older than 5 years. Hidden reserved in affiliated companies and equity investments are usually disallowed.

From the equity allowable by § 53c paragraph 3 number 3-5 the following amounts should be subtracted, since they cannot be counted on in case of losses:

- goodwill
- capitalized expenditure for setting up or expanding the business
- investments in affiliated companies in the financial sector as well as receivables from profit-sharing certificates and subordinate obligations against such companies.

8.2.4 Solvency II

8.2.4.1 Goals

The decision in June 1999 by the European Commission to fundamentally rework the financial supervision of insurance companies, and to adapt this to changed circumstances (see [5]), marks the beginning of the Solvency II project. This decision is to be seen, on the one hand, against the background of the evolution of supervision of banking toward Basel II, on the other hand as explained by the changes in the insurance market resulting from deregulation, increased competition and the development of new financial instruments.

Phase 1 of the project was devoted to taking stock of the situation and ended in 2003 with the publication of general considerations and recommendations for the reshaping of the supervision system (see [9]). Phase 2 had the goal of working out principles and concrete implementation specifications.

Solvency II has the following goals:

- setting up a principles-oriented, EU-wide harmonized, supervisory system that ensures neutrality in competition, encourages the development of an efficient insurance market, and provides adequate protection of the interests of policy holders.
- risk-sensitive capital requirements depending on the actual risk profile of the individual insurance company
- improvement of risk management and of the systems of internal control, and strengthening corporate governance

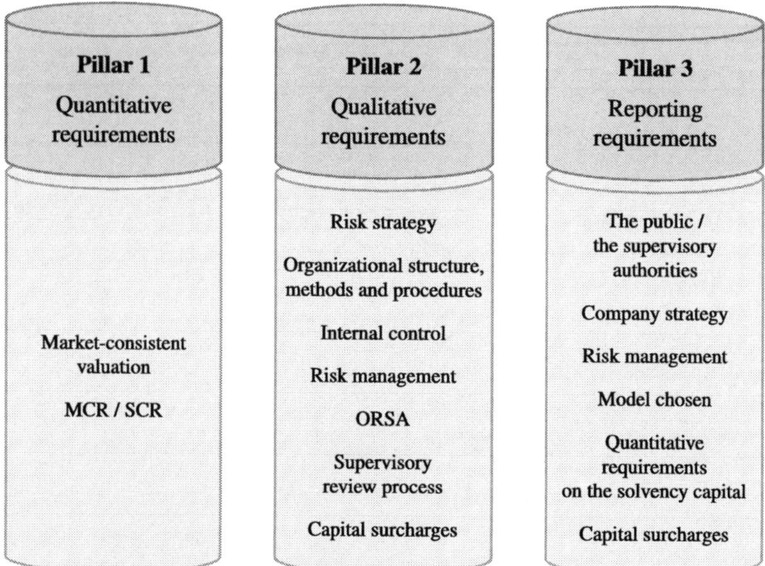

Fig. 8.1 The three pillars of the Solvency II architecture

- incentives for the development of internal models for the measurement and control of risks
- compatibility with international accounting standards.

8.2.4.2 Architecture

Like Basel II, Solvency II has an architecture based on 3 pillars (Fig. 8.1).

The first pillar defines requirements on capital resources. This includes capital requirements, principles for determining technical provisions, as well as regulations about the investment of assets.

Capital requirements are determined using the standard model, or using an internal model, which includes all relevant risks and which is certified by the supervisory authorities. The main risk categories are market risk, credit risk and operational risk, as well as insurance risks.

There are two levels of capital requirements falling below which triggers progressive regulatory measures. The SCR (solvency capital requirement) is calibrated to the maximal tolerable probability of ruin within a year, while the MCR (minimal capital requirement) represents a minimum amount. Falling short of the MCR means the supervisory authority will take the company off the market; falling short of the SCR means measures will be discussed intended to restore the necessary capital. While the SCR should be designed in a way as sensitive to risk as possible, the MCR should be as easily calculated as possible with an eye to possible legal enforcement.

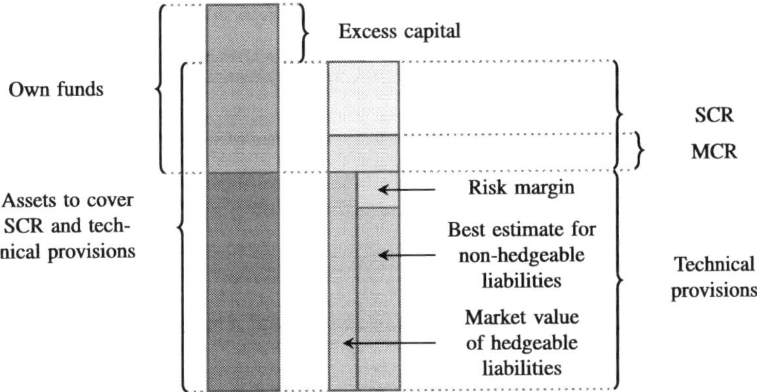

Fig. 8.2 Economic balance sheet

The second pillar regulates the supervisory review process. Since even generous amounts of capital can be quickly consumed by bad management, the supervisory review process should ensure an appropriate risk management and a functioning system of internal controls. Further components of the second pillar are early warning indicators and stress tests, to allow a timely reaction of the company to adverse developments. If there are weaknesses in the risk management system the regulators can require further capital be put up.

The third pillar has the goal to integrate market discipline. The information asymmetry concerning the financial situation of the company should be reduced by disclosure requirements, so that the various interested parties can form a satisfactory picture, and by their reactions contribute to ensuring the company manages its risks adequately.

8.2.4.3 Key Parts of Pillar 1

The starting point for pillar one is the economic balance sheet (Fig 8.2).

A characteristic of the economic balance sheet is the valuation of assets and liabilities at market values. As there is no market value for technical provisions which cover non-hedgeable liabilities, at first the present value, discounted with the risk-free interest curve, of the future payment stream under realistic assumptions (best estimate) is to be calculated. In addition a risk margin is to be determined that corresponds to the present value of the future costs of capital, which a knowledgeable investor would require for providing the necessary risk capital in the indefinite future. The sum of the present value and the risk margin then serves as a proxy for the market value of those technical provisions which cover non-hedgeable liabilities.

Determining the SCR (solvency capital requirement) is based on the value at risk with a confidence level of 99.5 % and a time horizon of 1 year. The computation can be done with a modularly constructed standard model[7] or follow an internal model certified by the regulators. Furthermore, there is the possibility for companies that develop internal models of using temporarily a partially internal model which includes elements of the standard model.

The standard model was already described in Sect. 4.6.2. Companies that wish to use an internal model in place of the standard model must preemptively have it certified by the supervisory authorities in accord with the following criteria:

- *Use test:* The use test is often held to be the most important category of certification. The company must have integrated the internal model so far into its company management that it is motivated to continually improve its quality. The model must be used to inform and to verify company decisions. Thus the internal model has to be consistent with the business model of the company. It must be organizationally deep and consistently integrated into risk management. What is more, the management must be able to demonstrate that it understands the model.
- *Internal supervision of the model:* The company's executive management is responsible for including the internal model in a comprehensive company-wide process of supervision and that there are adequate resources available for this. It is also responsible for the strategic direction of the model. Risk management is responsible for detailed supervision of the model. In particular part of its charge is the supervision of the design, implementation, testing and validation of the model. Input data have to be accurate, complete and sufficient.
- *Statistical quality standard:* The model must be technically correct and also the methodology must correspond to the present state of the art. All assumptions upon which the methodology of the model is based must be justified.
- *Calibration standard:* The model must be calibrated so that the calculated risk capital is equivalent to a 99.5 % VaR.
- *Validation:* The company must have its own validation guidelines. The guidelines cover both the validation of the calculation engine and all qualitative and quantitive processes that affect the model. Under this rubric fall at least data, methodology, assumptions, expert opinions, documentation, IT systems, model supervision, and application tests. Model results must be reconciled with past experience, and the robustness of the methods of the model must be ensured.
- *Documentation:* The company must maintain comprehensive and complete documentation for the model, the methodology at its base, and its applications. In doing so the boundaries of its validity must be explicitly drawn.
- *External models and data:* The company must document the role played by external data and models, and the extent of their use.

Finally pillar 1 specifies the *Prudent Person Plus Approach* which stipulates that assets are invested according to the principles of a knowledgeable and prudent manager, who carefully assesses the risk and returns of investments. The "Plus" denotes that the principles have been supplemented with quantitative investment requirements and a system of limits, which take account of risks which are not covered

[7] This standard model is often referred to as the "standard formula".

in the SCR. A concrete manifestation of the "Plus" in this principle can be seen in the "tiers system" for categorizing assets, which is modeled after the tiers system of Basel II.

8.2.4.4 Key Parts of Pillar 2

Pillar 2 is targeted toward effective risk management, which is closely enmeshed with company management.

The Supervisory Review Process (SRP) is intended to ensure that insurance companies satisfy certain minimum requirements as to risk management and internal control procedures. In addition, the top management must be sufficiently qualified and able to take on the responsibility for the company (Fit and Proper criteria).

Pillar 2 regulates the methods of the SRP, the levels of intervention by regulators and the regulatory powers.

In pillar 2 capital add-ons can be defined, in order to

- increase capital requirements because a standard model is used which insufficiently reflects the risk profile or because an internal model is used that has weaknesses that need to be compensated for, or
- to penalize weaknesses in the risk management process, internal controls and strategies.

However, before settling upon add-ons the insurance company and the supervisory authorities would first investigate measures to remove the weaknesses.

The SRP takes place as a dialog between the regulators and the insurance company. Ideally it has already been anticipated by the company in its Own Risk and Solvency Assessment (ORSA). The objectives of the ORSA are a better understanding of the company's own risks, the continuous monitoring of its risk profile, and to ensure that its solvency requirements can be satisfied in the future. The core elements of ORSA are:

- regular analysis of the risk profile taking into account possible future adverse developments
- estimation of the solvency requirements on the basis of the risk profile, individual risk tolerance and the business policy, where this is not a duplication of the SCR calculation, but a purely economic consideration
- ensuring the constant fulfillment of the SCR and the technical provisions
- analysis of differences between the risk profile and the assumptions of the SCR
- interlocking of risk management with strategic management and the integration into the decision process

The ORSA should be part of the risk culture of the company, and carried out at least yearly, as well as immediately after significant changes in the risk profile. All relevant risks are to be included, also including those not counted in the SCR. The ORSA is to be conducted from an entirely economic point of view, and, in particular, must use valuation methods consistent with the market. ORSA requires adequate documentation. There is the general principle of proportionality according to which the scope and intensity of the ORSA should be in accord with the individual situation of the company.

8.2.4.5 Minimal Requirements of Risk Management (MaRisk)

In anticipation of pillar 2 of Solvency 2 the German legislature introduced in 2008 § 64a "Business Organization" of the VAG, which along with an orderly administration and bookkeeping specified a suitable risk management process as a prerequisite for the orderly management of a business. § 64a VAG mentions as key elements a risk strategy in harmony with the management of the business, suitable regulations about structures and procedures, an internal management and control system, as well as internal auditing.

The BaFin made the specification in § 64a VAG concrete in the Circular 3/2009 (VA) in the form of a minimum requirement for risk management (MaRisk). MaRisk unites quantitative and qualitative risk management and meshes risk management with internal management.

The starting point is the company's individual overall risk profile, which results from the business strategy and the risk strategy of the company. MaRisk asks that all relevant risks be considered. The risk categories used must include at least: insurance risk, market risk, credit risk, operational risk, liquidity risk, concentration risk, strategic risk, reputation risk. The company's executive management has to specify the earning targets and capital targets. Using economic valuation approaches and its own risk appetite a risk-bearing capacity concept should be developed. This concept states the amount of available capital and the portion that is used to cover the economic risk capital. For this the solvency requirements constitute a lower bound, i.e. they only serve as a constraint. The risk-bearing capacity concept is made operational with a system of limits (see below).

Implementing a risk management system according to MaRisk poses the following requirements on the organization of a business:

- *Organizational structure.* The overall responsibility for risk management lies with the executive management, and this responsibility cannot be delegated. Executive management has to develop a risk strategy that suits the company's business strategy and to fix the risk tolerance. Starting from the resulting overall risk profile, executive management has to see to it that a risk culture is established which facilitates the implementation and continued improvement of a functioning risk management system. The organizational structure has to be geared towards the business strategy and permit a clear separation between operational units that carry risk and the independent risk control function (IRCF). Internal audit is also independent and answers directly to the executive management.
- *Operational structure.* All business procedures and processes that are relevant to risk have to be clearly defined. According to MaRisk the risk-relevant business procedures include at least product development, controlling, investment management, calculation of technical provisions, and reinsurance. All responsibilities

have to be clearly assigned to those who carry out the functions, and adequate communication and reporting structures have to be put in place. In respect of risks which are taken on adequate resources (e.g., qualified personnel, technical facilities) must be provided. The company's incentive system must be consistent with the risk strategy, and disincentives (e.g., exclusively maximizing short-term profit) must be avoided. The organization of the business needs to be continually improved.

- *Internal management and control system.* A risk-bearing capacity concept is to be developed on the basis of the risk strategy and the resulting overall risk profile. This concept determines the available capital to cover risks. It also states what portion of the available capital is used to cover economic risk capital, and to which business areas the economic capital is allocated. The internal management and control system has to make the risk-bearing capacity concept operational through a limit system. The limit system ensures that the overall risk profile of the company is managed in accordance with the risk tolerance prescribed by executive management. The internal management and control system monitors the whole risk management process, and ensures functional internal communication structures and risk reporting. Risk reporting provides information about the overall risk profile, limit utilization, and, in the form of ad hoc reports, about radical changes relevant to risk. Thus the internal management and control system is actively involved in developing a risk culture in the company.
- *Internal audit.* Internal audit is concerned with the whole organization of the company and with risk management. Internal audit has an unlimited right to examine. It is independent and answers directly to the executive management. It is allowed to outsource the internal audit to an external auditor or to outsource it to an affiliated company.
- *Outsourcing.* Activities and risks that are outsourced or subject to a service contract must still be monitored and appropriately managed by the outsourcing company. In particular, outsourcing or service contracts must not interfere with the ability of executive management to adequately manage or control these activities. The rights of the supervisory authorities to perform checks or audits must not be affected.
- *Documentation.* Reasonable documentation must ensure that all essential information on the risk management system, the internal management and control system, the organizational and operational structures as well as changes in the course of time, are given in a form that can be reproduced and checked by a knowledgeable third party.
- *Emergency planning.* The goal of emergency planning is to develop suitable measures that preserve the most important company processes (e.g., IT) in the face of a crisis or a catastrophe.

The requirements of MaRisk are applied in accordance with the *Proportionality Principle*. This means that the level of detail of the requirements to be implemented should be kept within an (economicly) reasonable proportion for the risk profile of the company.

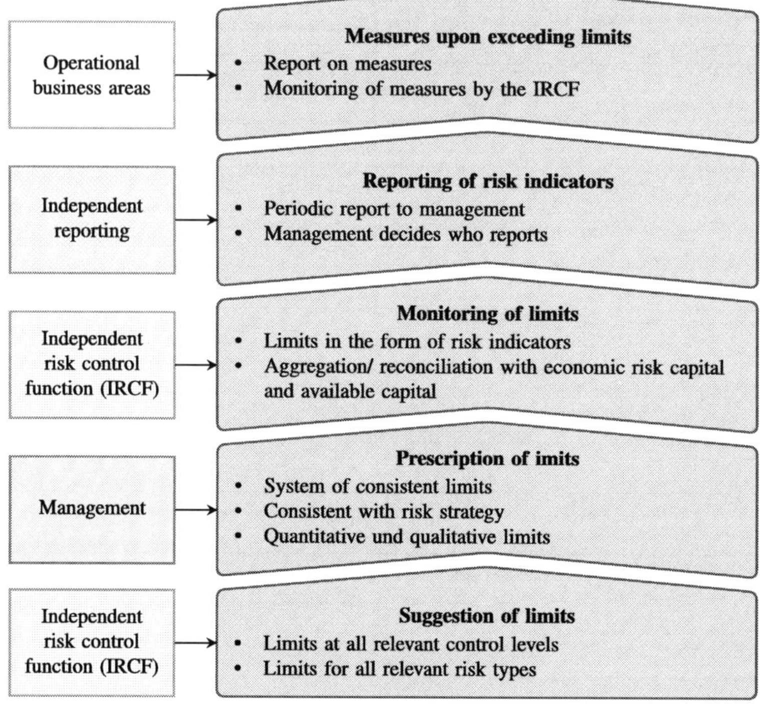

Fig. 8.3 The limit system according to MaRisk

MaRisk prescribes for the implementation of the risk-bearing capacity concept a *system of limits* (see Fig. 8.3). In agreement with the business strategy one deduces from the economic risk capital limits that can be applied at various levels (e.g., organizational areas, products, tariffs, risk types). It is necessary that this limit system is fine enough to directly address the most important organizational areas and all the risk types. The utilization of limits with regard to their capacity has to be monitored through quantitative or qualitative risk controlling. Responsibility for the definition and monitoring of limits lies with the independent risk control function (IRCF). Regular reports must be made about the respecting of limits. The limits have to be interpretable, and operationally implementable, by the corresponding addressees (e.g., allocated risk capital, volume of new business, limits on underwriting, share quotas, maximum credit risk with respect to some counterparty). A planned business transaction that will exceed a limit must either be disallowed, or made consistent with the risk strategy by some previously defined reaction (e.g., an authorized increase in the limits for a unit if the company has sufficient risk capital, reduction of other limits, risk transfer). Clear responsibilities in risk management and regular reporting are prerequisites for a functioning limit system.

Because MaRisk bases risk management on concrete business and risk strategies, and interlocks it with management of the company, it can be seen that MaRisk is an extension of the traditional regulatory view of risk to an economicly based view of value-oriented risk management.

References

1. Bundesanstalt für Finanzdienstleistungsaufsicht (BaFin), *Rundschreiben*, vol. 4/2005(VA) (2005). (p. 322)
2. C. Campagne, Standard minimum de solvabilité applicable aux entreprises d'assurances. Report of the OECD, March 1961. Reprint: Het Verzekerings-Archief 1971 (4), pp. 1–75 (p. 320)
3. Conference of Insurance Supervisory Services of the Member States of the European Union. Solvency of insurance undertakings, April 1997. Müller-Report (p. 321)
4. Der Bundesminister der Finanzen, Verordnung über die Kapitalausstattung von Versicherungsunternehmen, zuletzt geändert am 10.12.2003. BGBl I 2478 (1983). Zuletzt geändert durch Art. 4 G v. 29.7.2009 I 2305 (p. 321)
5. E. Commission, The review of the overall financial position of an insurance undertaking (Solvency II review). MARKT/2095/99-EN (December 1999) (pp. 321, 325)
6. European Commission, Directive 2002/12/EC of the European Parliament and of the Council of March 5 amending Council Directive 79/267/EEC as regards the solvency margin requirements for life insurance undertakings (2002) (p. 321)
7. European Commission, Directive 2002/13/EC of the European Parliament and of the Council of March 5 amending Council Directive 73/239/EEC as regards the solvency margin requirements for non-life insurance undertakings (2002) (p. 321)
8. European Commission, Directive 2002/83/EC of the European Parliament and of the Council of November 5 concerning life insurance (2002) (p. 321)
9. European Commission, Towards a new committee architecture in the insurance sector (including reinsurance and occupational pensions): Presentation of commission roadmap. MARKT/2505/03 (2003) (p. 325)
10. European Communities, First council directive of 24 July 1973 on the coordination of laws, regulations and administrative procedures relating to the taking-up and pursuit of the business of direct insurance other than life assurance. Off. J. Eur. Communities, L **228**(3) (1973) (pp. 319, 321)
11. European Communities, First council directive of 5 March 1979 on the coordination of laws, regulations and administrative procedures relating to the taking-up and pursuit of the business of direct life insurance. Off. J. Eur. Communities, L **63**(1) (1979) (pp. 319, 321)
12. International Association of Insurance Supervisors, Principles on capital adequacy and solvency. http://www.iaisweb.org/ (January 2002) (p. 319)
13. International Association of Insurance Supervisors, Glossary of terms. http://www.iaisweb.org/ (February 2007) (p. 319)
14. T. Pentikainen, On the net extension and solvency of insurance companies. Skand. Aktuarietidskr. **35**, 71–92 (1952) (p. 318)

Appendix A
The Capital Asset Pricing Model (CAPM)

The CAPM has historical significance, but it is only very restrictedly suitable for modern risk management because of its considerable simplifications.

We assume that the market consists in $n+1$ financial instruments $i \in \{0, \dots, n\}$. Denote by P_i the "normed" investment in the financial instrument i that has (at time 0) the value 1. We represent P_i by the $(1+i)$-th unit vector in $\mathbb{R}^{n+1}$. The return (at time 1) from this investment is a random variable that we denote by R_i. we make the following economic assumptions:

(i) For the financial instrument 0 the return R_0 is constant. P_0 is thus a *risk-free investment*. On competitive grounds there can be at most one risk-free financial instrument.
(ii) For all $i > 0$ we have $\operatorname{var}(R_i) > 0$. The investments $P_1, \dots, P_n$ are called *risky*.
(iii) The returns on financial instruments are positively correlated, but not perfectly, $\operatorname{corr}(R_i, R_j) \in [0, 1[$ for all i, j.
(iv) There are neither transaction costs nor personal income tax.
(v) Any investment can be arbitrarily divided.
(vi) There is perfect competition: No single investor can influence the share prices by buying or selling.
(vii) Each investor bases his decisions only upon the expectation values and variances of the returns of all possible portfolios.

Definition A.1 A *portfolio* $P(x) = \sum_{i=0}^{n} x^i P_i$ consists in $x^i \geq 0$ shares in the investment P_i for each $i \in \{0, \dots, n\}$. We denote the return on $P(x)$ by $R(x)$.

A *pure portfolio* $P(x)$ is a portfolio of investments with $x^0 = 0$. We denote the set of all pure portfolios by $\mathscr{P}_{\text{pure}}$.

A *normed portfolio* is a portfolio $P(x)$ with $\sum_{i=0}^{n} x^i = 1$. We denote the set of all normed portfolios by $\mathscr{P}^{\text{norm}}$.

Remark A.1 Obviously the value of a normed portfolio is 1.

M. Kriele, J. Wolf, *Value-Oriented Risk Management of Insurance Companies*, EAA Series, DOI 10.1007/978-1-4471-6305-3, © Springer-Verlag London 2014

For $P(x), P(y) \in \mathscr{P}^{\text{norm}}$ let

$$P(x) \prec P(y) \quad \Leftrightarrow \quad \begin{cases} \mathrm{E}(R(x)) < \mathrm{E}(R(y)), & \sigma(R(x)) \geq \sigma(R(y)) \\ \text{or} \\ \mathrm{E}(R(x)) \leq \mathrm{E}(R(y)), & \sigma(R(x)) > \sigma(R(y)). \end{cases}$$

If $P(x) \prec P(y)$, then (under the fundamental assumptions of CAPM) the normed portfolio $P(y)$ is to be preferred over the normed portfolio $P(x)$. The optimal boundary $\partial \mathscr{P}^{\text{norm}}$ of $\mathscr{P}^{\text{norm}}$ consists in the portfolios

$$\partial \mathscr{P}^{\text{norm}} = \left\{ P(x) \in \mathscr{P}^{\text{norm}} : \text{there is no } P(y) \in \mathscr{P}^{\text{norm}} \text{ with } P(x) \prec P(y) \right\}$$

Since $[0,1]^{n+1} \cap \{x : \sum_{i=1}^{n} x^i = 1\}$ is compact, $\partial \mathscr{P}^{\text{norm}}$ is well-defined.

The optimal boundary $\partial \mathscr{P}^{\text{norm}}_{\text{pure}}$ of $\mathscr{P}^{\text{norm}}_{\text{pure}} = \mathscr{P}^{\text{norm}} \cap \mathscr{P}_{\text{pure}}$ is analogously defined.

Definition A.2 Let $P(x)$ be a normed portfolio. Then

$$p(x) = \left(\sigma\big(R(x)\big), \mathrm{E}\big(R(x)\big)\right) \in \mathbb{R}^2$$

is a point in the *risk-return diagram* of the capital market.

The *capital market line* is the set

$$\left\{ p(x) \in \mathbb{R}^2 : P(x) \in \partial \mathscr{P}^{\text{norm}} \right\}$$

in the risk-return diagram.

The *efficient frontier* is the set

$$\left\{ p(x) \in \mathbb{R}^2 : P(x) \in \partial \mathscr{P}^{\text{norm}}_{\text{pure}} \right\}.$$

in the risk-return diagram.

We say that $P(x)$ lies on the capital market line or the efficient frontier if

$$(\sigma\big(R(x)\big), \mathrm{E}\big(R(x)\big)$$

is a point on the capital market line or the efficient frontier.

The capital market line describes the investment portfolios that are optimal in the market.

Theorem A.1 *The efficient frontier is concave.*

Proof Let $P(x)$ and $P(y)$ be pure normed portfolios on the efficient frontier. We can assume, without loss of generality, that $\sigma(R(y)) > \sigma(R(x))$ and $\mathrm{E}(R(y)) > \mathrm{E}(R(x))$. It suffices to show that for $a \in]0,1[$

$$\frac{\mathrm{E}(R(ay + (1-a)x) - R(x))}{\sigma(R(ay + (1-a)x)) - \sigma(R(x))} > \frac{\mathrm{E}(R(y) - R(x))}{\sigma(R(y)) - \sigma(R(x))} \tag{A.1}$$

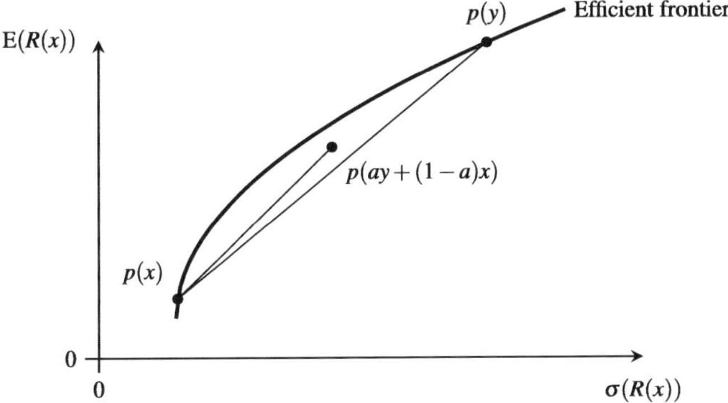

Fig. A.1 Proof of Theorem A.1. The points $p(x)$, $p(y)$, $p(ay + (1 - a)x)$ represent the pure normed portfolios $P(x)$, $P(y)$, $P(ay + (1 - a)x)$ in the risk-return diagram

holds (see Fig. A.1). Since $\operatorname{corr}(P_i, P_j) \in [0, 1[$ for all $i \neq j$, it follows that

$$\operatorname{cov}\big(P(y), P(x)\big) < \sigma(y)\sigma(x)$$

and thus

$$\begin{aligned} \operatorname{var}\big(R\big(ay + (1-a)x\big)\big) &< a^2 \operatorname{var}\big(R(y)\big) + 2a(1-a)\sigma\big(R(y)\big)\sigma\big(R(x)\big) \\ &\quad + (1-a)^2 \operatorname{var}\big(R(x)\big) \\ &= \big(a\sigma\big(R(y)\big) + (1-a)\sigma\big(R(x)\big)\big)^2. \end{aligned} \tag{A.2}$$

In particular, we have $\sigma(R(ay + (1 - a)x)) < \sigma(R(y))$ and (A.1) is equivalent to

$$\begin{aligned} 0 &< \big(\mathrm{E}\big(R\big(ay + (1-a)x\big) - R(x)\big)\big)\big(\sigma\big(R(y)\big) - \sigma\big(R(x)\big)\big) \\ &\quad - \big(\mathrm{E}\big(R(y) - R(x)\big)\big)\big(\sigma\big(R\big(ay + (1-a)x\big)\big) - \sigma\big(R(x)\big)\big) \\ &= \mathrm{E}\big(R(y) - R(x)\big)\big(a\big(\sigma\big(R(y)\big) - \sigma\big(R(x)\big)\big) \\ &\quad - \big(\sigma\big(R\big(ay + (1-a)x\big)\big) - \sigma\big(R(x)\big)\big)\big) \\ &= \mathrm{E}\big(R(y) - R(x)\big)\big(a\sigma\big(R(y)\big) + (1-a)\sigma\big(R(x)\big) - \sigma\big(R\big(ay + (1-a)x\big)\big)\big). \end{aligned}$$

The last inequality is equivalent to (A.2) and therefore satisfied. □

Theorem A.2 *There is a unique portfolio* $P(x_M) \in \mathscr{P}^{\text{norm}}_{\text{pure}}$, *such that the capital market line consists of the portfolios* $P(y) \in \mathscr{P}^{\text{norm}}$ *with*

$$\mathrm{E}\big(R(y)\big) = R_0 + \frac{\mathrm{E}(R(x_M)) - R_0}{\sigma(R(x_M))}\sigma\big(R(y)\big).$$

Proof Each normed portfolio $P(y)$ has the form $P(y) = aP_0 + (1-a)P(x)$, where $P(x)$ is a pure normed portfolio and $a \in [0, 1]$ holds. We have

$$\mathrm{E}\big(R(y)\big) = aR_0 + (1-a)\,\mathrm{E}\big(R(x)\big) = R_0 + (1-a)\big(\mathrm{E}\big(R(x)\big) - R_0\big), \tag{A.3}$$

where we used $\mathrm{E}(R_0) = R_0$. Since aR_0 is a constant random variable, we obtain

$$\mathrm{var}\big(R(y)\big) = \mathrm{var}\big(aR_0 + (1-a)R(x)\big) = (1-a)^2\,\mathrm{var}\big(R(x)\big)$$

resp. $\sigma(R(y)) = (1-a)\sigma(R(x))$. By solving this equation for $1-a$ and putting that in (A.3), we see that in the risk-return diagram the points $p(y)$ of the portfolios $P(y)$, that are for fixed x a combination of $P(x)$ and P_0, lie on a line. This line is given by

$$\mathrm{E}(R(y) = R_0 + \frac{\mathrm{E}(R(x)) - R_0}{\sigma(R(x))}\,\sigma\big(R(y)\big).$$

In particular, the point $(0, R_0)$ lies on the line and does not depend on $P(x)$. Therefore each line which arises in this way from a pure normed portfolio $P(x)$ goes through $(0, R_0)$. Let $j > 0$ be the financial instrument for which the return on the normed investment P_j has the smallest standard deviation $\sigma(R_j) > 0$, and let i be the financial instrument for which the return on the normed investment P_i has the largest expected value $\mathrm{E}(R_i)$. Then we have for the slope $\theta(x)$ of the line generated by $P(x)$

$$\theta(x) = \frac{\mathrm{E}(R(x)) - R_0}{\sigma(R(x)) - \sigma(R_0)} = \frac{\mathrm{E}(R(x)) - R_0}{\sigma(R(x))} \le \frac{R_i - R_0}{\sigma(R_j)} < \infty.$$

Therefore there exists

$$\theta_{\max} = \sup\big\{\theta(x)\colon\ P(x) \in \mathscr{P}^{\mathrm{norm}}_{\mathrm{pure}}\big\} < \infty.$$

Since $[0, 1]^{n+1} \cap \{x\colon x^0 = 0, \sum_{i=0}^{n} x^i = 1\}$ compact and $x \mapsto \theta(x)$ is continuous, there exists a sequence $\{x_k\}_{k\in\mathbb{N}}$ converging to a vector x_M, so that

(i) $P(x^k)$ and $P(x_M)$ are pure normed portfolios,
(ii) $\theta(x_M) = \theta_{\max}$,
(iii) $P(x_M)$ lies on the capital market line.

See Fig. A.2. Because $P(x_M)$ lies on the capital market line, the capital market line is tangent to the efficient frontier. $P(x_M)$ is unique, because the tangents to a concave curve each touch the curve at a single point. □

Definition A.3 $P(x_M)$ is *the market portfolio*. We write $R_M = R(x_M)$.

Since in CAPM each investor makes decisions exclusively from the expected values and variances of the possible portfolios, it is optimal for each investor to invest in a mixture of the market portfolio and risk-free investments. It follows that

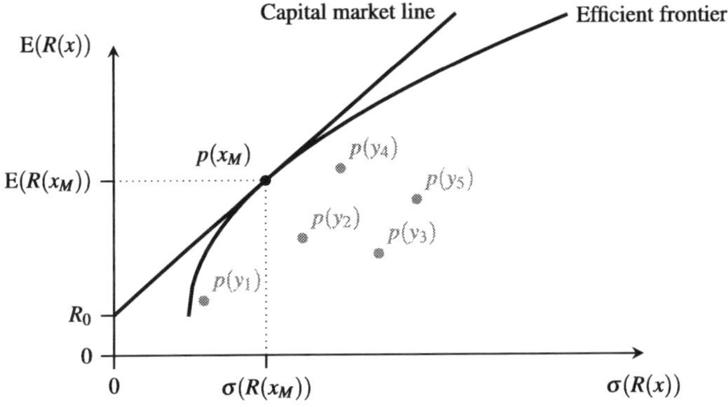

Fig. A.2 The capital market line and the efficient frontier. $p(x_M) = (\sigma(R(x_M)), \mathrm{E}(R(x_M)))$ is the point in the risk-return diagram generated by the market portfolio $P(x_M)$. The points $p(y_1), \dots, p(y_5)$ represent some pure normed portfolios and thus lie beneath the efficient frontier

each investor holds the same relative proportion of financial instruments $1, \dots, n$. In particular, the capital market portfolio is a multiple of the entire pure capital market.

This consequence of the assumptions of the CAPM does not occur in the real capital market, which shows that the CAPM is often too simplistic to be applied.

Corollary A.1 *For each normed portfolio $P(y) \in \mathscr{P}^{\mathrm{norm}}$ there holds*

$$\mathrm{E}\big(R(y)\big) = R_0 + \frac{\operatorname{cov}(R(y), R_M)}{\operatorname{var}(R_M)}\big(\mathrm{E}(R_M) - R_0\big).$$

Proof We use the same relations as in the proof of Theorem A.2. For the market portfolio $P(x_M)$ the slope $x \mapsto \theta(x)$ has maximum under the condition $\sum_{i=1}^n x_M^i = 1$. The coefficients x_M^i can be determined as Lagrange multipliers. To do this we must find the stationary points of

$$\big(x^1, \dots, x^n, \lambda\big) \mapsto \theta(x) - \lambda\left(\sum_{i=1}^n x^i - 1\right) = \frac{\mathrm{E}(R(x)) - R_0}{\sigma(R(x))} - \lambda\left(\sum_{i=1}^n x^i - 1\right).$$

Taking the derivative with respect to x_k at x_M gives

$$\begin{aligned}\lambda &= \frac{\partial}{\partial x^k}\left(\frac{\mathrm{E}(R(x)) - R_0}{\sigma(R(x))}\right)_{|x_M} \\ &= \frac{\partial}{\partial x^k}\left(\frac{\sum_{i=1}^n x^i\,\mathrm{E}(R_i) - R_0}{\sqrt{\sum_{i,j=1}^n x^i x^j \operatorname{cov}(R_i, R_j)}}\right)_{|x_M}\end{aligned}$$

$$= \frac{\mathrm{E}(R_k)\sigma(R_M) - (\mathrm{E}(R_M) - R_0)\sum_{i=1}^n x_M^i \operatorname{cov}(R_i, R_k)/\sigma(R_M)}{\sigma(R_M)^2}$$

$$= \frac{\mathrm{E}(R_k)\operatorname{var}(R_M) - (\mathrm{E}(R_M) - R_0)\operatorname{cov}(R_M, R_k)}{\sigma(R_M)^3}. \tag{A.4}$$

By multiplying this equation by x_M^k and summing over k, taking into account the constraint $\sum_{k=1}^n x_M^k = 1$, we obtain

$$\lambda = \frac{\mathrm{E}(R_M)\operatorname{var}(R_M) - (\mathrm{E}(R_M) - R_0)\operatorname{var}(R_M)}{\sigma(R_M)^3} = \frac{R_0}{\sigma(R_M)}.$$

From this follows using (A.4)

$$\frac{R_0}{\sigma(R_M)} = \frac{\mathrm{E}(R_k)\operatorname{var}(R_M) - (\mathrm{E}(R_M) - R_0)\operatorname{cov}(R_M, R_k)}{\sigma(R_M)^3}$$

$$\Leftrightarrow \quad R_0 \operatorname{var}(R_M) = \mathrm{E}(R_k)\operatorname{var}(R_M) - \big(\mathrm{E}(R_M) - R_0\big)\operatorname{cov}(R_M, R_k)$$

$$\Leftrightarrow \quad \big(\mathrm{E}(R_k) - R_0\big)\operatorname{var}(R_M) = \big(\mathrm{E}(R_M) - R_0\big)\operatorname{cov}(R_M, R_k),$$

which shows the assertion of the corollary for the special case $P(y) = P_k$, $k > 0$. Trivially, the assertion also holds for P_0. The general case for a normed portfolio $P(y)$ now directly follows by linearity of expectation values and the bilinearity of the covariance if we take into account $\sum_{i=0}^n y^i = 1$. □

Definition A.4 The β of a normed portfolio $P(y)$ is given by

$$\beta(y) = \frac{\operatorname{cov}(R(y), R_M)}{\operatorname{var}(R_M)}.$$

With this definition the expected value $\mathrm{E}(R(y))$ has the form

$$\mathrm{E}\big(R(y)\big) = R_0 + \beta(y)\big(\mathrm{E}(R_M) - R_0\big).$$

This relationship is sometimes considered the basic message of the CAPM.

Definition A.5 The *systematic risk of the portfolio* $P(y)$ is

$$\operatorname{corr}\big(R(y), R_M\big)\sigma\big(R(y)\big).$$

The *non-systematic risk of the portfolio* $P(y)$ is $(1 - \operatorname{corr}(R(y), R_M))\sigma(R(y))$.

Definition A.5 can be motivated by the following example.

Example A.1 We assume that the return on a financial instrument i depends linearly on the return R_M of the market portfolio:

$$R_i = \alpha_i + \beta_i R_M + \varepsilon_i, \tag{A.5}$$

where ε_i is a stochastic noise term with $\mathrm{cov}(\varepsilon_i, R_M) = 0$. α_i and β_i are constants, where without loss of generality α_i is chosen so that $\mathrm{E}(\varepsilon_i) = 0$. It follows that

$$\begin{aligned}
\mathrm{var}(R_j) &= \mathrm{cov}(\alpha_j + \beta_j R_M + \varepsilon_j, \alpha_j + \beta_j R_M + \varepsilon_j) \\
&= \mathrm{cov}(\beta_j R_M + \varepsilon_j, \beta_j R_M + \varepsilon_j) \\
&= (\beta_j)^2 \,\mathrm{var}(R_M) + \mathrm{var}(\varepsilon_j) \\
&= \mathrm{corr}(R_j, R_M)^2 \big(\sigma(R_j)\big)^2 + \mathrm{var}(\varepsilon_j).
\end{aligned}$$

Thus the standard deviation of the investment P_j with a vanishing ε_j is exactly the systematic risk of the investment P_j.

The non-systematic risk can be diversified away, if one chooses the market portfolio, since

$$\begin{aligned}
\mathrm{var}(R_M) &= \sum_{j=1}^{n} x_M^j \,\mathrm{cov}(R_j, R_M) \\
&= \sum_{j=1}^{n} x_M^j \,\mathrm{corr}(R_j, R_M)\sigma(R_j)\sigma(R_M).
\end{aligned}$$

Definition A.6 The *risk premium* of the portfolio P is $\mathrm{E}(R(y)) - R_0$.

From

$$\beta(y) = \frac{\mathrm{corr}(R(y), R_M)\sigma(R(y))}{\sigma(R_M)}$$

it follows that $\beta(y)$ is the ratio of the systematic risk and the market risk. Furthermore it follows that the risk premium of $P(y)$ is exactly $\beta(y)$ times the risk premium of the market portfolio.

Appendix B
R-Code for the SST Calculation Using the delta-Gamma-Model

The following code has been used for the calculations in Sect. 4.6.1.4:

```
########################################################
### Calculation of the SST-ratio following the SST
##  standard model for a simple example
########################################################

## library for multivariate normal distributions
library(mvtnorm)
time.start.user <-  proc.time()[["user.self"]]

########################################################
### Input data
########################################################

## repeatable random numbers
set.seed(1)

## number of Monte Carlo simulations
N <-  100000
## safety level
alpha <-  0.01
## risk free spot rates
return.spot.chf <- c(0.00155, 0.00013, -0.00011)
return.stock.chf <- 0.05

# cost of capital for market value margin
coc <- 0.06
## mortality of the insured
q <-c(0.02,0.02,0.02)
```

M. Kriele, J. Wolf, *Value-Oriented Risk Management of Insurance Companies*, EAA Series, DOI 10.1007/978-1-4471-6305-3, © Springer-Verlag London 2014

```
## insured sum, survival to the end of period
IS <- c(0, 0,1000)

## nominal values, i.e. cashflow
## -- assumption: no default risk
assets.bond.gov.chf <- c(0, 300, 500)

## value at the beginning of period 1
assets.stock.chf <- 300

## expected increase per period of value
stock.chf.incr <- 0.05

### standard deviations provides by FINMA (2012)
SD <- c(  0.0060292643195745          # spot.chf[1]
        , 0.00605990680656078         # spot.chf[2]
        , 0.00633152336734695         # spot.chf[3]
        , 0.150517788474455           # stock.chf
        , 0.05                        # q
        )

### correlation matrix provided by FINMA (2012)
###                   spot.chf[1], spot.chf[2]
###                   , spot.chf[3], stock.chf, q
Corr <- matrix( c(  1.000000000, 0.721560179
                  , 0.545556472, 0.402239567, 0
                  , 0.721560179, 1.000000000
                  , 0.953194547, 0.433388849, 0
                  , 0.545556472, 0.953194547
                  , 1.000000000, 0.413041139, 0
                  , 0.402239567, 0.433388849
                  , 0.413041139, 1.000000000, 0
                  , 0.000000000, 0.000000000
                  , 0.000000000, 0.000000000, 1
                  ), ncol = 5)

### Scen: Data frame that describes the scenarios
##        to be added.
###       The following columns are necessary:
###       -- Target:      0 or 1 depending on whether
###                       the particular scenario is
###                       included
###       -- Probability: Probability for this scenario
###       -- Delta_RTK:   Effect of this scenario on the
###                       risk bearing capital RTK)
```

```
Scen <-
    list( Name = c( "SZ01: Equity drop -60%"
             ,      "SZ03: Stock mkt crash (87)"
             ,      "SZ04: Nikkei crash (89/90)"
             ,      "SZ05: Europ. FX crisis (92)"
             ,      "SZ06: US interest crisis (94)"
             ,      "SZ07: Russian crisis/LTCM (98)"
             ,      "SZ08: Stock mkt crash (00/01)"
             ,      "SZ11: Financial crisis (08)"
             )
         , Target = c( 1, 1, 1, 1, 1, 1, 1, 1 )
         ##                 SZ01   SZ03   SZ04   SZ05
         ##     SZ06   SZ07   SZ08   SZ11
         , Probability = c(0.001, 0.001, 0.001, 0.001
             , 0.001, 0.001, 0.001, 0.001)
         , Delta.x = rbind(
             ## #spot.1   #spot.2  #spot.3   #stock #q
             c(  0.00000,  0.00000, 0.00000, -0.6000, 0)
             ,c(-0.00155, -0.00013, 0.00000, -0.2323, 0)
             ,c( 0.01563,  0.01098, 0.01177, -0.2643, 0)
             ,c(-0.00155, -0.00013, 0.00000, -0.0580, 0)
             ,c( 0.01109,  0.01406, 0.01509, -0.1852, 0)
             ,c(-0.00155, -0.00013, 0.00000, -0.2841, 0)
             ,c(-0.00155, -0.00013, 0.00000, -0.3567, 0)
             ,c(-0.00155, -0.00013, 0.00000, -0.3881, 0)
             )
         , Delta_RTK = c()
         )

########################################################
### Functions
########################################################

### Expected Shortfall for an (unsorted) vector of
### random outcomes
### -- In this implementation the expected shortfall
###    is taken with respect to the left tail. (In the
###    literature it is often taken with respect to
###    the right tail.)
ES <- function(x,perc){
    VaR <- quantile(x,perc,names=FALSE)
    sum(x[x<VaR])/(length(x)*perc)
}

### survival to the end of the period
```

```
Survival <- function(q) cumprod(1-q)

### discount to the beginning of the first period
Discount <-function(s)  1/(1+s)^(1:length(s))

### valuation of government bonds
bond.gov.value <-  function(ZB,date,s){
    T <- length(s)
    if (length(s) != length(ZB)){
        return(NA)
    }else{
        return(  sum(Discount(s)[date:T]*ZB[date:T])
               / if (date == 1) 1
               else Discount(s)[date-1]
               )
    }
}

# risk bearing capital at the beginning of period "year"
value.assets <- function(x,year){
return(  bond.gov.value(  assets.bond.gov.chf
                        , year
                        , x[i.spot.chf]
                        )
       + (1+stock.chf.incr)^(year-1)
       * assets.stock.chf
       * x[i.stock.chf]
       )
}

### valuation of insurance polices
### -- only survival benefit
### -- observe that we do not condition on the
###    policyholder's being alive at time year.
value.liabilities <- function(x,year){
    T <- length(IS)
    return(  sum(Discount(x[i.spot.chf])[year:T]
                * Survival(x[i.q]*q)[year:T]
                * IS[year:T]
                )
          / if (year == 1) 1
          else Discount(x[i.spot.chf])[year-1]
          )
}
```

```
RTK <- function(x, year){
    return(  value.assets(x,year)
           - value.liabilities(x,year) )
}


### positive sensitivity for component i of the
### risk factors x
x.up <- function(x,h,additive,i){
    if (additive[i]) x[i] <- x[i]+h[i]
    else x[i] <- x[i] * (1+h[i])
    return(x)
}
### negative sensitivity for component i of the
### risk factors x
x.down <- function(x,h,additive,i){
    if (additive[i]) x[i] <- x[i]-h[i]
    else x[i] <- x[i] * (1-h[i])
    return(x)
}

### Correction for the fact that some sensitivities
### are multiplicative
Delta <- function(x,h,additive){
    Delta.x <- NULL
    for (i in 1:length(x)){
         if (additive[i]) Delta.x[i] <- h[i]
         else Delta.x[i] <- x[i] * h[i]
     }
    return(Delta.x)
}

quad.approx <- function(x, D.RTK, DD.RTK){
    return(t(x) %*% D.RTK + 0.5*t(x) %*% DD.RTK %*% x)
}

rand.Delta.RTK <- function(N, risk.index){
    Cov <- (  SD[risk.index] %*% t(SD[risk.index])
            * Corr[risk.index, risk.index]
            )
    return( apply(  rmvnorm( N
                           , rep(0,length(risk.index))
                           , Cov
                           )
                  , 1
```

```
                    , quad.approx
                    , D.RTK = delta[risk.index]
                    , DD.RTK =
                    Gamma[risk.index,risk.index]
                    )
            )
}

### Aggr.Scen: Aggregate Scenarios to randomly
###            generated values rand
### -- We use the same approximation as in the shift
###     method, namely, two different scenarios cannot
###     happen in the same year.
Aggr.Scen <- function(rand,Scen){
    ## rand: random values for the distribution
    ##       without added scenarios
    ## Scen: Data frame that describes the scenarios
    ##       to be added.
    N.Scen <- length(Scen$Delta_RTK)
    N.rand <- length(rand)
    k <- 0
    for (j in 1:N.Scen){
        N.Adjust <- floor(  N.rand
                          * Scen$Target[j]
                          * Scen$Probability[j]
                          )
        if (N.Adjust>0){
            for (l in 1:N.Adjust){
                rand[k+l] <-
                    (  rand[k+l]
                     + min(0, Scen$Delta_RTK[j])
                     )
            }
            k <- k+N.Adjust
        }
    }
    return(rand)
}

########################################################
### Calculations
########################################################

### Risk factors, sensitivities, and indicators whether
### the sensitivity is additive
```

```
x <- NULL
h <- NULL
additive <- NULL
i.spot.chf <- (length(x)+1):( length(x)
                              +length(return.spot.chf))
h <- c(h, rep(0.01,length(i.spot.chf)))
additive <- c(additive, rep(TRUE,length(i.spot.chf)))
x <- c(x,return.spot.chf)
i.stock.chf <-  (length(x)+1):(length(x)+1)
h <- c(h,0.1)
additive <- c(additive, FALSE)
x <- c(x,1)
i.q <- (length(x)+1):(length(x)+1)
h <- c(h, 0.1)
additive <- c(additive, FALSE)
x <- c(x,1)

### expected risk bearing capital at the end of period 1
### and RTK at the beginning of the projection
RTK.end <- RTK(x,2)
RTK.start <- RTK(x,1)

### positive and negative linear sensitivities of the
### risk bearing capital
d.RTK.plus <- NULL
d.RTK.minus <- NULL
for (i in 1:length(x)){
  d.RTK.plus[i] <- RTK(x.up(x,h,additive,i),2)
  d.RTK.minus[i] <- RTK(x.down(x,h,additive,i),2)
}

### positive and negative quadratic sensitivities of the
### risk bearing capital at the end of period 2
dd.RTK.plus.plus <-
    matrix(  rep(NA,length(x)*length(x))
           , ncol=length(x)
           )
dd.RTK.plus.minus <-
    matrix(  rep(NA,length(x)*length(x))
           , ncol=length(x)
           )
dd.RTK.minus.plus <-
    matrix(  rep(NA,length(x)*length(x))
           , ncol=length(x)
           )
```

```
dd.RTK.minus.minus <-
    matrix( rep(NA,length(x)*length(x))
          , ncol=length(x)
          )
for (i in 1:length(x)){
    for (j in 1:length(x)){
        dd.RTK.plus.plus[i,j] <-
            RTK(x.up( x.up( x, h, additive, i)
                    , h , additive, j)
                , 2)
        dd.RTK.plus.minus[i,j] <-
            RTK(x.up( x.down( x, h, additive, i)
                    , h, additive, j)
                , 2)
        dd.RTK.minus.plus[i,j] <-
            RTK(x.down( x.up( x, h, additive, i)
                      , h, additive, j)
                , 2)
        dd.RTK.minus.minus[i,j] <-
            RTK(x.down( x.down( x, h, additive, i)
                      , h, additive, j),2)
    }
}

### numerical derivative and second derivative of the
## risk bearing capital
### at the end of period 2
Delta.x <- Delta(x,h,additive)
delta <- (d.RTK.plus-d.RTK.minus)/(2*Delta.x)
Gamma <- ( dd.RTK.plus.plus - dd.RTK.plus.minus
         + dd.RTK.minus.minus - dd.RTK.minus.plus
         ) / (4 * Delta.x %*% t(Delta.x))


rand.market <-
    rand.Delta.RTK(N,c(i.spot.chf,i.stock.chf))
rand.insurance <-
    rand.Delta.RTK(N,c(i.q,i.q))
rand.market.insurance <-
    rand.Delta.RTK(N,c(i.spot.chf,i.stock.chf, i.q))

for (j in 1:length(Scen$Target)){
    Scen$Delta_RTK[j] <-
        quad.approx(Scen$Delta.x[j,],delta,Gamma)
}
```

```
rand.market.insurance.scen <-
    Aggr.Scen(rand.market.insurance, Scen)

risk.capital <-
    list(  insurance = ES(rand.insurance, alpha)
         , market = ES(rand.market, alpha)
         , market.insurance =
         ES(rand.market.insurance, alpha)
         , market.insurance.scen =
         ES(rand.market.insurance.scen, alpha)
         )

### market value margin for non-hedgeable risks
### -- we assume that the following risks are
###    non-hedgeable:
###    -- both risk from extreme scenarios and
###    -- biometric risk is non-hedgeable
### -- we assume that the following risks are hedgeable:
###    -- investment risks for bonds of
###       duration <= 15 years
###    -- investment risk for stocks
### -- for simplicity we approximate future risk capital
###    using the market value of liabilities.
###    This is an approximation.
cost.of.capital <- 0
for (i in 1:length(IS)){
 cost.of.capital[i] <-
     (  coc
      * value.liabilities(x,i)
      / value.liabilities(x,1)
      * (  (risk.capital$market.insurance.scen
            - risk.capital$market.insurance)
         + risk.capital$insurance)
      )

}
MVM <- sum(Discount(return.spot.chf) * cost.of.capital)

### observe that the function RTK already incorporates
### expected gains
target.capital <- (  risk.capital$market.insurance.scen
                   - return.spot.chf[1] * RTK.start
                   + MVM
                   ) / (1+return.spot.chf[1])
```

```
SST.ratio <- -RTK.start/target.capital

#########################################################
### Output
#########################################################

Results <- list()

## placeholder for the 1st entry in Results
Results$calc_start<- numeric(0)

## placeholder for the 2nd entry in Results
Results$calc_time_in_s <- numeric(0)

Results$number.simulations <- N
Results$alpha <- alpha
Results$RTK.start <- RTK.start
Results$RTK.end <- RTK.end
Results$ES.insurance <- risk.capital$insurance
Results$ES.market <- risk.capital$market
Results$ES.market.insurance <-
    risk.capital$market.insurance
Results$ES.market.insurance.scen <-
    risk.capital$market.insurance.scen
Results$MVM <- MVM
Results$target.capital <- target.capital
Results$SST.ratio <- SST.ratio
time.end.user <-  proc.time()[["user.self"]]

Results$calc_start <-
    format(Sys.time(), "%Y-%m-%d_%H:%M:%S")
Results$calc_time_in_s <-
    time.end.user-time.start.user

print(t(data.frame(Results)))
```

Appendix C
R-Script for the Scenario-Based Solvency II SCR Computation

The computations in Example 4.13 were made with the following script.

```
tiny <- 1E-6
# Our three scenarios
BE <- 1
UP <- 2
DOWN <- 3

# Maturity t, rel. change sup(t), re. change sdown(t)
# Durations shorter than 1 year have the same relative
# changes as duration 1
Shocks <- rbind(c(1, 70/100, -75/100),
                c(2, 70/100, -65/100),
                c(3, 64/100, -56/100),
                c(4, 59/100, -50/100),
                c(5, 55/100, -46/100),
                c(6, 52/100, -42/100),
                c(7, 49/100, -39/100),
                c(8, 47/100, -36/100),
                c(9, 44/100, -33/100),
                c(10, 42/100, -31/100),
                c(11, 39/100, -30/100),
                c(12, 37/100, -29/100),
                c(13, 35/100, -28/100),
                c(14, 34/100, -28/100),
                c(15, 33/100, -27/100),
                c(16, 31/100, -28/100),
                c(17, 30/100, -28/100),
                c(18, 29/100, -28/100),
                c(19, 27/100, -29/100),
                c(20, 26/100, -29/100),
```

M. Kriele, J. Wolf, *Value-Oriented Risk Management of Insurance Companies*, EAA Series, DOI 10.1007/978-1-4471-6305-3, © Springer-Verlag London 2014

```
                    c(21, 26/100, -29/100),
                    c(22, 26/100, -30/100),
                    c(23, 26/100, -30/100),
                    c(24, 26/100, -30/100),
                    c(25, 26/100, -30/100),
                    c(30, 25/100, -30/100))

T <- 8

Bonds <- c(1200, 500, 300, 100,    0,    0,    0,   0)
Repl <-  c( 500, 450, 300, 250, 200, 150, 100, 50)
RF_BE <- c( 0.5, 1.0, 1.2, 1.5, 2.0, 2.2, 2.3, 2.3)/100
Stat_Interest <- 0.5 / 100

## statutory provisions at the beginning of the period
## are equal to the statutory provisions at the end of
## the previous period
StatProv_BoP <- vector("numeric",T)
for (t in 1:T){
  StatProv_BoP[t] <-
    sum( Repl[t:T] / (1+Stat_Interest)^(1:(T-t+1)) )
}

# Initialization of vectors and matrices
Assets <- c()
Bonus <- matrix(NA,3,T)
Liabs <- rep(NA,3)
NAV <- rep(NA,3)
D_NAV <- rep(NA,3)
ADJ_Liabs <- rep(NA,3)
ADJ_NAV <- rep(NA,3)
ADJ_D_NAV <- rep(NA,3)

# risk free rate for the three scenarios: BE, UP, DOWN
RF <- rbind(RF_BE,
            RF_BE * (1+Shocks[1:T,UP]),
            pmax(RF_BE +
                 pmin(RF_BE * Shocks[1:T,DOWN],
                      -0.01),
                 0) )

# The policyholders are credited 90% of risk free rate
for (i in c(BE,UP,DOWN)){
  Bonus[i,] <-
     pmax(0, 0.9 * RF[i,] - Stat_Interest) *
```

```
            StatProv_BoP
}

for (i in c(BE,UP,DOWN)){
  Assets[i] <- sum(Bonds / cumprod(1+RF[i,]))
  # Without risk mitigation we use the best estimate
  # claims in all scenarios
  Liabs[i] <-
    sum((Repl+Bonus[BE,]) / cumprod(1+RF[i,]))
  NAV[i] <- Assets[i]-Liabs[i]
  ADJ_Liabs[i] <-
    sum((Repl+Bonus[i,]) / cumprod(1+RF[i,]))
  ADJ_NAV[i] <- Assets[i]-ADJ_Liabs[i]
}
for (i in c(UP,DOWN)){
  D_NAV[i] <- max(0,-(NAV[i]-NAV[BE]))
  ADJ_D_NAV[i] <- max(0,-(ADJ_NAV[i]-ADJ_NAV[BE]))
}

ADJ_SCR <- max(ADJ_D_NAV[UP],ADJ_D_NAV[DOWN],0)

# tiny rather than 0 to avert rounding problems
if (ADJ_SCR < tiny){
  SCR <- 0
}else if (abs(ADJ_SCR-ADJ_D_NAV[UP]) < tiny){
  SCR <- D_NAV[UP]
}else{
  SCR <- D_NAV[DOWN]
}

FDB <- sum(Bonus[BE,]/cumprod(1+RF[BE,]))
ADJ <- -min(SCR-ADJ_SCR,FDB)
SCR_Final=SCR+ADJ


Results <- data.frame(row.names = c("BE","UP","DOWN"),
                      Assets,
                      Liabs,
                      NAV,
                      D_NAV,
                      ADJ_Liabs,
                      ADJ_NAV,
                      ADJ_D_NAV
                      )
names(Results) = (c("Assets",
```

```
                    "Liabs",
                    "NAV",
                    "D_NAV",
                    "ADJ_Liabs",
                    "ADJ_NAV",
                    "ADJ_D_NAV"))
print(round(Results,0))
SCR_Result <- data.frame(SCR,ADJ_SCR)
names(SCR_Result) <- c("SCR","ADJ_SCR")
print(SCR_Result)
print(FDB)
print(ADJ)
print(SCR_Final)
```

Appendix D
R-Script for the Solvency II SCR Computation for XYZ Inc in Example 4.14

The computation consists in 3 scripts,

D_SII_NL_Input.r, D_SII_NL_Calc.r and D_SII_NL_Output.r,

that are input one after another. The results appear directly in the R console.

D.1 Input Definition

The file D_SII_NL_Input.r defines the input data in Example 4.14:

```
### Global constants
tiny <- 1E-6
# gross: before reinsurance, net: after reinsurance
gross <- 1; net <- 2

ConfLevel <- 0.995

# Following assignments make indexing more comfortable
NoIns <- 3
F <- 1 # Fire
L <- 2 # Liability
T <- 3 # Theft
PY <- 1 # Past Year
CY <- 2 # Current Year
NY <- 3 # Next Year

Ins_Name <- c("Fire", "Liability", "Theft")
Ins_No <- c(
            NA, #  1
            NA, #  2
            NA, #  3
```


M. Kriele, J. Wolf, *Value-Oriented Risk Management of Insurance Companies*, EAA Series, DOI 10.1007/978-1-4471-6305-3,

```
             F, #  4
             L, #  5
            NA, #  6
            NA, #  7
            NA, #  8
             T, #  9
            NA, # 10
            NA, # 11
            NA  # 12
            )

#  Row Index: Lob_Name
#  Column index: Time (PY, CY, NY)

# Gross written premium for PY and CY
Prem_written_gross <- rbind(c( 500, 600),
                            c( 250, 300),
                            c(  50, 100))

# Mean and variation coefficient of the cost per claim
# gross of reinsurance
Mean_Y_gross <- c(0.1, 0.1, 0.1)
VC_Y_gross <- c( 0.5, 0.6, 0.7)

# unearned premiums reserve for PY, CY, NY
# (net values)
UPR <- rbind(c( 50, 70, 75),
             c( 20, 40, 45),
             c(  5, 12, 15))

#Costs <- c( 0.05, 0.05, 0.05)

# Proportional reinsurance
ReP_Q <- c( 0.25, 0.20, 0.20)

# Non-proportional reinsurance
ReNP_a <- c( 0.2,    0,    0)
ReNP_h <- c( 0.5,    0,    0)

# Reserve pattern
t_runoff <- 6
beta <-  array(NA, c(NoIns,t_runoff))
beta_temp <- rbind(c(1, 0.6, 0.1, 0.0, 0.0, 0.0),
                   c(1, 0.9, 0.6, 0.2, 0.1, 0.0),
                   c(1, 0.4, 0.0, 0.0, 0.0, 0.0))
```

```
for (k in 1:NoIns) {
  beta[k,] <- beta_temp[k,]/sum(beta_temp[k,])
}

# Reserve and premium reserve: end of previous year
# (= beginning of current year)
ResPY_undiscounted <- c(200,150,20)
PremResPY <- c(0,0,0)

# Result from lapse scenarios: BE, Up, Down
NAV_BE <- 3000
NAV_Up <- 2800
NAV_Down <- 3100

### Quantities provided by EIOPA

# Risk free interest curve
Interest <- c(0.03, 0.031, 0.0315, 0.032, 0.032, 0.032)

# gross variation coefficient for premium risk
VC_Prem_gross <- c( 0.100, 0.150, 0.130)
# net variation coefficient for reserve risk
VC_Res <- c( 0.110, 0.110, 0.150)

#  Data for the catastrophe risk factor model
CatFactor <- c(1.75,
               1.13,
               1.20,
               0.30,
               1.75,
               1.00,
               0.40,
               0.85,
               1.39,
               0.40,
               2.50,
               2.50,
               2.50)
CatSet <-  list( c(2,4),
                 c(2,4),
                 c(2,4),
                 c(2),
                 c(4),
                 c(3),
                 c(1),
```

```
                  c(5),
                  c(6),
                  c(9),
                  c(10),
                  c(11),
                  c(12)
                )
# Sets of indices that reflect the structure of
# the SCR Cat square root formula
CatSetA <- c(1,2,3,5)
CatSetB <- c(11)
CatSetC <- c(4,7,8,9,10,12)
CatSetD <- c(6,13)

# Event i:              1    2    3    4    5    6
#                       7    8    9   10   11   12   13
CatSplit <- rbind(c(0.1, 0.2, 0.3, 0.0, 0.4, 0.0,
                    0.0, 0.0, 0.0, 0.0, 0.0, 0.0, 0.0),
                  c(0.0, 0.0, 0.0, 0.0, 0.0, 0.0,
                    0.0, 1.0, 0.0, 0.0, 0.0, 0.0, 0.0),
                  c(0.0, 0.0, 0.0, 0.0, 0.0, 0.0,
                    0.0, 0.0, 0.0, 1.0, 0.0, 0.0, 0.0))

# Correlation matrices

# Correlation: premium risk and reserve risk
# same correlation for every LoB
corr_PremRes <- rbind(c( 1.0, 0.5),
                      c( 0.5, 1.0))

# Premium and reserve risk,
# correlation: lines of business
#                           F      H      D
corr_LoB_LoB <- rbind(c( 1.00,  0.25,  0.50), # F
                      c( 0.25,  1.00,  0.50), # H
                      c( 0.50,  0.50,  1.00)) # D

# Correlation: premium/reserve risk and lapse risk
corr_PremRes_Lapse <- rbind(c(1,0),
                            c(0,1))
# Cat risk,
# correlation: lines of business
#                    F  H  D
corr_Cat <- rbind(c( 1, 0, 0),   #  F
                  c( 0, 1, 0),   #  H
```

```
                    c( 0, 0, 1))  #  D


# Correlation: premium/reserve/lapse risk and cat risk
corr_PremResLapse_Cat <- rbind(c(1.00, 0.25),
                               c(0.25, 1.00))
```

D.2 Computing the SCR

The file D_SII_NL_Calc.r contains the actual computation:

```
## Earned Premium
Prem <- array(NA, c(NoIns,2))
Prem_written<- array(NA, c(NoIns,2))
for (t in PY:CY){
  Prem_written[,t] <- (1-ReP_Q)*Prem_written_gross[,t]
  Prem[,t] <- Prem_written[,t]+UPR[,t]-UPR[,t+1]
}

# calculation of the variation coefficient
# for the net premium risk
s <-  sqrt(log(1 + VC_Y_gross^2))
m <- log(Mean_Y_gross) - s^2 / 2

aq <- ReNP_a/(1-ReP_Q)
hq <- ReNP_h/(1-ReP_Q)

Mean_Y <-(1-ReP_Q) * (  exp(m+s^2/2) - hq
                        - aq * plnorm(aq,m,s)
                        + (aq+hq) * plnorm(aq+hq,m,s)
                        + exp(m+s^2/2)
                          * plnorm(aq,m+s^2,s)
                        - exp(m+s^2/2)
                          * plnorm(aq+hq,m+s^2,s)
                        )
VC_Y <- sqrt((1-ReP_Q)^2 *
               (
                (exp(2*m+2*s^2)
                 - 2*hq*exp(m+s^2/2) + hq^2)
               - aq^2 * plnorm(aq,m,s)
               + (aq^2-hq^2) * plnorm(aq+hq,m,s)
               + 2*hq*exp(m+s^2/2)
                 * plnorm(aq+hq,m+s^2,s)
               + exp(2*m+2*s^2)
```

```
                  * plnorm(aq,m+2*s^2,s)
                - exp(2*m+2*s^2)
                  * plnorm(aq+hq,m+2*s^2,s)
                ) / Mean_Y^2 -1
               )
VC_Prem <- VC_Prem_gross * sqrt((1+VC_Y^2) / exp(s^2))

#Calculation of volume measured for premium and reserve
V_Prem <- c()
for (k in 1:NoIns){
   V_Prem[k] <- max(Prem_written[k,PY],
                    Prem_written[k,CY],
                    Prem[k,CY])+PremResPY[k]
 }

DiscountFactor <- 1
V_Res <- array(0,c(NoIns))
for (t in 1:t_runoff){
  DiscountFactor <- DiscountFactor/(1+Interest[t])
  V_Res <- ( V_Res +
             ResPY_undiscounted
             *beta[,t] * DiscountFactor )
}

# Calculation of the variation coefficient
# for the total premium reserve risk
sd_Prem <- V_Prem * VC_Prem
sd_Res <- V_Res * VC_Res
sd_PremRes <- c()
for (k in 1:NoIns){
  sd_PremRes[k] <- sqrt(c(sd_Prem[k],sd_Res[k])
                   %*% corr_PremRes %*%
                   c(sd_Prem[k],sd_Res[k]))
}
VC_PremRes <- sd_PremRes / (V_Prem+V_Res)

Mean_PremRes_total <- sum(V_Prem) + sum(V_Res)
sd_PremRes_total <- sqrt( sd_PremRes
                          %*% corr_LoB_LoB %*%
                          sd_PremRes )
VC_PremRes_total <- sd_PremRes_total/Mean_PremRes_total

# Calculation of SCR components and their aggregation
SCR_PremRes <- (  Mean_PremRes_total
                / sqrt( 1 + VC_PremRes_total^2 )
```

```
                   * exp(sqrt(log(1 + VC_PremRes_total^2))
                         * qnorm(ConfLevel,0,1))
                   )
SCR_Lapse <- max(NAV_BE-NAV_Up,NAV_BE-NAV_Down)
SCR_PremResLapse <- sqrt(c(SCR_PremRes,SCR_Lapse)
                         %*% corr_PremRes_Lapse %*%
                         c(SCR_PremRes,SCR_Lapse) )


Cat_P_Events <- t(CatSplit) %*% Prem_written_gross[,CY]
Cat_Loss <- Cat_P_Events*CatFactor
SCR_Cat <- sqrt( (sqrt(sum(Cat_Loss[CatSetA]^2))
                   + sum(Cat_Loss[CatSetB]))^2
                 + sum(Cat_Loss[CatSetC]^2)
                 + sum(Cat_Loss[CatSetD])^2
                 )

SCR_PremResLapseCat <- sqrt(c(SCR_PremResLapse,SCR_Cat)
                        %*% corr_PremResLapse_Cat %*%
                        c(SCR_PremResLapse,SCR_Cat) )
```

D.3 Output from the Computation

The most important input quantities and computed values are output to the file `D_SII_NL_Output.r`:

```
print("Company inputs:")
dIns <- data.frame(row.names = c("F","L","T"),
                   Prem_written_gross,
                   UPR,
                   Mean_Y_gross,
                   VC_Y_gross,
                   ReP_Q,
                   ReNP_a,
                   ReNP_h,
                   ResPY_undiscounted,
                   PremResPY)
names(dIns) = (c("Prem_written_gross[,PY]",
                 "Prem_written_gross[,CY]",
                 "UPR[,PY]",
                 "UPR[,CY]",
                 "UPR[,NY]",
                 "Mean_Y_gross",
```

```
                    "VC_Y_gross",
                    "ReP_Q",
                    "ReNP_a",
                    "ReNP_h",
                    "ResPY_undiscounted",
                    "PremResPY")
                  )
print(dIns)

dInsCat <- data.frame(row.names = c("F","L","T"),
                      CatSplit)
names(dInsCat) = (c("lambda_1","lambda_2","lambda_3",
                      "lambda_4","lambda_5","lambda_6",
                      "lambda_7","lambda_8","lambda_9",
                      "lambda_10","lambda_11",
                      "lambda_12","lambda_13")
                    )
print(dInsCat)

print("")
print("Regulatory inputs:")

dEIOPA <- data.frame(row.names = c("F","L","T"),
                         VC_Prem_gross,
                         VC_Res,
                         corr_LoB_LoB)

names(dEIOPA) = (c("VC_Prem_gross",
                     "VC_Res",
                     "corr_LoB_LoB[,F]",
                     "corr_LoB_LoB[,L]",
                     "corr_LoB_LoB[,T]")
                   )
print(dEIOPA)

dEIOPACat <- data.frame(CatFactor,t(CatSet))
names(dEIOPACat) = (c("CatFactor",
                        "A_1","A_2","A_3","A_4","A_5",
                        "A_6","A_7","A_8","A_9","A_10",
                        "A_11","A_12","A_13")
                      )
print(dEIOPACat)


print("");print("Results")
```

```
dg <- 2

dResults <-  data.frame(row.names = c("F","L","T"),
                        round(Prem_written,dg),
                        round(Prem,dg),
                        round(m,dg),
                        round(s,dg),
                        round(Mean_Y,dg+1),
                        round(VC_Y,dg),
                        round(VC_Prem,dg+1),
                        round(V_Prem,dg),
                        round(V_Res,dg),
                        round(sd_Prem,dg),
                        round(sd_Res,dg),
                        round(VC_PremRes,dg+1))
names(dResults) = (c(
                     "Prem_written[,PY]",
                     "Prem_written[,CY]",
                     "Prem[,PY]",
                     "Prem[,CY]",
                     "m",
                     "s",
                     "Mean_Y",
                     "VC_Y",
                     "VC_Prem",
                     "V_Prem",
                     "V_Res",
                     "sd_Prem",
                     "sd_Res",
                     "VC_PremRes")
                   )
print(dResults)

print(paste("Mean_PremRes_total: ",
            round(Mean_PremRes_total,dg), sep=""))
print(paste("VC_PremRes_total: ",
            round(VC_PremRes_total,dg), sep=""))
print(paste("SCR_PremRes: ",
            round(SCR_PremRes,dg), sep=""))
print(paste("SCR_Lapse: ",
            round(SCR_Lapse,dg), sep=""))
print(paste("SCR_PremResLapse: ",
            round(SCR_PremResLapse,dg), sep=""))
print(paste("SCR_Cat: ",
```

```
              round(SCR_Cat,dg), sep=""))
print(paste("SCR_PremResLapseCat: ",
              round(SCR_PremResLapseCat,dg), sep=""))
```

Appendix E
R-Script of the Simplified Economic Capital Model

The computation consists in 3 scripts,

`E_EC_Input.r`, `E_EC_Calc.r` and `E_EC_Output.r`,

that constitute the economic capital model for XYZ Inc. They are input one after another. The results appear directly in the R console.

E.1 Input Definition

The file `E_EC_Input.r` defines the input data from Sect. 7.2.3.1:

```
# This will give reproducible results:
set.seed(2)

### Global constants
tiny <- 1E-6
# gross: before reinsurance, net: after reinsurance
gross <- 1; net <- 2
# The BU[[1]] represents Investments
# The first insurance BU has therefore index 2
FirstIns <- 2

strSep = ",";
strDec = ".";

# Global Projection Parameters
NoScen <- 10000
StartCapital <- 1400
FixedCosts <- 20
RiskFreeRate <- 0.03
ConfLevel <- 0.99
```

M. Kriele, J. Wolf, *Value-Oriented Risk Management of Insurance Companies*, EAA Series, DOI 10.1007/978-1-4471-6305-3, © Springer-Verlag London 2014

```
BU_Name <- c( "Investment",
             "Fire",
             "Indemnity",
             "Theft")
BU_Premium <- c( NA, 600, 300, 100)
BU_LossRatio <- c( NA, 0.75, 0.75, 0.75)
BU_CostRatio <- c( 0.005, 0.05, 0.05, 0.05)
BU_VarKoeff <- c( NA, 0.5, 0.6, 0.7)
BU_Investments_Mean <- 0.05
BU_Investments_Sd <- 0.02
BU_ReCeded <- c( 0.00, 0.25, 0.20, 0.20)
# Reinsurance commission = negative costs:
BU_ReCosts <- c( 0.00, -0.06, -0.06, -0.06)
BU_KendalTau <- c( 0.0, 0.0, 0.0, 0.3, 0.2, 0.6)
```

E.2 Computation of the Economic Capital

The file `E_EC_Calc.r` contains the actual computation:

```
require(copula)

NoBU <- length(BU_Name) # Number of business units
                        # including investments

# Structure for all business units, investments
# are always BU[[1]]
BU <- NULL
for (j in 1:NoBU){
  # investments: different handling because of
  # different distributions
  if (j==1){
    tempDistrName <- "norm"
    tempDistrPar <- list(mean=BU_Investments_Mean,
                         sd=BU_Investments_Sd)
  }else{
    tempDistrName <- "lnorm"
    tempSdlog <- sqrt(log(1+BU_VarKoeff[j]^2));
    tempMeanlog <- (log(BU_LossRatio[j]*BU_Premium[j])
                     -tempSdlog^2/2);
    tempDistrPar <- list(meanlog=tempMeanlog,
                         sdlog=tempSdlog)
  }
```

```
  NextBU <- list(Name = BU_Name[j],
                 Premium=BU_Premium[j],
                 LossRatio=BU_LossRatio[j],
                 CostRatio = BU_CostRatio[j],
                 Distr = list(Name=tempDistrName,
                              NoArg = 2,
                              Par = tempDistrPar),
                 Q = list(Ceded=BU_ReCeded[j],
                          Costs=BU_ReCosts[j]))
  BU <- c(BU,list(NextBU))
}
NextBU <- NULL # Not needed any more

# Set-up of copula using functionality from the
# copula package
cop <- normalCopula(param =  sin( pi/2 * BU_KendalTau),
                    dim = NoBU,
                    dispstr = "un");
copMargin <- NULL
copMarginPar <- NULL
for (i in 1:NoBU){
    copMargin <- c(copMargin,BU[[i]]$Distr$Name)
    copMarginPar <- c(copMarginPar,
                      list(BU[[i]]$Distr$Par))
}
rDistribution <- rMvdc(NoScen,
                       mvdc(cop,
                            copMargin,
                            copMarginPar)
                       )

##### We will now calculate the profit for each run

Profit <- list(BU = array(NA,dim=c(NoScen,NoBU,2)),
               total = array(NA,dim=c(NoScen,2)),
               BU.mean = array(NA,dim=c(NoBU,2)),
               total.mean = array(NA,dim=c(2)))

##### Premium ------------------------------------
#Premium <- NULL;
# BU index that corresponds to insurance index:
bu <- NULL
# Insurance index that corresponds to BU index:
ins <- NULL
for (j in FirstIns:NoBU){
```

```
  ins[j] <- j-FirstIns+1
  bu[ins[j]] <- j
}
NoIns <- length(bu)

Premium <- array(NA,dim=c(NoIns,2));
for (j in 1:NoIns){
 Premium[j,gross] <- BU[[bu[j]]]$Premium;
 Premium[j,net] <- (Premium[j,gross]
                     * (1-BU[[bu[j]]]$Q$Ceded));
}

# Investment assets
Assets <- array(NA,dim=c(2));
for (gn in gross:net){
  Assets[gn] <- StartCapital+sum(Premium[1:NoIns,gn])
}

##### Costs ---------------------------------------
Costs <-  NULL;
Costs <- array(NA,dim=c(NoBU,2));

for (gn in gross:net){
  Costs[1,gn] <-  BU[[1]]$CostRatio * Assets[gn];
}
for (j in 1:NoIns){
  for (gn in gross:net){
    Costs[bu[j],gn] <- (BU[[bu[j]]]$CostRatio
                        * Premium[j,gross])
  }
  Costs[bu[j],net] <- (Costs[bu[j],net]
                       + BU[[bu[j]]]$Q$Costs
                         * (Premium[j,gross]
                            -Premium[j,net]))
}

##### Profit ---------------------------------------
for (gn in gross:net){
  # rDistribution[,1] represents the relative
  # return from the investment
  Profit$BU[,1,gn] <- ((rDistribution[,1]
                        -RiskFreeRate)
                       *Assets[gn]-Costs[1,gn])
}
for (j in 1:NoIns){
```

```
  Profit$BU[,bu[j],gross] <- ((1+RiskFreeRate)
                              *Premium[j,gross]
                              - rDistribution[,bu[j]]
                              - Costs[bu[j],gross])
  Profit$BU[,bu[j],net] <- ((1+RiskFreeRate)
                            *Premium[j,net]
                            - (1-BU[[bu[j]]]$Q$Ceded)
                            *rDistribution[,bu[j]]
                            - Costs[bu[j],net])
}
for (gn in gross:net){
  for (j in 1:NoBU){
    Profit$BU.mean[j,gn] <- (sum(Profit$BU[,j,gn])
                             /NoScen)
  }
}

for (gn in gross:net){
  Profit$total[,gn] <-  (RiskFreeRate * StartCapital
                         - FixedCosts)
  for (j in 1:NoBU){
    Profit$total[,gn] <- (Profit$total[,gn]
                          + Profit$BU[,j,gn])
  }
  Profit$total.mean[gn] <- (sum(Profit$total[,gn])
                            /NoScen)
}

#### We will now calculate the economic capital
#### and the risk-adjusted return

EcoCap <- list(BU = array(NA,dim=c(NoBU,2)),
               total = array(NA,dim=c(1,2)))

RORAC <- list(BU = array(NA,dim=c(NoBU,2)),
              total = array(NA,dim=c(1,2)))

NotDefined <- -999

ES <- function(profit, perc){
  # Calculation of the expected shortfall
  quantiles <- profit[order(profit,c(1:length(profit)),
                            decreasing=FALSE)]
  result <- NA;
  if(!is.na(perc)){
```

```
    if (0 < perc & perc < 1){
      result <- sum(quantiles[1:ceiling(
                          (1-perc)*length(quantiles))])
      result <- -(result
                  / ceiling((1-perc)
                            *length(quantiles)))
    }else{
      warning(paste(" perc should be in ]0,1[,",
                    " perc = ", perc, sep=""))
    }
  }
  result
}

### Calculation of all economic capital values
for (gn in gross:net){
  for (j in 1:NoBU){
    EcoCap$BU[j,gn] <- ES(Profit$BU[,j,gn], ConfLevel)
  }
  EcoCap$total[gn] <-  ES(Profit$total[,gn], ConfLevel)
}

### Calculation of all RAROC values
for (gn in gross:net){
  for (j in 1:NoBU){
    if (EcoCap$BU[j,gn] <= tiny){
      RORAC$BU[j,gn] <- NotDefined
    }else{
      RORAC$BU[j,gn] <- (Profit$BU.mean[j,gn]
                         /EcoCap$BU[j,gn])
    }
  }
  if (EcoCap$total[gn] <= tiny){
    RORAC$total[gn] <- NotDefined
  }else{
    RORAC$total[gn] <- (Profit$total.mean[gn]
                        /EcoCap$total[gn])
  }
}
```

E.3 Output from the Computation

The expected return, economic capital and RORAC are output to `E_EC_Output.r`:

```
# Check that it works:
print("Order of business units:")
for (j in 1:NoBU){
  print(paste(j, ": ", BU[[j]]$Name, sep=""))
}
print(paste("Gross: ", gross,
            ", Net ", net, sep=""))

print("Profit of business units: ")
print(Profit$BU.mean)
print("Total profit:")
print(Profit$total.mean)

print("Economic capital for business units: ")
print(EcoCap$BU)
print("Total economic capital:")
print(EcoCap$total)

print("RORAC for business units: ")
print(RORAC$BU)
print("Total RORAC:")
print(RORAC$total)
```

Index

M. Kriele, J. Wolf, *Value-Oriented Risk Management of Insurance Companies*,
EAA Series, DOI 10.1007/978-1-4471-6305-3, © Springer-Verlag London 2014

Printed in Poland
by Amazon Fulfillment
Poland Sp. z o.o., Wrocław